Current Progress in Hydrometeorological Observations and Modeling

Current Progress in Hydrometeorological Observations and Modeling

Edited by Alfredo Mason

STATES
ACADEMIC PRESS
www.statesacademicpress.com

States Academic Press,
109 South 5th Street,
Brooklyn, NY 11249, USA

Visit us on the World Wide Web at:
www.statesacademicpress.com

ISBN: 978-1-63989-742-1

Cataloging-in-Publication Data

Current progress in hydrometeorological observations and modeling / edited by Alfredo Mason.
p. cm.
Includes bibliographical references and index.
ISBN 978-1-63989-742-1
1. Hydrometeorology. 2. Meteorology. 3. Hydrology. 4. Atmospheric science. I. Mason, Alfredo.
GB2803.2 .C87 2023
363.349--dc23

Table of Contents

Preface

The branch of study that focuses on the energy, chemistry, physics, and water fluxes in the atmosphere and Earth's surface is known as hydrometeorology. Hydrometeorological hazards originate due to extreme meteorological and climatic occurrences having atmospheric, hydrological, or oceanographic origin. A few examples of hydrometeorological hazards include typhoons, hurricanes, floods, drought, heatwaves, cold spells, and coastal storm surges. Hydrometeorological monitoring systems are required to develop adequate preparedness and mitigation strategies. The hydrometeorological phenomena can be modeled in three major ways, namely, nowcasting, numerical weather prediction, and statistical techniques. Nowcasting refers to the weather forecasting on a very short mesoscale period. This book traces the progress of hydrometeorology, and highlights some of its key concepts and applications. It unfolds the innovative aspects of hydrometeorological observations and modeling, which will be crucial for the progress of this field in the future. The book is appropriate for students as well as for experts seeking detailed information in the area of hydrometeorology.

Significant researches are present in this book. Intensive efforts have been employed by authors to make this book an outstanding discourse. This book contains the enlightening chapters which have been written on the basis of significant researches done by the experts.

Finally, I would also like to thank all the members involved in this book for being a team and meeting all the deadlines for the submission of their respective works. I would also like to thank my friends and family for being supportive in my efforts.

Editor

1

Hydrometeorological Ensemble Forecast of a Highly Localized Convective Event in the Mediterranean

Luca Furnari, Giuseppe Mendicino and Alfonso Senatore *

Department of Environmental Engineering, University of Calabria, 87036 Rende (Cosenza), Italy; luca.furnari@unical.it (L.F.); giuseppe.mendicino@unical.it (G.M.)

* Correspondence: alfonso.senatore@unical.it

Abstract: The uncertainties that affect hydrometeorological modelling chains can be addressed through ensemble approaches. In this paper, a convection-permitting ensemble system was assessed based on the downscaling of all members of the ECMWF ensemble prediction system through the coupled atmospheric-hydrological WRF-Hydro modelling system. An exemplary highly localized convective event that occurred in a morphologically complex area of the southern Italian coast was selected as a case study, evaluating the performance of the system for two consecutive lead times up to the hydrological forecast on a very small (11.4 km^2) catchment. The proposed approach accurately downscales the signal provided by the global model, improving up to almost 200% the quantitative forecast of the accumulated rainfall peak in the area affected by the event and supplying clear information about the forecast uncertainty. Some members of the ensemble simulations provide accurate results up to the hydrological scale over the catchment, with unit peak discharge forecasts up to 3 $m^3 \cdot s^{-1} \cdot km^{-2}$. Overall, the study highlights that for highly localized convective events in coastal Mediterranean catchments, ensemble approaches should be preferred to a classic single-based simulation approach, because they improve the forecast skills and provide spatially distributed information about the forecast uncertainty, which can be particularly useful for operational purposes.

Keywords: atmospheric-hydrological forecasting chains; convection-permitting models; ensemble forecasting systems; ECMWF ensemble prediction system; lead time; WRF-Hydro; southern Italy; Mediterranean; coastal systems; convective storms; uncertainty

---

1. Introduction

The coupled weather-hydrological forecasting chains, whose development has undergone a decisive acceleration in recent years, represent potentially invaluable tools for the management of emergencies and civil protection purposes [1] since they allow both the forecast of atmospheric events and the estimate of their hydrological impact on the ground. Nevertheless, both the main components of such complex systems, i.e., the atmospheric and the hydrological models, are subject to so many uncertainties [2,3] that an approach based on single simulations is most often not enough, but it needs to be supported or even replaced by a probabilistic approach based on multiple forecasts (e.g., [4]). The awareness of the high uncertainty lying behind the weather forecasting emerged soon [5], leading to the development of global ensemble prediction systems like, e.g., the Ensemble Prediction System (EPS), which has been operational since 1992 [6] at the European Centre for Medium-Range Weather Forecasts (ECMWF).

The need to estimate in greater detail the extent and intensity of the predicted atmospheric phenomena and the related impact on the ground has been answered by the growing availability of mesoscale models able to perform the dynamical downscaling of forecasts provided by the global models, giving rise to the development of high-resolution ensemble forecasting systems operating at convection-permitting resolutions [7]. Many studies based on such systems focus on meteorological variables, above all precipitation (e.g., [8–13]). Some other studies incorporate specific aspects (e.g., [14] perturbed the physical parameterizations; [15] and [16] addressed initial conditions perturbations). Nevertheless, increasing examples of coupled ensemble atmospheric-hydrological systems are available, where the Quantitative Precipitation Forecast (QPF) is the main input of the hydrological models (e.g., [17–21]).

Localized highly convective episodes are particularly challenging in terms of both QPF and, consequently, the hydrological response, especially in those regions characterized by a complex orography, where errors of a few kilometers in localizing a given precipitation event, which is still a valuable result in terms of the weather forecast, can provide a rather different hydrological scenario, shifting the flood forecast from one catchment to another. The Mediterranean Basin is a morphologically complex region of the world typically affected by convective episodes, and, in fact, it has been identified as a hot spot of climate change [22], where the intensity and frequency of high-impact weather events are bound to increase [23,24]. In this area, the increasingly complex challenge of efficient and effective hydrometeorological forecasting is also addressed through convection-permitting ensemble forecasting systems (e.g., [25–28]).

The Calabria region at the southern tip of the Italian peninsula, placed in the middle of the Mediterranean Basin, is particularly prone to high-impact episodes, producing noteworthy ground impact with damages and, not seldom, loss of human lives [29]. In this area, the effects of the large gradients between the cooler atmosphere and the warmer sea surface, typical of summer (mainly because of isolated cut-off lows) and autumn (generally connected to frontal systems) seasons, are amplified by the steep orography close to the coast, leading to intense convective precipitation very often characterized by orographic enhancement.

This study aimed to assess the potential benefit of an ensemble approach on the predictability of highly localized convective events in Calabria by analyzing them up to the hydrological-scale impact with a case study that occurred in summer 2015. The analysis was performed using the WRF-Hydro [30] modelling system, composed by the Weather Research and Forecast (WRF) mesoscale model [31], allowing high-resolution (in the order of 10^3 m) convection-permitting downscaling of the boundary conditions provided by 50 members of the ECMWF EPS for two consecutive lead times, and its hydrological extension providing detailed information about the impact on the ground at a resolution in the order of 10^2 m, which is suitable to medium and small-size catchments' hydrological modelling. Therefore, a case study concerning a complete hydro-meteorological ensemble forecasting chain is proposed, where the uncertainty is accounted only for the atmospheric modelling, and then cascaded throughout the forecasting chain. This choice is explained both by the fact that the meteorological input is generally supposed to be the main source of uncertainty in short-range hydrological forecasts (even though this assumption is debatable, according to [32]) and by the purpose of clearly highlighting, for the sake of clarity, only the effects of the meteorological uncertainty, evaluating also its magnitude with different forecast lead times.

The described research followed and aimed to integrate a previous analysis already performed for the same event [33], which showed the limits of a single-based simulation approach, even supported by variational data assimilation techniques, for the accurate predictability of both the meteorological and hydrological features. Specifically, among the different mesoscale model boundary conditions investigated in [33], the most suitable were selected and used as a starting point for the analysis shown in this paper. Therefore, the choice of the summer 2015 event is justified by a combination of features that makes it almost unique in the analyzed region, since it is: (1) A relatively recent high-impact event (with almost 250 mm of rain in 24 h); (2) a highly localized event, whose signal was recorded by only one weather station of the dense Calabrian monitoring network; and (3) an already well-studied event, simulated with varying boundary conditions and with/without data assimilation techniques. These features allow it to be considered as an exemplary case study, from which general conclusions, applicable to similar events, can be drawn.

In the following, Section 2 describes the study area and the weather event analyzed and provides details about the modelling approach and the validation strategy. Section 3 shows the results of the coupled meteo-hydrological analysis, while Section 4, finally, discusses the main findings achieved, summarizes the main conclusions, and provides insights for future research.

2. Materials and Methods

2.1. Study Area and Modelling Approach

Calabria is the southernmost Italian peninsula (Figure 1), surrounded by the Tyrrhenian Sea (western coast) and the Ionian Sea (southern and eastern coast). Despite its proximity to the sea, more than 40% of its total area of about 15,000 km^2 is higher than 500 m above the sea level (a.s.l.). The orographic features play a fundamental role in the development of high-impact events, hitting especially (but not exclusively) the eastern coast [34–39].

The limited area model (LAM) used in this study is the Weather Research and Forecasting—Advanced Research (WRF-ARW, [31]) version 3.7.1, with two one-way nested domains (Figure 1), whose main properties are shown in Table 1. The same table also reports the physics parameterization of the mesoscale model (which is the same used by [33,40]), together with information about the initial and boundary conditions (ICs and BCs, respectively) provided by the high-resolution simulation and both the ensemble control and member simulations of the reference Global Circulation Model (GCM). Considering the problems related to the spatial interpolation of the Sea Surface Temperature (SST) variable arisen with the ECMWF-EPS, which were also found in previous studies [33,39], Table 1 also provides information about the lower (SST) initial and boundary conditions used.

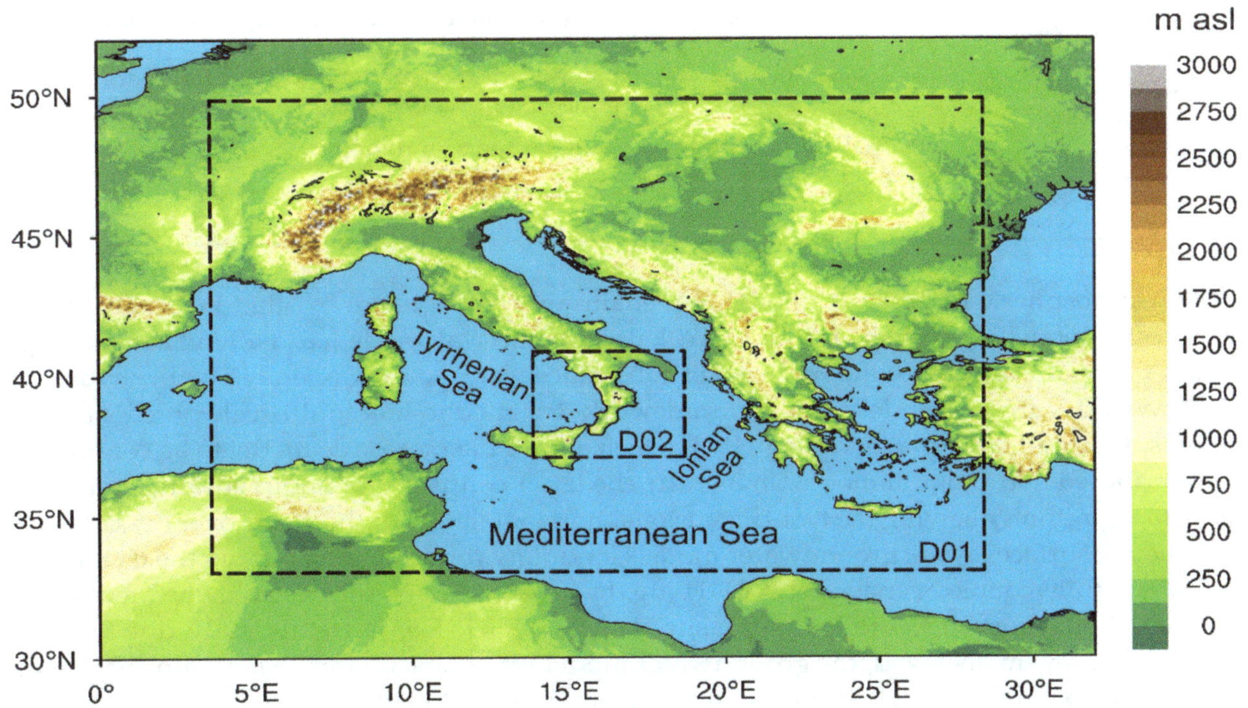

Figure 1. WRF computational domains (dashed lines). The outermost domain D01 is centered over the central Mediterranean Sea, and the innermost D02 is centered over the Calabria region, whose northern border is highlighted with a black bold line.

Table 1. Main features of the simulations performed in this study.

WRF Physics Parameterization	
Component	*Scheme*
Microphysics	Lin-Purdue [41]
PBL	MJY [42]
Shortwave Radiation	Dudhia [43]
Longwave Radiation	RTTM [44]
Land Surface Model	Unified NOAH [45]
Surface Layer	Eta Similarity [46]
Cumulus	Kain-Fritsch (D01) [47]
WRF Domains Space and Time Resolutions	
D01	10 km (205 × 187 grid points), 60 s
D02	2 km (200 × 200 grid points), 12 s
Vertical layers	44 terrain-following layers above the surface and 4 layers in the soil
WRF Initial and Boundary Conditions (ICs and BCs)	
GCM ICs and BCs	ECMWF-EPS, reference IFS Cycle 41r1 [48]
HRES:	High-resolution simulation at 16 km resolution
LRES_CON:	Ensemble control simulation at 32 km resolution
LRES_ENS:	50 perturbations are applied to the initial conditions of the GCM, based on a multivariate Gaussian sampling technique [49]
Lower BCs (SST)	Native SST fields replaced with the Medspiration L4 Ultra-High Resolution SSTfnd product as a daily mean at a resolution of 0.022° [50,51]. The *sst_update* and the *sst_skin* [52] options were also activated, which permit dynamic SST and allow the system to simulate the daily SST cycle, respectively
WRF-Hydro Main Features	
Land Surface Model	Unified NOAH, 2 km resolution
Active modules	Subsurface, surface and channel water routing
Input from WRF	Precipitation and pressure on the ground, air temperature and humidity, wind speed and solar radiation (1 h time step)
Resolution	200 m (2000 × 2000 grid points), disaggregation factor with respect to the atmospheric model of 1/10
Initialization Time and Range of the Simulations	
0000 UTC	0000 UTC 11 August 2015 (48 h range)
1200 UTC	1200 UTC 11 August 2015 (36 h range)

The atmospheric-hydrologic modelling chain was completed by the WRF Hydrological (WRF-Hydro) modelling system version 3.0 [30], used in one-way mode (i.e., without considering any feedback to the atmosphere of the surface and subsurface water routing operated by the hydrological model). WRF-Hydro is a model coupling framework designed to link multi-scale process models of the atmosphere and terrestrial hydrology [53], which allows the mesoscale model to be connected to a semi-distributed hydrological model. In its turn, the latter couples a monodimensional land-surface scheme to algorithms allowing the two-dimensional horizontal flow of the water over the land surface and in the subsurface layers, groundwater models, and flow routing models along the channel. Both one-way and fully-coupled versions of WRF-Hydro have been successfully tested in the Mediterranean region with both short-term and long-term simulations (e.g., [54–56]). In this study, due to the short duration of the event analyzed, the groundwater model was not activated. More information about the WRF-Hydro modelling details for this study is provided in Table 1.

2.2. Case Study

The analyzed case study concerns intense and rather localized precipitation that occurred in the first hours of 12 August 2015 and affected almost only a small area in the north-eastern side of the region (Figure 2a), corresponding to the municipalities of Corigliano Calabro and Rossano [57]. The synoptic analysis shows that the event was due to a cut-off low generated over the Atlantic, which reached the central Mediterranean some days after, entailing the advection of the convective cells that developed over the Ionian Sea towards the impacted area, corresponding broadly to the coast and the first slopes of the Sila Plateau (Figure 2b). Figure 2a shows the spatial distribution of rainfall observations on the ground provided by the Centro Funzionale Multirischi—ARPACAL of the Calabria Region and detected by the radar network managed by the National Civil Protection for 24 h between 1800 UTC on 11 August 2015 and 1800 UTC on 12 August 2015. It is noteworthy that the Corigliano rain gauge recorded 246.6 mm in 24 h, of which 223.2 mm in only 12 h, corresponding to a return-time period of 151 years, and 167.4 mm in 6 h, corresponding to a return-time period of 179 years (Figure 2c). The starting times selected for the two ensemble simulations performed in the study were 0000 UTC and 1200 UTC on 11 August 2015 (Table 1), allowing lead times of about one day and half-day, respectively, to the time when the highest rainfall intensity started.

Among the several small and very small catchments affected by the event, the most relevant hydrological impact concerned the Citrea creek, a small coastal catchment extending for only 11.4 km^2 (Figure 2b). The length of the creek is 9.6 km, while the basin has an average elevation of 236 m a.s.l. (maximum elevation of 785 m a.s.l.), an average slope of 9.6%, and a lag time estimated between 15 and 20 min. According to the Corine Land Cover 2018 project [58], about 28% of the catchment area is covered by artificial surfaces, 49% by agriculture areas, and the remaining 23% by forest and semi-natural areas. Unfortunately, neither discharge nor hydrometric observations are available for the Citrea creek; therefore, the hydrological model was not calibrated and the default parameterization was chosen, which, however, was enough for the aims of this study, whose objective was to explore the added value of an ensemble forecasting chain to the predictability of the event rather than estimating specific hydrological quantities, such as, e.g., peak flow.

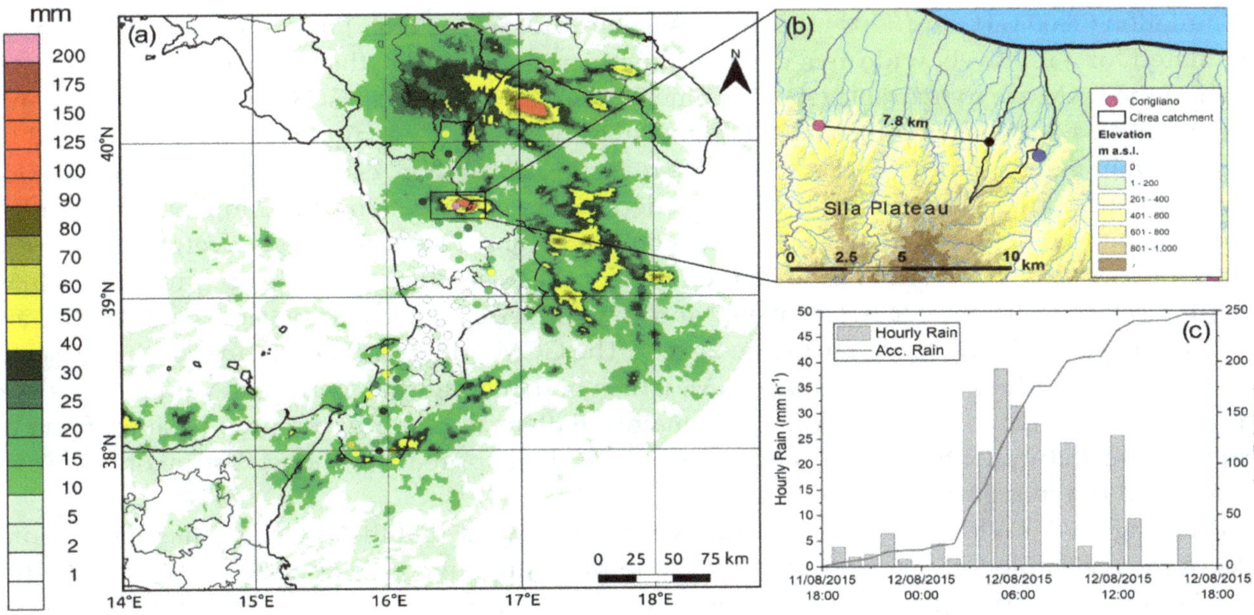

Figure 2. (**a**) The accumulated precipitation observed by the regional monitoring network (dots), overlying the meteorological radar observations during the interval between 1800 UTC, 11 August 2015 and 1800 UTC, 12 August 2015; regional borders are shown together with the coastlines; (**b**) zoom of the study area highlighting both the local orography and river network, which shows a distance lower than 8 km between the Citrea creek catchment and the Corigliano rain gauge; the location of the municipality of Rossano (old city) is indicated with a blue point; (**c**) cumulative and hourly observed hyetograph (mm) at the Corigliano rain gauge.

2.3. Validation Strategy

Given the peculiarity of the event, which essentially interested only one rain gauge, the performances of the 51 (50 ensemble members and the control run) weather simulations for each of the two ensemble forecasts were assessed, developing a tailored approach.

Concerning the spatial pattern of the simulated precipitation, the rainfall fields produced by the ensemble members were accumulated over 24 h, performing later over the whole domain D02 a cell-by-cell statistical analysis, which led to maps of the mean values, standard deviation, and different percentiles. Such analysis allowed, in particular, the spatial variability of uncertainty to be highlighted.

Concerning the comparison with observed rainfall in the analyzed area (i.e., with the observations at the Corigliano rain gauge), since radar data underestimate observations and the event is extremely localized, the analysis focused on a square box with an 80 km side (i.e., 1600 grid points), centered on the Corigliano rain gauge and including the Citrea creek catchment. Within this box, for each weather simulation, the maximum cumulative value over 24 h and its position within the domain D02 was identified and, for the relative grid point, the rainfall time series was extracted.

Finally, concerning the evaluation of the hydrological impact, for each ensemble member, the averaged precipitation amount over the Citrea creek catchment was assessed. Furthermore, the discharge series calculated by the WRF-Hydro model, starting from initial conditions provided by the ERA-Interim [59] reanalysis, were evaluated in terms of the peak flow time and hydrograph shape.

3. Results

3.1. Meteorological Ensembles

Though its increasingly higher detail, the ECMWF Integrated Forecasting System (IFS) does not provide an adequate spatial resolution (both as high-resolution and ensemble forecasts) for capturing highly localized convective episodes, such as that analyzed. Table 2 shows for the two starting times of the simulation (0000 UTC and 1200 UTC on 11 August 2015, respectively) that, if the precipitation rate is evaluated for a relatively wide area equal to the innermost domain, the amounts estimated by the GCM and the LAM are comparable (starting time 0000 UTC) or even the GCM amount is significantly higher than the LAM (starting time 1200 UTC). Nevertheless, if the analysis is tightened around the Corigliano rain gauge, the GCM cannot provide realistic information, estimating rainfall peaks equal about from one-third to one-half of those provided by the mesoscale model and from one-tenth to one-seventh of that observed (with starting time 0000 UTC and 1200 UTC, respectively). In particular, in both the forecasts, only 2 out of the 50 members of the ensembles estimate accumulated rainfall amounts to close to 100 mm·24 h^{-1}, while 29 and 15 members, with a starting time of 0000 UTC and 1200 UTC, respectively, do not reach 30 mm·24 h^{-1}. The GCM underestimations are tightly connected to its coarser resolution compared to the LAM, which both averages precipitation values over larger grid areas (i.e., from 2^2 to 16^2 or 32^2 km^2) and, mainly, flattens the complex orography simplifying the morphological features of the area. Therefore, a downscaling approach, either statistical or (such as in this case study) dynamical, is unavoidable.

Focusing on the WRF downscaled forecasts, it is noteworthy that the average value of the rainfall peaks of the ensemble forecast near the Corigliano rain gauge increase of about 30% from the first (0000 UTC) to the second (1200 UTC) forecast, while the high-resolution and low-resolution control simulations peaks slightly decrease.

Figures 3 and 4 relate to forecast starting times of 0000 UTC and 1200 UTC, respectively, and provide outlooks on the overall behavior of the downscaled ensemble simulations. Concerning the forecast starting at 0000 UTC, the low-resolution control simulation (Figure 3a) forecasts some relevant rainfall peaks both over the Ionian sea (at about the coordinates 39° N, 18° E; about 130 km from the centroid of the Citrea creek) and over the land, on the Tyrrhenian coast (about 39° N, 16° E; about 70 km from the Citrea creek) and on the northern (near the location where the event actually occurred) and southern Ionian coastline. In the latter location (about 38.5° N, 16.5° E; about 110 km from the

Citrea creek), the highest rainfall peak over the land (>100 mm·24 h^{-1}) is forecasted, not corresponding, however, to the observations (Figure 2a). The ground monitoring network highlights noteworthy accumulated precipitation (>40 mm), besides the Corigliano area, only along the southern Tyrrhenian coast and at the southernmost bottom of the peninsula. The control simulation of the ensemble forecast starting at 1200 UTC (Figure 4a) does not provide more detailed information, showing a bit drier pattern over the northern Ionian coast. Both in the 0000 UTC and 1200 UTC forecasts, each of the 50 perturbed simulations shows rainfall patterns like the respective low-resolution control simulations, with localized peaks especially in the eastern side of the domain. However, while in the 0000 UTC forecast, on average, no particularly intense rainfall is expected (the pattern in Figure 3b is explained by the fact that the locations of the rainfall peaks forecasted by the ensemble simulations are rather scattered), some more detailed information is given by Figure 4b, relating to the 1200 UTC forecast, where some rainfall peaks are more evident. The highest mean values in the 0000 UTC forecast are found along the Tyrrhenian coastline where the low-resolution control simulation erroneously forecasts rainfall rates of about 70–80 mm·24 h^{-1}. The same peak (more intense) is found in the 1200 UTC forecast, together with another one in the south of the region. The standard deviation maps (Figures 3c and 4c) are more interesting since they account directly for the forecast uncertainty. Over the land, the highest standard deviation values are found roughly in the same locations where the low-resolution control simulations forecast the highest peaks, i.e., on the Tyrrhenian coast (about 39° N, 16° E; about 70 km from the Citrea creek), the southern Ionian coast (about 38.5° N, 16.5° E; about 110 km from the Citrea creek), and where the analyzed event occurred, i.e., in the Corigliano area. The high standard deviation entails higher rainfall rates in the percentile maps. As an example, the 80th percentile map in Figure 3d shows, together with a couple of incorrectly forecasted peaks (on the Tyrrhenian and southern Ionian coasts, respectively), that intense rainfall is correctly forecasted both in the extreme south of the peninsula and, mainly, in the Corigliano area, even though at rates still considerably lower than observations. Figure 4b confirms the same behavior in the 1200 UTC forecast, with higher rainfall amounts.

Table 2. Average accumulated precipitation in the innermost domain and rainfall peak in an 80 km-side square box centered on the Corigliano rain gauge in the period 1800 UTC, 11 August 2015—1800 UTC, 12 August 2015, forecasted by ECMWF IFS and WRF. Units are in (mm·24 h^{-1}). HRES: high-resolution simulation; LRES_CON: low-resolution ensemble control simulation; LRES_ENS: low-resolution ensemble simulations.

Variable	ECMWF IFS			WRF		
	HRES	LRES_CON	LRES_ENS	HRES	LRES_CON	LRES_ENS
Starting time 11 August 2015 0000 UTC						
Average accumulated precipitation in D02	9.3	8.5	10.6 ± 3.4	9.5 [1]	6.6	6.7 ± 2.0
Rainfall peak near Corigliano	22.9	27.4	31.0 ± 22.2	73.7 [1]	80.3	87.2 ± 35.4
Starting time 11 August 2015 1200 UTC						
Average accumulated precipitation in D02	12.7	12.1	12.0 ± 2.7	7.0	7.0	7.3 ± 1.1
Rainfall peak near Corigliano	39.3	35.1	41.4 ± 18.7	63.7	70.9	112.9 ± 40.0

[1] results from [33].

Therefore, the spatial analysis of the high-resolution ensemble forecasts over the whole region highlights that, though some false alarms are identified, a signal of the upcoming extreme event on the Corigliano area is detected.

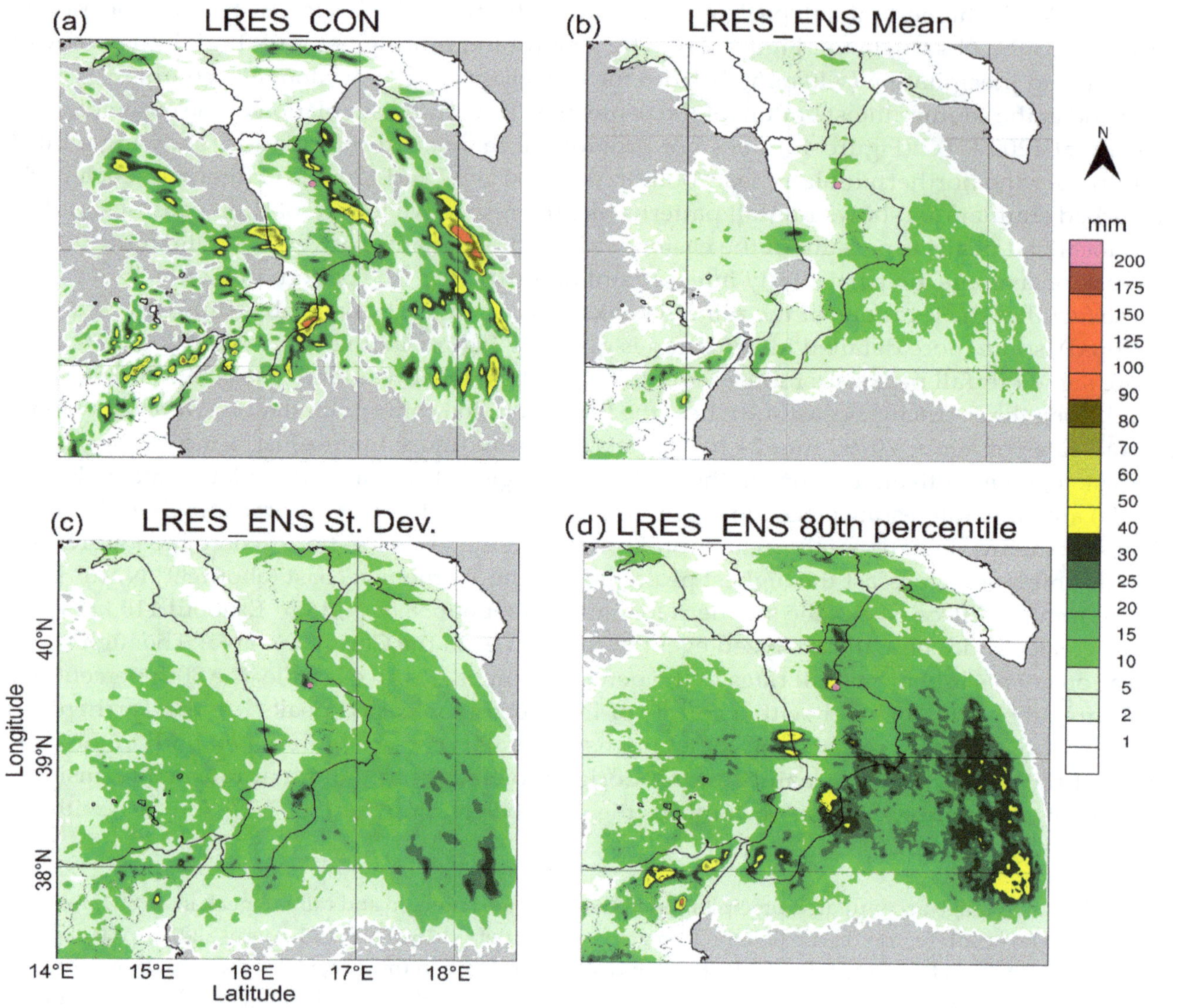

Figure 3. Accumulated precipitation (mm) simulated by WRF using the ECMWF EPS boundary conditions (simulation start at 0000 UTC, 11 August 2015) during the 24-h-long period 1800 UTC, 11 August 2015—1800 UTC, 12 August 2015: (**a**) low-resolution control simulation; (**b**) cell-by-cell mean of the 50 perturbed forecasts, (**c**) same as (**b**), but for standard deviation; (**d**) same as (**b**), but for the 80th percentile. The colored dot represents the Corigliano gauge station.

Focusing on the Corigliano area, both the performances in space and time of the ensemble forecasts are assessed through the graphs in Figures 5 and 6. Specifically, Figure 5 summarizes the location, the 24-h accumulated value, and its time correlation with observations for each of the rainfall peaks simulated by the 50 perturbed simulations of both the 0000 UTC and 1200 UTC forecasts, together with the respective low-resolution control and the high-resolution simulations, within an 80 km-side squared box centered on the Corigliano rain gauge. To avoid detecting insignificant peaks, a lower threshold of 30 mm was assumed. Almost all the simulations forecast noteworthy rainfall peaks within the considered squared area. The peaks are detected at the border of the box (since they belong to rainfall clusters having their maximum values outside the box itself) only for nine (5) simulations in the 0000 UTC (1200 UTC) forecast. These peaks are represented in the circles mainly gathered on the bottom-left sides of both Figure 5a,b. As it is clear from Table 2, all simulations underestimate the observations (their amount is 35.4 ± 14.4% and 45.8 ± 16.2% of the rain gauge value on average, for the 0000 UTC and 1200 UTC forecasts, respectively), yet they are quite correlated (Pearson's correlation coefficient $r = 0.85$ for the 0000 UTC and $r = 0.91$ for the 1200 UTC forecast). In both the forecasts, the low-resolution control and the high-resolution simulations do not show particularly high performances, especially concerning the rainfall amount. Since the low-resolution control simulation is equal to the high-resolution except resolution, the distances between the peaks related to the high-resolution and the low-resolution control simulations highlight that the effect of the boundary conditions is not negligible. On the other hand, the relatively low performances of the high-resolution forecasts (especially with the 0000 UTC starting time) highlight that, in this case, the higher resolution of the boundary conditions (i.e., 16 vs. 32 km) does not provide a clear benefit to the forecast accuracy. This result can be related to the chaotic nature of the precipitation pattern in the convective event, where other sources of uncertainty overwhelm the potentially positive effect of higher resolution boundary conditions. It is noteworthy that the precipitation variability in the innermost domain is highly enhanced by the resolution of the mesoscale model. For example, the correlation coefficient r between the accumulated rainfall peak around Corigliano and the averaged accumulated rainfall in the whole domain is equal to 0.68 (0.42) for the EPS but only to 0.02 (0.11) for the downscaled ensemble in the 0000 UTC (1200 UTC) forecast.

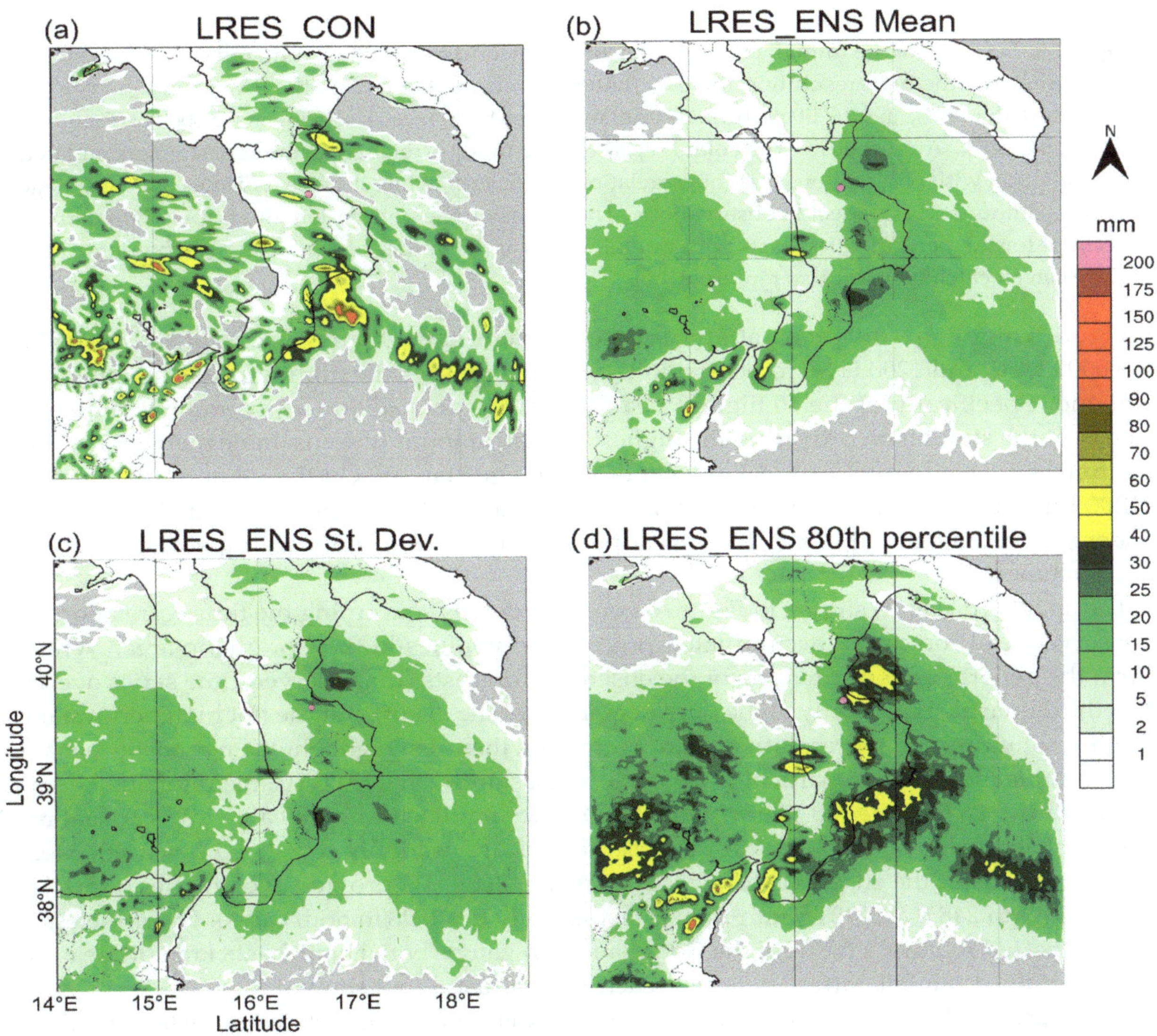

Figure 4. Accumulated precipitation (mm) simulated by WRF using the ECMWF EPS boundary conditions (simulation start at 1200 UTC, 11 August 2015) during the 24-h-long period 1800 UTC, 11 August 2015—1800 UTC, 12 August 2015: (**a**) low-resolution control simulation; (**b**) cell-by-cell mean of the 50 perturbed forecasts, (**c**) same as (**b**), but for standard deviation; (**d**) same as (**b**), but for the 80th percentile. The colored dot represents the Corigliano gauge station.

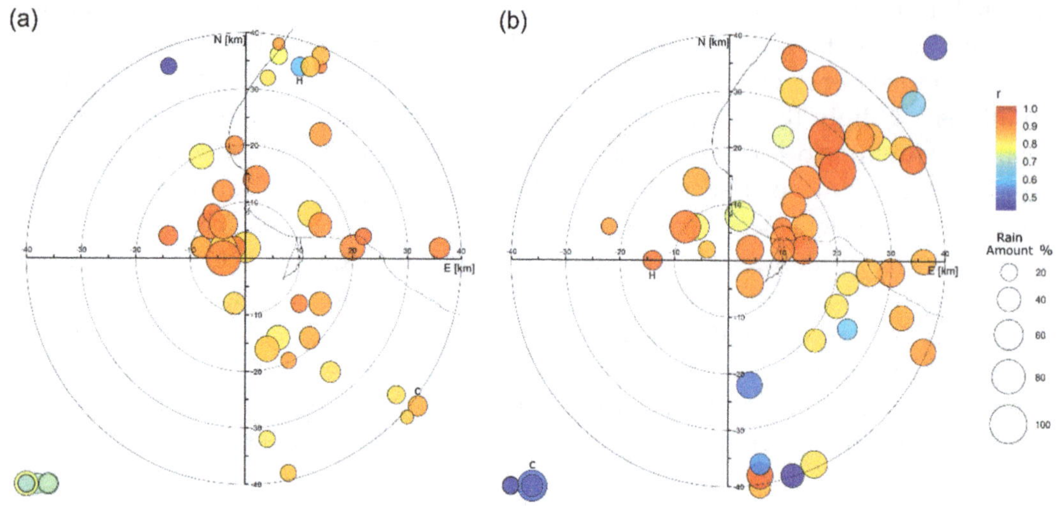

Figure 5. Location of the rainfall peaks within an 80 km-side squared box centered on the Corigliano rain gauge for the high-resolution forecast (letter "H" inside the circle), the low-resolution control forecast (letter "C"), and the 50 perturbed simulations, for the forecasts starting at (**a**) 0000 UTC and (**b**) 1200 UTC on 11 August 2015. The colors represent the time correlation of the series of accumulated precipitation with observations (r is the correlation coefficient), while the size indicates the percentage rainfall amount with respect to observations. The circles on the bottom-left and upper-right corners account for the perturbed forecasts with rainfall peaks at the border of the boxes. The borders of the Citrea creek catchment are also represented.

Figure 5 highlights different patterns for the locations of the highest accumulated rainfall peaks in the 0000 UTC and 1200 UTC forecasts. In the former, most of the highest peaks are quite close to Corigliano. Specifically, of the 10 highest peaks, 8 are less than 15 km away and 6 less than 10 km. The peak of the simulation providing the highest rainfall amount (the ensemble member no. 36, with 204.8 mm·24 h^{-1}) is located only 4 km west of the gauge station. On the other hand, the peaks of the 1200 UTC forecast (Figure 5b) are generally biased towards the north-east quadrant (52% of the members, including 10 out of the 12 highest peaks) but some of them are capable of affecting the Citrea catchment.

The spaghetti graphs shown in Figure 6 goes into details concerning the time correlation of both the observed and forecasted accumulated precipitation values in the Corigliano area. While they confirm the general underestimation of the ensemble forecasts (the observed time series overlay the first to third quartile bands only at the very beginning of the event) and the not high performances of the high-resolution and low-resolution control forecasts, they also highlight the reasonable behavior of some ensemble members. In particular, for the forecast starting at 0000 UTC, the rainfall time series of the member no. 36 (i.e., the red line with the highest accumulated rainfall value in Figure 6a) reaches an accumulated value equal to 83% of the observations and is highly correlated ($r = 0.92$). Members no. 34 and no. 39 of the ensemble forecast starting at 1200 UTC provide even higher rainfall peaks (Figure 6b), with 248.0 and 221.5 mm·24 h^{-1}, respectively ($r = 0.98$ in both cases). Nevertheless, these peaks are located relatively far from the Corigliano rain gauge, both about 26–28 km north-east, hence the very localized rainfall pattern does not reach the mainland (not shown).

The overall high indices of the member no. 36 of the ensemble forecast starting at 0000 UTC suggest a further more detailed analysis of this simulation. Figure 7 shows its forecasted 24-h accumulated precipitation map from 1800 UTC, 11 August 2015 to 1800 UTC, 12 August 2015. The rainfall spatial pattern is slightly different from that detected by the low-resolution control simulation (Figure 3a). The precipitation amount increases considerably in the eastern side of the domain, over the Ionian Sea, while it reduces over the western side. This behavior is a consequence of several differences in the atmospheric features caused by the boundary conditions provided by the EPS member no. 36: More humid and warmer air is advected over the Ionian Sea, with integrated water vapor values of up to 50 kg·kg^{-1} and higher potential temperature (up to about 292 K at 850 hPa) during the night (maps not shown), entailing higher convective available potential energy (CAPE) values (over 3500 J·kg^{-1} offshore the Ionian coast). Overall, the forecast provided by the ensemble member no. 36 looks more coherent than the low-resolution control simulation with radar observations (e.g., the model can simulate the rainfall peak detected by the radar in the northern part of the Ionian Sea shown in Figure 2a).

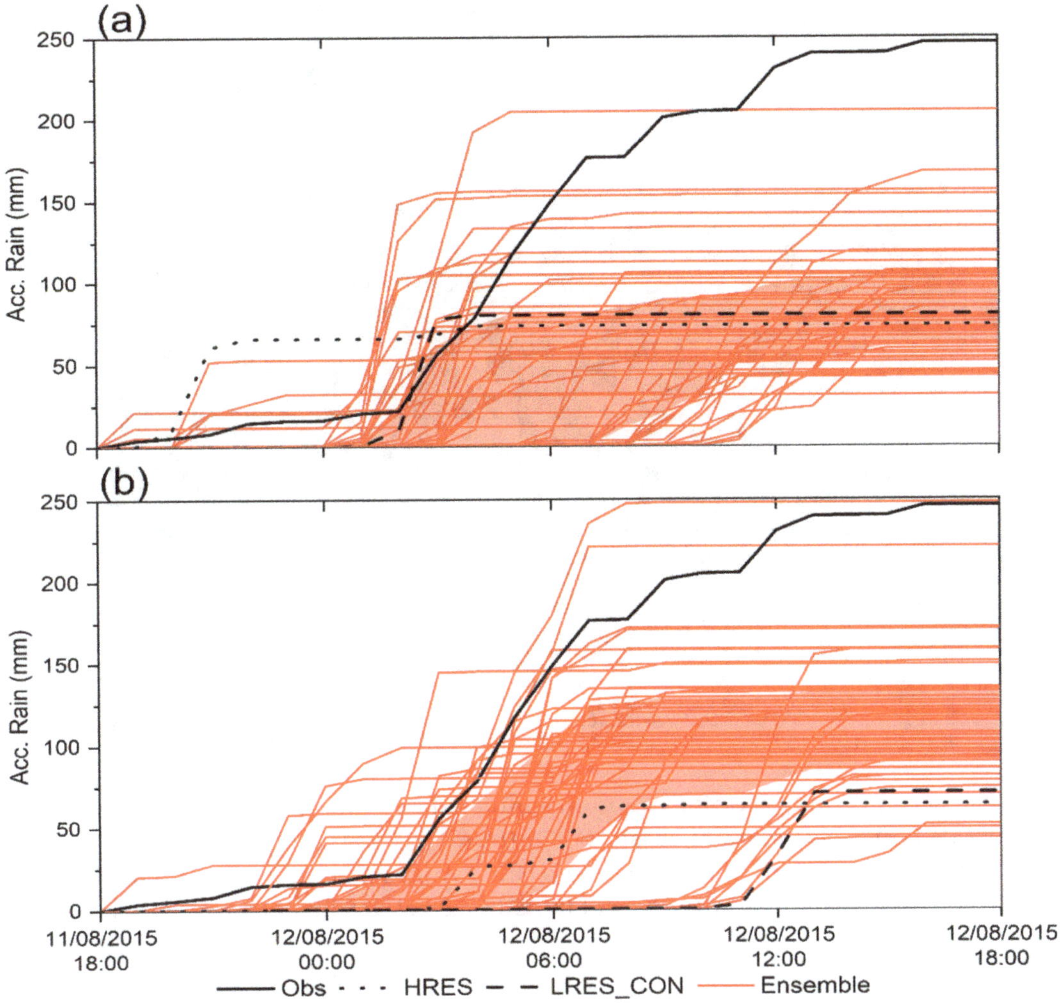

Figure 6. Spaghetti graphs of the accumulated rainfall peaks around the Corigliano rain gauge shown in Figure 5: (**a**) forecasts starting at 0000 UTC, 11 August 2015; (**b**) forecasts starting at 1200 UTC, 11 August 2015. The light red bands highlight the intervals between the 1st and the 3rd quartiles. The dotted line indicates the HRES simulation, while the dashed line is the low-resolution EPS control simulation.

Over the land, the rainfall overestimation in the central-southern Ionian coast, already highlighted in the ensemble low-resolution control simulation, is more widespread, while an opposite behavior is observed for the overestimation in the Tyrrhenian coast (i.e., the extent of the area with over-forecasted precipitation is reduced compared to the low-resolution control simulation). The main change, however, occurs in the Corigliano area, where the scattered rainfall clusters simulated in the low-resolution control simulation are gathered into a more intense one, which is centered almost exactly on the rain gauge station. Nevertheless, the rainfall cluster is so highly concentrated in space that it does not include the Citrea creek catchment. Indeed, in the analyzed 24-h interval, the ensemble member no. 36 forecasts only about 1 mm of averaged precipitation over the Citrea creek catchment area. The high performance of the precipitation forecast at the Corigliano rain gauge does not lead to an improved hydrological simulation in a catchment only 8 km away from the rain gauge itself.

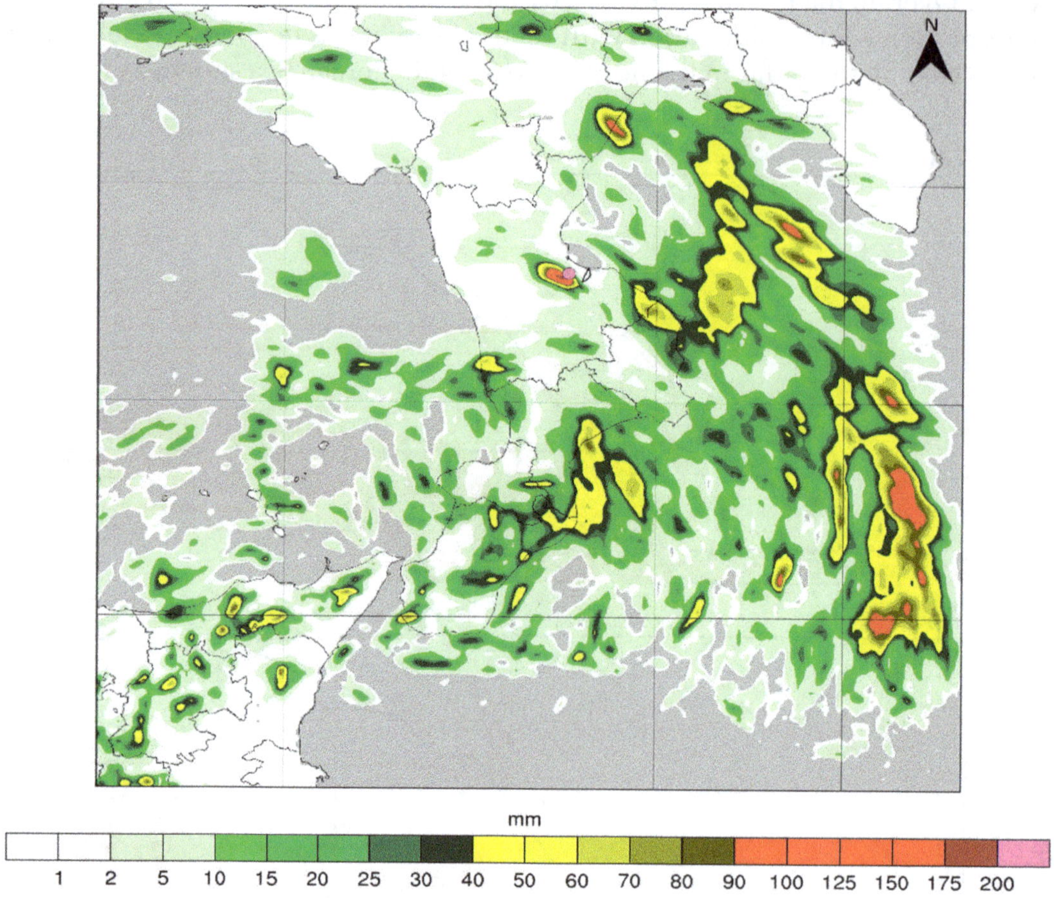

Figure 7. 24-h accumulated precipitation (mm) simulated by WRF using the boundary conditions from the ECMWF EPS perturbed forecast no. 36 (from 1800 UTC, 11 August 2015 to 1800 UTC, 12 August 2015; starting time of the forecast 0000 UTC, 11 August 2015). The colored dot highlights the Corigliano's rain gauge station. The borders of the Citrea creek catchment are also represented.

3.2. *Hydrological Ensembles*

Figure 8 shows the hourly accumulated rainfall averaged over the Citrea creek catchment and the resulting hydrographs achieved with the WRF-Hydro model using the ensemble low-resolution control simulation and the 50 perturbed ensemble simulations in the 0000 UTC and 1200 UTC forecasts. Both the highest 24-h accumulated rainfall and the highest hourly rainfall intensity in the 0000 UTC forecast are provided by the ensemble member no. 16 (58.8 and 39.8 mm·h^{-1} between 0200 and 0300 UTC, respectively). The 24-h rainfall peak of this simulation within the 80 km-side squared box centered on the Corigliano rain gauge is 112.6 mm·24 h^{-1} and is localized not far from the Citrea creek (about 12 km to the east and 8 to the north of the Corigliano station). In the 1200 UTC forecast, the ensemble member no. 47 provides the highest accumulated rainfall amount (78.0 mm·24 h^{-1}) and almost the highest hourly intensity (57.4 mm·h^{-1} between 0600 and 0700 UTC), only slightly exceeded by the member no. 21 (57.8 mm·h^{-1}, again between 0600 and 0700 UTC). The 24-h rainfall peaks of both these simulations are rather far from the Citrea creek (from 25 to 30 km).

Notwithstanding the likely underestimation of the accumulated precipitation (it is worth reminding that the Corigliano rain gauge recorded 223.2 mm in only 12 h), the hourly intensity peak of about 40 mm·h^{-1} of the ensemble member no. 16 (0000 UTC forecast) is comparable to that observed

(Figure 2c) and is timely, while the highest hourly intensities of the ensemble members of the 1200 UTC forecast are even higher (almost 60 mm·h^{-1}) but late.

The ensemble member no. 16 of the 0000 UTC forecast leads to a low peak flow (about 3.4 m^3·s^{-1}, Figure 8a), even with a sufficiently high simulated rainfall intensity peak. On the other hand, in the ensemble forecast starting at 1200 UTC, much higher values are reached (up to 30 m^3·s^{-1}, Figure 8b), which, though lower than the peak flow of about 90 m^3·s^{-1} estimated by the ARPACAL Agency [57] (and it makes sense, given the overall underestimation of the accumulated precipitation), is enough for triggering early warning measures for that specific catchment.

Even neglecting the problems due to the missed calibration, which can influence the peak discharge value, it is noted that the aforementioned simulations provide almost all isolated rainfall peaks, producing a rather impulsive response of the catchment. On the other hand, the observed hyetograph (Figure 2c), though relative to a site slightly out of the catchment, highlights five consecutive hours (from 0200 to 0700 UTC) with an average intensity over 30 mm·h^{-1}, suggesting that no simulation can correctly reproduce the persistence of the precipitation event over the Corigliano area. Furthermore, even with comparable hyetographs, the spatial distribution of the simulated precipitation within the catchment area should also have been carefully evaluated, given the sharp differences in land use between the upper (mainly rural) and lower (urbanized) areas of the catchment, entailing rather different hydrological responses.

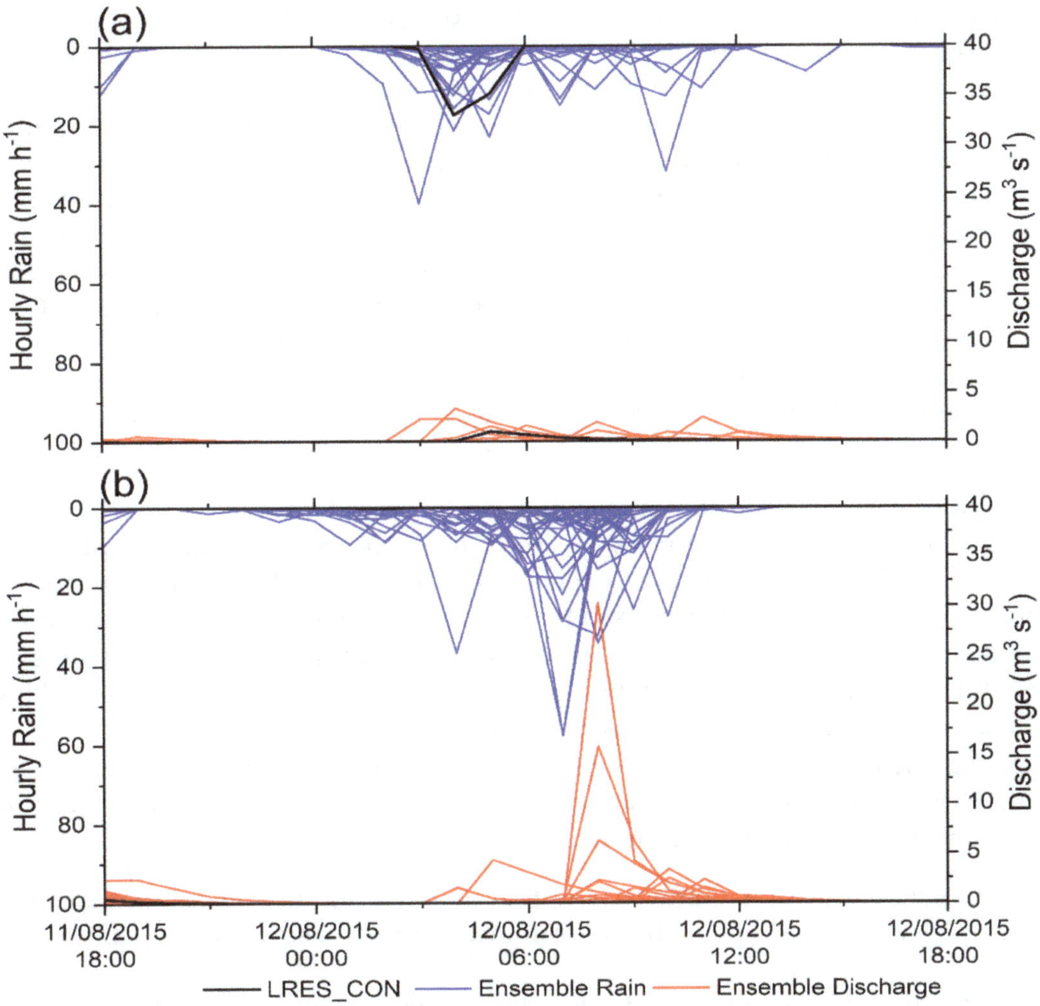

Figure 8. Hourly accumulated rainfall averaged over the Citrea creek catchment with the ensemble low-resolution control forecast and the 50 perturbed ensemble forecasts, and resulting hydrographs: (**a**) forecasts starting at 0000 UTC, 11 August 2015; (**b**) forecasts starting at 1200 UTC, 11 August 2015.

Overall, none of the members of the ensembles seems able to contribute to completely reliable hydrological forecasts at the resolution of the small catchment analyzed, even though many of them can warn about the probability of a significant hydrological impact considering a wider area, and some of them can even forecast a significant flow increase for the analyzed 11.4 km^2 wide catchment.

4. Discussion and Conclusions

The objective of a quantitatively accurate assessment of the hydrometeorological impact at the resolution of the small and very small coastal Mediterranean catchments is still a too demanding challenge for the current level of development of forecasting techniques. This awareness leads the Italian National Civil Protection agency to a strategy based on the definition of relatively wide warning areas, whose extents are indicative of the maximum resolution to go for assumed by operational agencies, to have reliable forecasts in an operating context. Specifically, the Calabria region is divided into eight warning areas almost 2000 km^2 wide on average, and the Corigliano area falls in that to the north-east.

The analyzed event has specific features that make its accurate forecast particularly difficult: Highly convective dynamics, very localized impact on the ground both in time and (mainly) in space, and a morphologically complex area with complex air–sea interactions. Based on this test case, we argue that, within the current operational framework based on warning areas of the order of 10^3 km^2, the dynamical downscaling approach based on an ensemble forecast strategy can provide further useful information for the correct detection of the hydrological effects even for such particular events. Most of the ensemble members amplify and localize the larger-scale signal provided by the GCM, supplying indications about the spots, within the warning area, where the impact on the ground is most probable. The output of the ensemble forecasts can be communicated in a relatively straightforward way by employing percentile maps, such as those shown in Figures 3d and 4d, which, together with the quantitative information, also provide a measure of the forecast uncertainty. Furthermore, in the analyzed test case, the ensemble forecast starting closer to the occurrence time of the event (i.e., the simulation starting at 1200 UTC with a lower lead time) provides more accurate predictions of rainfall intensity and, partially, of its location (Figure 5b), reaching in few but significant cases also an accurate hydrological forecast in the small catchment under analysis (Figure 8b).

Regarding the main aim of this study, i.e., the assessment of the benefit of an ensemble approach on the predictability of a highly localized convective event (and related hydrological impact) in the Mediterranean area, the analyzed test case leads to some generalizable conclusions, i.e.,:

1. In orographically complex areas, prone to high-impact very localized weather events, such as it is the case for most of the mountainous Mediterranean coasts, an ensemble approach should be preferred to single-based simulations, even if at the cost of a higher calculation burden, given its capability of improving forecast skills in terms of both rainfall intensity and location;
2. The higher information content offered by an ensemble system can be managed through percentile maps, which facilitate the interpretation of the forecast uncertainty in space and highlight the sub-zones, within the warning areas, most likely subject to risk; and
3. Such management of the forecast uncertainty can be very useful for operational purposes, being capable, in principle, to support civil protection actions that, though activated in the whole warning area, can start to prepare more targeted actions for specific sub-zones.

Furthermore, this case study demonstrates that by reducing the forecast horizon, the ensemble forecast can be very accurate up to the hydrological impact also on a coastal catchment about 10 km^2 wide.

The experiment performed gives hints also about the limits of the proposed approach. From an operational point of view, it is noteworthy that, according to the results of the ensemble simulation, the Corigliano area is highlighted as a possible hotspot of the event together with a couple of other well-localized areas, where, however, false alarms would have been issued. Furthermore, the local morphology, entailing a subdivision of the territory into many small basins, does not allow in any case to attribute a completely reliable rainfall rate (and the corresponding hydrological response) to the catchment that was most affected. From a scientific point of view, instead, it is interesting that the accuracy of the forecast is not enhanced by the higher-resolution (16 km vs. 32 km) boundary conditions provided by the high-resolution ECMWF IFS simulation, such as Figure 6 highlights. This result has to be connected to the small-scale and intrinsically chaotic properties of the convective event, which can hide possible positive effects of improved initial and/or boundary conditions [33] and whose predictability can be addressed only in terms of the average behavior [60]. Convective storms forecasting is still a significantly evolving research field, which involves both theoretical and observational aspects [61].

The research presented in this paper shows an assessment strategy outlined for a specific, though exemplary, event. Many options are possible for strengthening the results achieved. Beyond taking into consideration more episodes, either convective or caused by frontal systems, which would expand the range of test cases and the related performance of the ensemble simulations, also going into details regarding the issue of predictability of a single event at different lead times is of particular interest from a civil protection point of view, given the importance of the time available to implement defense measures. Furthermore, the investigation on uncertainty can be expanded by considering also the issues connected to the initialization of the hydrological models, among which the soil moisture conditions are included (e.g., [17,62–64]). All these research lines, however, must deal with a higher computational burden, which could represent a strong constraint to the development of such systems in operational contexts. From this point of view, for example, the ensemble forecast starting at 1200 UTC could be useless in practice, due to the high number of simulations to perform in a very short time before the occurrence of the event. Therefore, the good results achieved up to the hydrological forecast in a very small catchment could have no positive impact in an operational modelling chain. To reduce this problem, looking for effective ensemble strategies (e.g., [65,66]) could help while waiting for the further increase of efficiency of parallel computing systems. On the other hand, complementary strategies for improving the forecast accuracy with a longer lead time can be based on both ensemble forecasts and data assimilation (e.g., [67,68]). Even the impact on the ground can be assessed through innovative approaches dealing with uncertainties, like [69], that is particularly useful for highly localized events.

Finally, while studying the potential of convection-permitting ensemble systems, the very rapid improvement of the GCMs providing the boundary conditions must be taken into account. For example, in 2016, the year after the occurrence of the analyzed event, both the ECMWF IFS and EPS resolutions increased a lot, in the framework of an overall forecasting system upgrade that provided other improvements as well. Therefore, for a state-of-the-art assessment of the predictability of complex convection-permitting hydro-meteorological (either single or probabilistic) forecasting systems, it makes sense to focus not only on the most up-to-date modelling systems but also, when possible, on recent events.

Author Contributions: Conceptualization, A.S. and G.M.; methodology and validation, L.F., G.M. and A.S.; formal analysis, L.F.; investigation, L.F.; writing—original draft preparation, A.S.; writing—review and editing, L.F., G.M. and A.S.; visualization, L.F. and A.S.; supervision, G.M. All authors have read and agreed to the published version of the manuscript.

Acknowledgments: Authors thank the Italian National Civil Protection "Centro Funzionale Centrale Rischio Meteo-idrogeologico e Idraulico" for providing radar data, and the "Centro Funzionale Multirischi" of the Calabrian Regional Agency for the Protection of the Environment (ARPACAL) for providing the observed precipitation data. This work is a contribution to the HyMeX international programme.

References

1. Pagano, T.C.; Wood, A.W.; Ramos, M.H.; Cloke, H.L.; Pappenberger, F.; Clark, M.P.; Cranston, M.; Kavetski, D.; Mathevet, T.; Sorooshian, S.; et al. Challenges of operational river forecasting. *J. Hydrometeorol.* **2014**, *15*, 1692–1707. [CrossRef]
2. Buizza, R. Chapter 2-Ensemble Forecasting and the Need for Calibration. In *Statistical Postprocessing of Ensemble Forecasts*; Vannitsem, S., Wilks, D.S., Messner, J.W., Eds.; Elsevier: Amsterdam, The Netherlands, 2018; pp. 15–48. ISBN 9780128123720.
3. Renard, B.; Kavetski, D.; Kuczera, G.; Thyer, M.; Franks, S.W. Understanding predictive uncertainty in hydrologic modelling: The challenge of identifying input and structural errors. *Water Resour. Res.* **2010**, *46*, W05521. [CrossRef]
4. Clark, P.; Roberts, N.; Lean, H.; Ballard, S.P.; Charlton-Perez, C. Convection-permitting models: A step-change in rainfall forecasting. *Meteorol. Appl.* **2016**, *23*, 165–181. [CrossRef]
5. Palmer, T. The ECMWF ensemble prediction system: Looking back (more than) 25 years and projecting forward 25 years. *Q. J. R. Meteorol. Soc.* **2019**, *145* (Suppl. 1), 12–24. [CrossRef]
6. Palmer, T.N.; Molteni, F.; Mureau, R.; Buizza, R.; Chapelet, P.; Tribbia, J. Ensemble Prediction. ECMWF Technical Memorandum No. 188. 1992. Available online: https://www.ecmwf.int/en/elibrary/11560-ensemble-prediction (accessed on 27 May 2020).
7. Evans, C.; Dyke, D.; Lericos, T. How Do Forecasters Utilize Output from a Convection-Permitting Ensemble Forecast System? Case Study of a High-Impact Precipitation Event. *Weather Forecast.* **2014**, *29*, 466–486. [CrossRef]
8. Baker, L.H.; Rudd, A.C.; Migliorini, S.; Bannister, R.N. Representation of model error in a convective-scale ensemble prediction system. *Nonlinear Process. Geophys.* **2014**, *21*, 19–39. [CrossRef]
9. Dey, S.R.; Plant, R.S.; Roberts, N.M.; Migliorini, S. Assessing spatial precipitation uncertainties in a convective-scale ensemble. *Q. J. R. Meteorol. Soc.* **2016**, *142*, 2935–2948. [CrossRef]
10. Nuissier, O.; Marsigli, C.; Vincendon, B.; Hally, A.; Bouttier, F.; Montani, A.; Paccagnella, T. Evaluation of two convection-permitting ensemble systems in the HyMeX Special Observation Period (SOP1) framework. *Q. J. R. Meteorol. Soc.* **2016**, *142*, 404–418. [CrossRef]
11. Schellander-Gorgas, T.; Wang, Y.; Meier, F.; Weidle, F.; Wittmann, C.; Kann, A. On the forecast skill of a convection-permitting ensemble. *Geosci. Model Dev.* **2017**, *10*, 35–56. [CrossRef]
12. Ma, S.; Chen, C.; He, H.; Wu, D.; Zhang, C. Assessing the Skill of Convection-Allowing Ensemble Forecasts of Precipitation by Optimization of Spatial-Temporal Neighborhoods. *Atmosphere* **2018**, *9*, 43. [CrossRef]
13. Gowan, T.M.; Steenburgh, W.J.; Schwartz, C.S. Validation of Mountain Precipitation Forecasts from the Convection-Permitting NCAR Ensemble and Operational Forecast Systems over the Western United States. *Weather Forecast.* **2018**, *33*, 739–765. [CrossRef]
14. Hally, A.; Richard, E.; Ducrocq, V. An ensemble study of HyMeX IOP6 and IOP7a: Sensitivity to physical and initial and boundary condition uncertainties. *Nat. Hazards Earth Syst. Sci.* **2014**, *14*, 1071–1084. [CrossRef]
15. Johnson, A.; Wang, X. A Study of Multiscale Initial Condition Perturbation Methods for Convection-Permitting Ensemble Forecasts. *Mon. Weather Rev.* **2016**, *144*, 2579–2604. [CrossRef]
16. Keresturi, E.; Wang, Y.; Meier, F.; Weidle, F.; Wittmann, C.; Atencia, A. Improving initial condition perturbations in a convection-permitting ensemble prediction system. *Q. J. R. Meteorol. Soc.* **2019**, *145*, 993–1012. [CrossRef]
17. Zappa, M.; Jaun, S.; Germann, U.; Walser, A.; Fundel, F. Superposition of three sources of uncertainties in operational flood forecasting chains. *Atmos. Res.* **2011**, *100*, 246–262. [CrossRef]
18. Calvetti, L.; Pereira Filho, A.J. Ensemble Hydrometeorological Forecasts Using WRF Hourly QPF and TopModel for a Middle Watershed. *Adv. Meteorol.* **2014**, *2014*, 484120. [CrossRef]
19. Golding, B.; Roberts, N.; Leoncini, G.; Mylne, K.; Swinbank, R. MOGREPS-UK Convection-Permitting Ensemble Products for Surface Water Flood Forecasting: Rationale and First Results. *J. Hydrometeors* **2016**, *17*, 1383–1406. [CrossRef]
20. Saleh, F.; Ramaswamy, V.; Georgas, N.; Blumberg, A.F.; Pullen, J. A retrospective streamflow ensemble forecast for an extreme hydrologic event: A case study of Hurricane Irene and on the Hudson River basin. *Hydrol. Earth Syst. Sci.* **2016**, *20*, 2649–2667. [CrossRef]

21. Olsson, J.; Pers, B.C.; Bengtsson, L.; Pechlivanidis, I.; Berg, P.; Körnich, H. Distance-dependent depth-duration analysis in high-resolution hydro-meteorological ensemble forecasting: A case study in Malmö City, Sweden. *Environ. Model. Softw.* **2017**, *93*, 381–397. [CrossRef]
22. Giorgi, F. Climate change hot-spots. *Geophys. Res. Lett.* **2006**, *33*, L08707. [CrossRef]
23. Scoccimarro, E.; Villarini, G.; Vichi, M.; Zampieri, M.; Fogli, P.G.; Bellucci, A.; Gualdi, S. Projected Changes in Intense Precipitation over Europe at the Daily and Subdaily Time Scales. *J. Clim.* **2015**, *28*, 6193–6203. [CrossRef]
24. Drobinski, P.; Silva, N.D.; Panthou, G.; Bastin, S.; Muller, C.; Ahrens, B.; Borga, M.; Conte, D.; Fosser, G.; Giorgi, F.; et al. Scaling precipitation extremes with temperature in the Mediterranean: Past climate assessment and projection in anthropogenic scenarios. *Clim. Dyn.* **2018**, *51*, 1237–1257. [CrossRef]
25. Vié, B.; Molinié, G.; Nuissier, O.; Vincendon, B.; Ducrocq, V.; Bouttier, F.; Richard, E. Hydro-meteorological evaluation of a convection-permitting ensemble prediction system for Mediterranean heavy precipitating events. *Nat. Hazards Earth Syst. Sci.* **2012**, *12*, 2631–2645. [CrossRef]
26. Davolio, S.; Miglietta, M.M.; Diomede, T.; Marsigli, C.; Montani, A. A flood episode in northern Italy: Multi-model and single-model mesoscale meteorological ensembles for hydrological predictions. *Hydrol. Earth Syst. Sci.* **2013**, *17*, 2107–2120. [CrossRef]
27. Hally, A.; Caumont, O.; Garrote, L.; Richard, E.; Weerts, A.; Delogu, F.; Fiori, E.; Rebora, N.; Parodi, A.; Mihalović, A.; et al. Hydrometeorological multi-model ensemble simulations of the 4 November 2011 flash flood event in Genoa, Italy, in the framework of the DRIHM project. *Nat. Hazards Earth Syst. Sci.* **2015**, *15*, 537–555. [CrossRef]
28. Corazza, M.; Sacchetti, D.; Antonelli, M.; Drofa, O. The ARPAL operational high resolution Poor Man's Ensemble, description and validation. *Atmos. Res.* **2018**, *203*, 1–15. [CrossRef]
29. Petrucci, O.; Salvati, P.; Aceto, L.; Bianchi, C.; Pasqua, A.A.; Rossi, M.; Guzzetti, F. The Vulnerability of People to Damaging Hydrogeological Events in the Calabria Region (Southern Italy). *Int. J. Environ. Res. Public Health* **2018**, *15*, 48. [CrossRef]
30. Gochis, D.J.; Yu, W.; Yates, D.N. The WRF-Hydro Model Technical Description and User's Guide, Version 3.0. NCAR Technical Document. 120 Pages. 2015. Available online: http://www.ral.ucar.edu/projects/wrf_hydro/ (accessed on 27 May 2020).
31. Skamarock, W.C.; Klemp, J.B.; Dudhia, J.; Gill, D.O.; Barker, D.M.; Duda, M.G.; Huang, X.-Y.; Wang, W.; Powers, J.G. *A Description of the Advanced Research WRF Version 3*; NCAR Tech. Note NCAR/TN-475+STR: Boulder, CO, USA, 2008; p. 113.
32. Cloke, H.L.; Pappenberger, F. Ensemble flood forecasting: A review. *J. Hydrol.* **2009**, *375*, 613–626. [CrossRef]
33. Senatore, A.; Furnari, L.; Mendicino, G. Impact of improved Sea Surface Temperature representation on the forecast of small Mediterranean catchments hydrological response to heavy precipitation. *Hydrol. Earth Syst. Sci.* **2020**, *24*, 269–291. [CrossRef]
34. Federico, S.; Bellecci, C.; Colacino, M. Numerical simulation of Crotone flood: Storm evolution. *Il Nuovo Cim.* **2003**, *26*, 357–371.
35. Federico, S.; Avolio, E.; Bellecci, C.; Lavagnini, A.; Colacino, M.; Walko, R.L. Numerical analysis of an intense rainstorm occurred in southern Italy. *Nat. Hazards Earth Syst. Sci.* **2008**, *8*, 19–35. [CrossRef]
36. Chiaravalloti, F.; Gabriele, S. Vibo Valentia flood and MSG rainfall evaluation. *Atmos. Res.* **2009**, *93*, 286–294. [CrossRef]
37. Gascón, E.; Laviola, S.; Merino, A.; Miglietta, M.M. Analysis of a localized flash-flood event over the central Mediterranean. *Atmos. Res.* **2016**, *182*, 256–268. [CrossRef]
38. Avolio, E.; Federico, S. WRF simulations for a heavy rainfall event in southern Italy: Verification and sensitivity tests. *Atmos. Res.* **2018**, *209*, 14–35. [CrossRef]
39. Avolio, E.; Cavalcanti, O.; Furnari, L.; Senatore, A.; Mendicino, G. Brief communication: Preliminary hydro-meteorological analysis of the flash flood of 20 August 2018 in Raganello Gorge, southern Italy. *Nat. Hazards Earth Syst. Sci.* **2019**, *19*, 1619–1627. [CrossRef]
40. Senatore, A.; Mendicino, G.; Knoche, H.R.; Kunstmann, H. Sensitivity of Modeled Precipitation to Sea Surface Temperature in Regions with Complex Topography and Coastlines: A Case Study for the Mediterranean. *J. Hydrometeors* **2014**, *15*, 2370–2396. [CrossRef]
41. Chen, S.H.; Sun, W.H. A One-dimensional Time Dependent Cloud Model. *J. Meteorol. Soc. Jpn.* **2002**, *80*, 99–118. [CrossRef]

42. Mellor, G.L.; Yamada, T. Development of a turbulence closure model for geophysical fluid problems. *Rev. Geophys.* **1982**, *20*, 851–875. [CrossRef]
43. Dudhia, J. Numerical Study of Convection Observed during the Winter Monsoon Experiment Using a Mesoscale Two-Dimensional Model. *J. Atmos. Sci.* **1989**, *46*, 3077–3107. [CrossRef]
44. Mlawer, E.J.; Taubman, S.J.; Brown, P.D.; Iacono, M.J.; Clough, S.A. Radiative transfer for inhomogeneous atmospheres: RRTM, a validated correlated-k model for the longwave. *J. Geophys. Res.* **1997**, *102*, 16663–16682. [CrossRef]
45. Tewari, M.; Chen, F.; Wang, W.; Dudhia, J.; LeMone, M.A.; Mitchell, K.; Ek, M.; Gayno, G.; Wegiel, J.; Cuenca, R.H. Implementation and verification of the unified NOAH land surface model in the WRF model. In Proceedings of the 20th Conference on Weather Analysis and Forecasting/16th Conference on Numerical Weather Prediction, Seattle, WA, USA, 12–16 January 2004; pp. 11–15.
46. Janjić, Z.I. The Step-Mountain Eta Coordinate Model: Further Developments of the Convection, Viscous Sublayer, and Turbulence Closure Schemes. *Mon. Weather Rev.* **1994**, *122*, 927–945. [CrossRef]
47. Kain, J.S. The Kain–Fritsch Convective Parameterization: An Update. *J. Appl. Meteorol.* **2004**, *43*, 170–181. [CrossRef]
48. ECMWF (European Centre for Medium-Range Weather Forecasts). IFS Documentation Cy41r1 Operational Implementation 12 May 2015, Part V: Ensemble Prediction System. 2015. Available online: https://www.ecmwf.int/sites/default/files/elibrary/2015/9212-part-v-ensemble-prediction-system.pdf (accessed on 27 May 2020).
49. Leutbecher, M.; Palmer, T.N. Ensemble forecasting. *J. Comput. Phys.* **2008**, *227*, 3515–3539. [CrossRef]
50. Merchant, C.J.; Filipiak, M.J.; Le Borgne, P.; Roquet, H.; Autret, E.; Piollé, J.-F.; Lavender, S. Diurnal warm-layer events in the western Mediterranean and European shelf seas. *Geophys. Res. Lett.* **2008**, *35*, L04601. [CrossRef]
51. Robinson, I.; Piolle, J.F.; Leborgne, P.; Poulter, D.; Donlon, C.; Arino, O. Widening the application of AATSR SST data to operational tasks through the Medspiration Service. *Remote Sens. Environ.* **2012**, *116*, 126–139. [CrossRef]
52. Zeng, X.; Beljaars, A. A prognostic scheme of sea surface skin temperature for modelling and data assimilation. *Geophys. Res. Lett.* **2009**, *32*, L14605.
53. Gochis, D.J.; Barlage, M.; Dugger, A.; FitzGerald, K.; Karsten, L.; McAllister, M.; McCreight, J.; Mills, J.; RafieeiNasab, A.; Read, L.; et al. *The WRF-Hydro Modeling System Technical Description, (Version 5.0)*; NCAR Technical Note: Boulder, CO, USA, 2018; p. 107. Available online: https://ral.ucar.edu/sites/default/files/public/WRF-HydroV5TechnicalDescription_update512019_0.pdf (accessed on 9 April 2020).
54. Yucel, I.; Onen, A.; Yilmaz, K.K.; Gochis, D.J. Calibration and Evaluation of a Flood Forecasting System: Utility of Numerical Weather Prediction Model, Data Assimilation and Satellite-Based Rainfall. *J. Hydrol.* **2015**, *523*, 49–66. [CrossRef]
55. Senatore, A.; Mendicino, G.; Gochis, D.J.; Yu, W.; Yates, D.N.; Kunstmann, H. Fully coupled atmosphere-hydrology simulations for the central Mediterranean: Impact of enhanced hydrological parameterization for short and long time scales. *J. Adv. Modeling Earth Syst.* **2015**, *7*, 1693–1715. [CrossRef]
56. Givati, A.; Gochis, D.; Rummler, T.; Kunstmann, H. Comparing One-Way and Two-Way Coupled Hydrometeorological Forecasting Systems for Flood Forecasting in the Mediterranean Region. *Hydrology* **2016**, *3*, 19. [CrossRef]
57. CFM (Centro Funzionale Multirischi della Calabria). Technical Report, Rapporto Speditivo di Evento Meteopluviometrico del 12 Agosto 2015 (In Italian). 2015. Available online: http://www.cfd.calabria.it//DatiVari/Pubblicazioni/rapporto%20di%20evento%2012%20agosto.pdf (accessed on 27 May 2020).
58. ISPRA. *Territorio. Processi e Trasformazioni in Italia*; ISPRA: Roma, Italy, 2018; p. 296. ISBN 978-88-448-0921-8. Available online: http://www.isprambiente.gov.it/it/pubblicazioni/rapporti/territorio.-processi-e-trasformazioni-in-italia (accessed on 27 May 2020).
59. Dee, D.P.; Uppala, S.M.; Simmons, A.J.; Berrisford, P.; Poli, P.; Kobayashi, S.; Andrae, U.; Balmaseda, M.A.; Balsamo, G.; Bauer, P.; et al. The ERA-Interim reanalysis: Configuration and performance of the data assimilation system. *Q. J. R. Meteorol. Soc.* **2011**, *137*, 553–597. [CrossRef]
60. Raymond, D.; Fuchs, Ž.; Gjorgjievska, S.; Sessions, S. Balanced dynamics and convection in the tropical troposphere. *J. Adv. Modeling Earth Syst.* **2015**, *7*, 1093–1116. [CrossRef]
61. Brooks, H.E.; Doswell, C.A., III; Zhang, X.; Chernokulsky, A.M.; Tochimoto, E.; Hanstrum, B.; de Lima Nascimento, E.; Sills, D.M.; Antonescu, B.; Barrett, B. A Century of Progress in Severe Convective Storm Research and Forecasting. *Meteorol. Monogr.* **2018**, *59*, 18.1–18.41. [CrossRef]

62. Edouard, S.; Vincendon, B.; Ducrocq, V. Ensemble-based flash-flood modelling: Taking into account hydrodynamic parameters and initial soil moisture uncertainties. *J. Hydrol.* **2018**, *560*, 480–494. [CrossRef]
63. Li, L.; Pontoppidan, M.; Sobolowski, S.; Senatore, A. The impact of initial conditions on convection-permitting simulations of a flood event over complex mountainous terrain. *Hydrol. Earth Syst. Sci.* **2020**, *24*, 771–791. [CrossRef]
64. Fersch, B.; Senatore, A.; Adler, B.; Arnault, J.; Mauder, M.; Schneider, K.; Völksch, I.; Kunstmann, H. High-resolution fully-coupled atmospheric–hydrological modeling: A cross-compartment regional water and energy cycle evaluation. *Hydrol. Earth Syst. Sci.* **2020**, *24*, 2457–2481. [CrossRef]
65. Ravazzani, G.; Amengual, A.; Ceppi, A.; Homar, V.; Romero, R.; Lombardi, G.; Mancini, M. Potentialities of ensemble strategies for flood forecasting over the Milano urban area. *J. Hydrol.* **2016**, *539*, 237–253. [CrossRef]
66. Roux, H.; Amengual, A.; Romero, R.; Bladé, E.; Sanz-Ramos, M. Evaluation of two hydrometeorological ensemble strategies for flash-flood forecasting over a catchment of the eastern Pyrenees. *Nat. Hazards Earth Syst. Sci.* **2020**, *20*, 425–450. [CrossRef]
67. Munsell, E.B.; Zhang, F. Prediction and uncertainty of Hurricane Sandy (2012) explored through a real-time cloud-permitting ensemble analysis and forecast system assimilating airborne Doppler radar observations. *J. Adv. Modeling Earth Syst.* **2014**, *6*, 38–58. [CrossRef]
68. Yussouf, N.; Dowell, D.C.; Wicker, L.J.; Knopfmeier, K.H.; Wheatley, D.M. Storm-scale data assimilation and ensemble forecasts for the 27 April 2011 severe weather outbreak in Alabama. *Mon. Weather Rev.* **2015**, *143*, 3044–3066. [CrossRef]
69. Lombardi, G.; Ceppi, A.; Ravazzani, G.; Davolio, S.; Mancini, M. From Deterministic to Probabilistic Forecasts: The 'Shift-Target' Approach in the Milan Urban Area (Northern Italy). *Geosciences* **2018**, *8*, 181. [CrossRef]

2

Hydrometeorological Drought Forecasting in Hyper-Arid Climates using Nonlinear Autoregressive Neural Networks

Abdullah A. Alsumaiei * and Mosaed S. Alrashidi

Civil Engineering Department, College of Engineering and Petroleum (COEP), Khaldiya Campus, Kuwait University, P.O. Box 5969, Safat 13060, Kuwait; mosaed.alrashidi@ku.edu.kw
* Correspondence: alsumaiei.a@ku.edu.kw

Abstract: Drought forecasting is an essential component of efficient water resource management that helps water planners mitigate the severe consequences of water shortages. This is especially important in hyper-arid climates, where drought consequences are more drastic due to the limited water resources and harsh environments. This paper presents a data-driven approach based on an artificial neural network algorithm for predicting droughts. Initially, the observed drought events in the State of Kuwait were tested for autocorrelation using the correlogram test. Due to the cyclic nature of the observed drought time series, nonlinear autoregressive neural networks (NARs) were used to predict the occurrence of drought events using the Levenberg–Marquardt algorithm to train the NAR models. This approach was tested for the forecasting of 12- and 24-month droughts using the recently developed precipitation index (*PI*). Four statistical measures were used to assess the model's performance during training and validation. The performance metrics indicated that the drought prediction was reliable, with Nash–Sutcliffe values of 0.761–0.878 during the validation period. Additionally, the computed R^2 values for model forecasts ranged between 0.784–0.883, which indicated the quality of the model predictions. These findings contribute to the development of more efficient drought forecasting tools for use by water managers in hyper-arid regions.

Keywords: drought; artificial neural network; precipitation index; hyper-arid climate; autocorrelation

---

1. Introduction

Drought events are climatic phenomena featuring substantial reductions in precipitation over a defined period. Droughts may take place in all climatic zones and generally deplete water reserves at a slow pace (typically between several months to several years), significantly impacting the moisture levels in the soil, lakes, streams, and groundwater aquifers. Moreover, droughts affect human activities, such as agriculture. The global demand for freshwater has increased enormously over the past few decades due to rapidly increasing human populations and the demands of the ever-expanding agricultural and industrial sectors. Climate change and increasing water pollution have contributed to the need for drought management, particularly as the frequency of extreme hydrological events has increased in recent times [1,2]. Additionally, drought impacts can influence both society and the economy [3,4]. Water quality can also be affected by drought, as the quantities of sediments, nutrients, and organic matter carried by surface streams can experience significant decreases as flow rates fall [5]. Therefore, quantitative drought impact assessments should be conducted to establish resilient water management plans that can secure water resources for the growing human population.

Climate change has major implications for the future of humanity and represents a significant challenge for the global population, potentially causing more extreme global events that have major socioeconomic repercussions [6–8]. Climate change has attracted increasing worldwide attention from governments and researchers, where significant efforts have been invested in identifying strategies for reversing its outcomes. Due to the global warming effect, the impact of climate change is expected to increase in the future, leading to variability in sea surface temperature (SST) anomalies. This variability in the SST will certainly lead to more intense droughts events [9,10]. The impacts of droughts on arid regions are particularly concerning, as conventional water resources are severely limited in such regions, which generally receive <250 mm per year of rainfall and whose water reserves are easily depleted. In these regions, groundwater is often the sole or major conventional water supply available, requiring efficient management (especially during water shortages). One method of addressing extreme climate change outcomes is through a risk management approach [11]. Unlike conventional methods, this takes a proactive approach that involves identifying and implementing solutions. Risk management can play a fundamental role in overcoming the problems associated with weather-related extremes and establishing systems that prevent further climate change [12].

Developing drought forecasting tools is a cornerstone component of risk management for water resource planning. In arid areas, drought forecast models help to evaluate how quickly groundwater can be recharged, providing a quantitative assessment of the anticipated impacts on water storage. Drought forecasting commonly uses data-driven stochastic methods. One of the most commonly used methods for this purpose involves generating autoregressive integrated moving average models (ARIMAs) [13–17]. These models are linear and have a limited capacity for forecasting nonlinear data. For the effective forecasting of nonlinear data, researchers have expanded the use of artificial neural networks (ANNs) with hydrological data over the past 25 years. In particular, the utility of ANNs in hydrological sciences includes forecasting drought events [18–27].

Khan et al. [28] utilized an ANN-based model to forecast hydrometeorological drought events in Malaysia's Langat River basin using wavelet transforms to preprocess raw precipitation data that were then used as the input of an ANN model to forecast droughts using the Standard Index of Annual Precipitation. Wavelet transforms effectively improved the ANN model predictions in comparison to non-processed precipitation data, leading the study to recommend the use of wavelet transforms for drought analysis purposes for more reliable results. However, the moist climatic region investigated in that study was very different from the harsh environmental conditions in arid regions, where the ability of wavelet transforms to improve ANN-based hydrological forecasts is limited [29].

Recent efforts in utilizing ANNs for operational hydrological drought forecasting were dedicated to developing reliable approaches for enhancing knowledge extraction from trained black box models. A recent study by Vidyarthi and Jian presented a novel approach for extracting operational forecasting outcomes for classifying droughts in the Indian subcontinent based on the Standardized Precipitation Index (SPI) [27]. The researchers relied on a decision tree support tool to establish simple rules for drought monitoring and forecasting. They showed that simple rules can be extracted from trained ANN models that can reliably aid policymakers in setting resilient operational decisions that can help with tolerating drought consequences. The researchers recommend that future studies should work on extensively developing ANN-based models that can be comprehensively utilized when managing water resource systems. Additionally, the researchers emphasized that the performance of their methods was excellent in the examined study area [27]. However, the global validity of ANN-based methods in other climatic regions should be thoroughly investigated before making generalized conclusions.

The cyclic nature of drought occurrences has prompted researchers to employ nonlinear autoregressive neural networks (NARs) in drought forecasting, as these combine the advantages of the ARIMA and ANN methods. For example, Belayneh et al. [30] used this approach to successfully forecast drought events in Ethiopia's Awash River basin using the SPI as a drought indicator. Despite the soundness of the SPI drought analysis capabilities, its computations require the normalization of precipitation data, including fitting to theoretical or empirical probability distributions. However, the goodness of fit is usually poor, especially for precipitation data in arid climates, where the time series include many null values, making the probabilistic distribution fitting highly problematic. Alsumaiei [31] has recently presented a precipitation index (*PI*) as a robust tool for drought analysis in arid climates and tested this *PI* in the arid State of Kuwait at 12- and 24-month time scales to assess the meteorological drought impacts on fresh groundwater reserves. That study showed that the *PI* provided similar drought assessment capabilities to the SPI but with the major advantage of not requiring precipitation data normalization, and thus providing a suitable drought analysis indicator for hyper-arid regions.

Data-driven techniques for forecasting droughts based on the utility of artificial neural networks have been shown to have powerful capabilities regarding making reliable forecasts. However, the implementation of NAR models in predicting drought occurrence in extremely harsh climatic conditions, such as those found in hot and dry areas, has not been thoroughly examined. The major objective of the current study was to bridge this gap. The study employed the *PI* to construct an NAR-based drought forecast model using historical precipitation records in Kuwait; we then validated the model when predicting 12- and 24-month droughts from 1992–2020. The correlogram test was used on the observed *PI* time series to determine the significant number of autoregressive terms in the NAR model. The results contribute to water resources risk management in arid and hyper-arid regions by developing a reliable drought forecasting tool that can provide early warnings of upcoming droughts, which can aid water managers in developing resilient water-supply plans in water-shortage scenarios.

2. Materials and Methods

2.1. Study Area

Kuwait is one of the smallest countries in the Arabian Gulf, with a total area of 17,818 km^2 and 195 km of coastline (Figure 1). The country has a hyper-arid climate and suffers from extremely dry and hot summer months (from May to August), where the temperatures range from 40–50 °C with an average annual precipitation of 115.7 mm [31,32]. During this season, increased evaporation rates lead to high water stress, while during June and July northwesterly winds can cause extreme sandstorms.

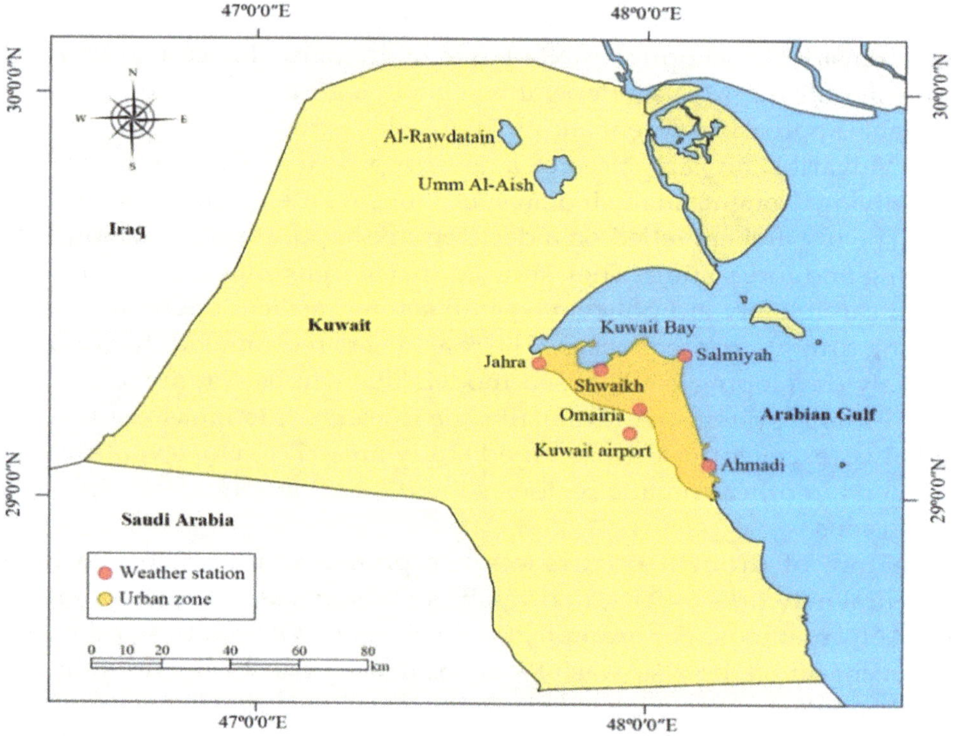

Figure 1. Locations of urban areas, weather stations, and major groundwater aquifers in Kuwait.

Like all western Arabian Gulf countries, Kuwait suffers from a lack of natural water resources due to the lack of surface water bodies, such as rivers and lakes. The availability of fresh groundwater is also limited. Natural sources of usable water exist in the Al-Rawdatain and Umm Al-Aish aquifers in northern Kuwait [33]. However, the water in the Umm Al-Aish aquifer was contaminated by an oil spill in 1990 [34,35]. Other aquifers are present in the southwestern part of the country, but the water is brackish and cannot be used without treatment. Moreover, the water demand has rapidly increased in recent years due to a high population growth rate from 1.25–8.6% [36]. Kuwait's oil reserves of >104 billion barrels mean that oil wealth currently supports the use of economically unviable water purification methods to supply its population and fertile land, including seawater desalination using distillation and reverse osmosis, along with wastewater treatment for reuse [37].

Such diminishing water sources and increasing water requirements make it necessary to accurately predict the drought status in Kuwait. Figure 2 presents annual and monthly rainfall data from 1992–2019 that were collected from the Kuwait International Airport (KIA). These are the only available meteorological records with continuous historical rainfall that can be used for drought analysis. As rainfall records from weather stations across other urban areas in Kuwait are spatially correlated [38], the KIA data can be used to analyze the drought status in Kuwait.

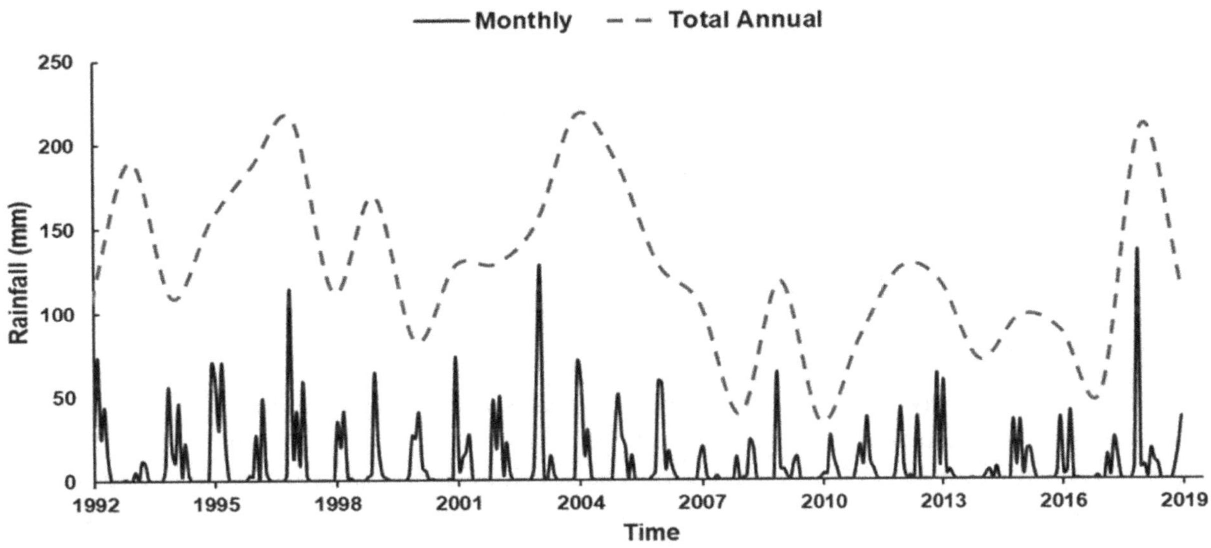

Figure 2. Total annual and monthly precipitation at the Kuwait International Airport (KIA) weather station from 1992–2019.

2.2. Drought Analysis Using the PI

The *PI* was presented by Alsumaiei [31] as a surrogate drought index for the SPI [39]. The *PI* is a meteorological drought index that resembles the error equation; it is computed using the following equation:

$$PI = \frac{P - \overline{P}}{\overline{P}}, \tag{1}$$

where P is the precipitation and $\overline{P}$ is the average precipitation. For example, extreme drought conditions ($P = 0$) produce a *PI* of −1, average precipitation produces a *PI* of 0, and twice the average precipitation produces a *PI* of 1. Table 1 lists drought categories according to their *PI* value. The definition of the original SPI was equivalent to a *Z*-score for a normal probability deviation, while the *PI* represents an evaluation of the precipitation deficit, which is divided by the long-term average in a way that relaxes the normalization requirements.

Two measures of the drought extent are important for achieving an objective drought assessment using this method. First, the cumulative magnitude of a drought is calculated using the drought severity S with PI values below zero for every consecutive occurrence:

$$S = -\sum_{i=1}^{x} PI_i, \tag{2}$$

where i is a drought occurrence and x is the total consecutive drought occurrences. In Equation (2), the negative sign is added for the sake of convenience such that S becomes positive. Although Table 1 defines a drought as occurring when $PI < -0.5$, the severity calculation is undertaken when PI is below 0. Second, the drought duration d represents the time during which severe drought takes place. Similar to severity, drought duration is calculated using PI values below 0. Droughts lasting for only one month are ignored, as they frequently occur before the start of the rainfall season due to the long zero-rainfall period during summer.

Table 1. Precipitation index (*PI*) drought classifications.

Drought Description	*PI* Value
Extremely wet	>1.0
Very wet	0.75 to 1.0
Moderately wet	0.5 to 0.75
Normal	0.5 to −0.5
Moderate drought	−0.5 to −0.75
Severe drought	−0.75 to −1.0
Extreme drought	−1.0

2.3. *Nonlinear Autoregressive Networks*

2.3.1. Network Structure

NARs are unique networks of ANNs, which were inspired by information processing in human neural systems. ANNs comprise simple processing units called nodes with specific neuron connections. During the training process, each neuron is allocated a weight to provide the best fit between the input and output data (targets). ANNs can be classified based on the direction in which the data flows [40]. Feedforward neural networks (FNNs) have a layered node arrangement in which the input layer receives data, which are then sent through hidden layers to a final output layer; the information flow is normally unidirectional. A special type of FNN is a recurrent neural network (RNN), in which the flow information is bidirectional due to the outputs being reused.

An NAR is a particular form of RNN used for describing processes based on lagged input/output variables. NARs are a combination of ANNs and autoregressive models (ARs), which are a widely employed statistical method for analyzing and modeling time series. Combining these elements in an NAR makes it capable of capturing nonlinear behavior within an autoregressive time series. Figure 3 demonstrates a conceptual illustration of an NAR.

A standard formulation for NAR models is:

$$y(t) = f\Big((y(t-1), y(t-2), \ldots, y\big(t-n_y\big)\Big), \tag{3}$$

where $y(t)$ is the target time series, n_y is the target delay term, and f is a nonlinear function, which is usually unknown. Thus, Equation (3) predicts the next time step value based on the past values according to target delay terms.

The number of hidden layers is calibrated to regress the black box function to the target time series; then, the input time series is passed through a specified number of hidden layers using a black box function. A specific algorithm is used to train the NAR network to find the optimal match between inputs and target variables.

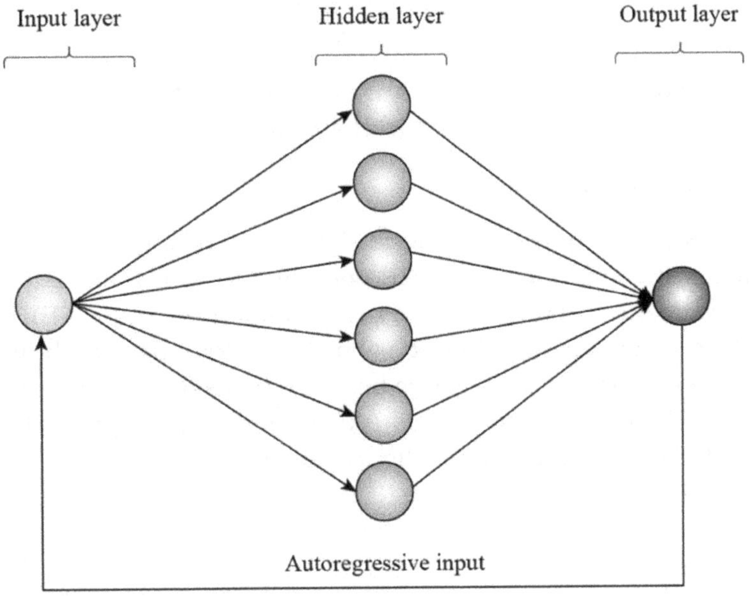

Figure 3. Conceptual illustration of a nonlinear autoregressive neural network (NAR).

2.3.2. Training Algorithms

The NAR training process employed the Levenberg–Marquardt (LM) backpropagation algorithm, which is a well-established training method that has been widely utilized in ANN-based hydrological models [41–45]. In this method, the input variable is passed through the hidden layers toward the output layer. After the error is computed, the input variable is recycled back to the input layer after being modified with a weight. This helps the LM algorithm effectively search for function minima and provides an optimized solution, making it suitable for training autoregressive networks.

2.4. Autocorrelation Analysis

A key element of autoregressive models is their employment of autocorrelation within the observed data when making forecasts, making it essential to conduct a correlogram test on the observed precipitation index time series to assess the level of correlation for adjacent values. The temporal separation period is known as the lag (τ). Initially, the observed time series of the precipitation index is standardized using Equation (4):

$$Z_j = \frac{PI_j - \mu}{\sigma}, \quad (4)$$

where *PI* is the precipitation index value at month j, μ is the precipitation index time series arithmetic average, and σ is the precipitation index time series standard deviation. The correlogram r is computed for the standardized data Z_j at lag τ according to Equation (5):

$$r_\tau = \frac{n}{n-\tau} \frac{\sum_{j=1}^{n-\tau}\left(Z_j - Z\right)\left(Z_{j+\tau} - \overline{Z}\right)}{\sum_{j=1}^{N}\left(Z_j - \overline{Z}\right)^2}, \quad (5)$$

where $\overline{Z}$ is the arithmetic mean and n is the number of values in the standardized series Z_j. The correlogram's lower and upper limits are -1 and $+1$, respectively. The confidence bounds

(CBs) are represented by the upper and lower lines drawn on the correlogram, in which the significance level is defined by the confidence lines with a significance level of $\alpha = 0.05$:

$$\mathrm{CB} = \pm \frac{z_{1-\alpha/2}}{\sqrt{n}}, \tag{6}$$

where z denotes the quartile function of the standard normal distribution. A significant autocorrelation exists within the *PI* time series if the computed correlogram exceeds the designated significance limits.

2.5. Model Validation and Statistical Evaluation

Validation of the NAR model was conducted using the chronological division of observed data into two subsets: training and validation. The NAR performance was assessed using four performance metrics: the Nash–Sutcliffe coefficient (NASH), the mean absolute error (MAE), the Pearson correlation coefficient, and the coefficient of determination (R^2). The latter demonstrates how strongly the observed and simulated targets are associated; it has a range between 0 (no statistical association) and 1 (an exact match between the simulated and observed targets). R^2 represents the square of the Pearson correlation coefficient (R); it does not directly measure prediction accuracy (as R does), but rather assesses how good a predictor can be built using the model. The R^2 for both the observed and forecasted precipitation index values were calculated using:

$$R^2 = \left(\frac{\sum_{i=1}^{n} \left(PIo_i - \overline{PIo}\right)\left(PIf_i - \overline{PIf}\right)}{\sqrt{\sum_{i=1}^{n} \left(PIo_i - \overline{PIo}\right)^2} \sqrt{\sum_{i=1}^{n} \left(PIf_i - \overline{PIf}\right)^2}} \right)^2, \tag{7}$$

where PIo_i is the observed precipitation index for time step i, PIf_i is the forecasted precipitation index for time step i, $\overline{PIo}$ is the mean for the observed precipitation index time series, $\overline{PIf}$ is the mean for the forecasted precipitation index time series, and n is the number of time steps within the validation period.

The MAE was used to quantify the variation between the forecasted targets and the observations in an objective fashion, as follows:

$$\mathrm{MAE} = \frac{1}{n} \sum_{i=1}^{n} |PIo_i - PIf_i|. \tag{8}$$

NASH is widely used within hydrological modeling [46]; it has a range from $-\infty$ to 1, with values <0 showing that the observed data arithmetic average is a superior predictor than the forecasted data. NASH was calculated as follows:

$$\mathrm{NASH} = 1 - \frac{\frac{1}{n} \sum_{i=1}^{n} \left(PIo_i - PIf_i\right)^2}{\frac{1}{m} \sum_{i=1}^{m} \left(PIo_i - \overline{PIo}\right)^2}, \tag{9}$$

where m denotes the number of time steps in the observation period.

Equation (9) ensured that the model's performance in the validation period was assessed by NASH relative to the variation from the mean for the full observation dataset. Using NASH to assess the model ensured adequate representation for the modeled system's variability at any point for NASH values > 0.

2.6. Drought Forecasting Procedures

The computational procedures for drought forecasting first computed the observed *PI* time series based on the precipitation data according to Equation (1). The computed *PI* time series was then tested for autocorrelation to determine the significant number of autocorrelation lags. The feedback delay term in Equation (3) was then selected according to the significant number of lags. The NAR model parameters (number of hidden layers and number of epochs) were then arbitrarily selected. The network parameters could be calibrated to optimize the model performance based on the statistical evaluation metrics. When acceptable statistical performance metrics were achieved, the constructed NAR model could then be used for the *PI* predictions. Figure 4 depicts the computational methodology used for the drought forecasting in this current study.

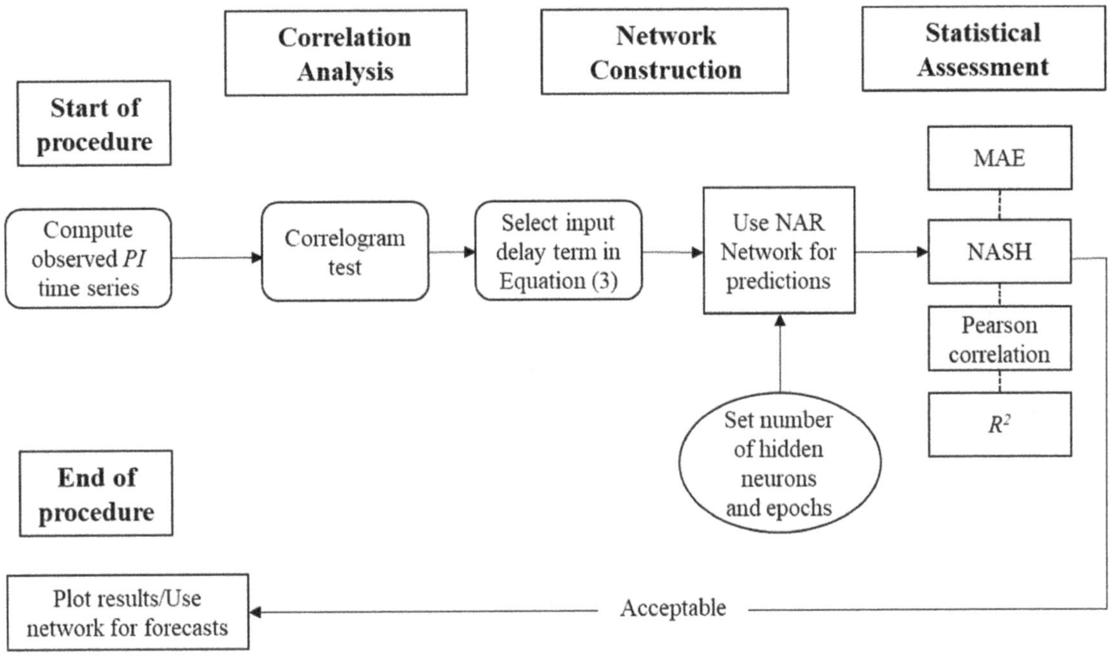

Figure 4. Flowchart illustrating the computational procedures for drought forecasting. MAE: mean absolute error, NASH: Nash–Sutcliffe coefficient.

3. Results and Discussion

3.1. NAR Drought Forecasting Model

The first step in developing autoregressive models is testing the target data for autocorrelation. Figure 5 shows the correlogram test results for the observed 12- and 24-month precipitation index time series for the study area. The figure boundaries designate the upper and lower significance boundaries, beyond which, significant autocorrelations exist. Determining the significant number of autocorrelation lags is essential for determining the number of autoregressive terms to use in Equation (3). Both time series clearly exhibited strong autocorrelations, with PI_{12} showing a significant autocorrelation up to 9 months and PI_{24} up to 35 months. Accordingly, the autoregression terms in the drought forecasting models were set to 9 and 35 months for PI_{12} and PI_{24}, respectively. The NAR model predicted the next time step drought index based on the past 9 and 35 months' values for the PI_{12} and PI_{24} time series, respectively. The relatively high number of autoregressive terms in the forecast model, especially at the 24-month scale, provided the model with an early detection capability for upcoming drought events. This feature will assist water managers in developing resilient water governance plans to cope with anticipated water-shortage scenarios.

The number of hidden layers and the number of epochs for the autoregressive neural network models were set to 10. The NARs were constructed using 10 epochs. The best NAR validation performance was achieved at epoch 4 (Figure 6). The observed drought indices' data series were calculated using the observed rainfall data at KIA station between October 1992 and April 2020 at 12- and 24-month scales. The observed drought data were used to construct the NAR models. The data were chronologically divided into subsets of 80 and 20% for training and validation purposes, respectively. Figure 7 shows the NAR modeling results for the drought indices at 12- and 24-month scales. The NAR models were found to efficiently represent the long-term variations in the drought index. The long-term trend and periodic drought behavior were well represented at the temporal scales examined.

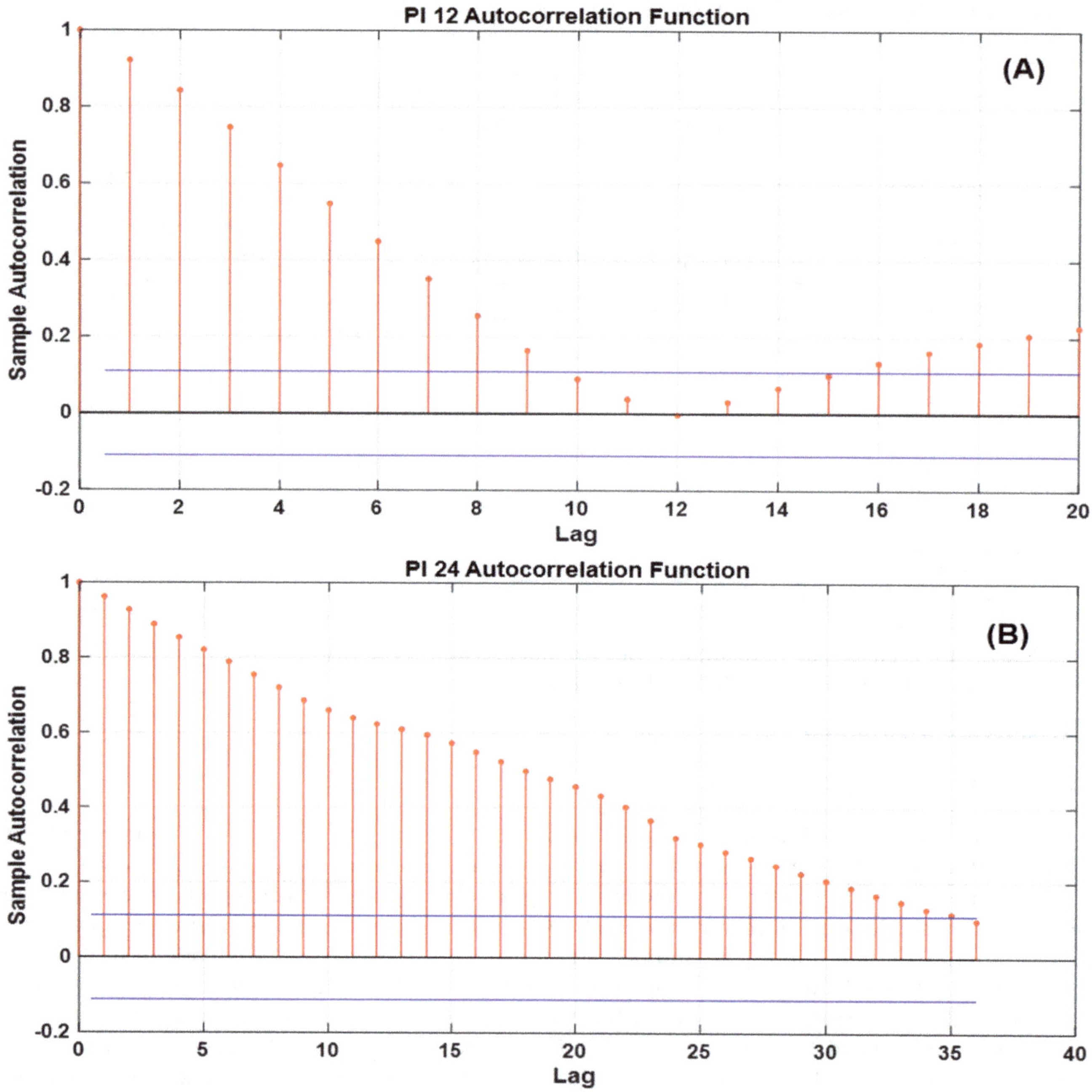

Figure 5. Correlogram test results for the (**A**) 12-month and (**B**) 24-month observed precipitation index time series for the study area.

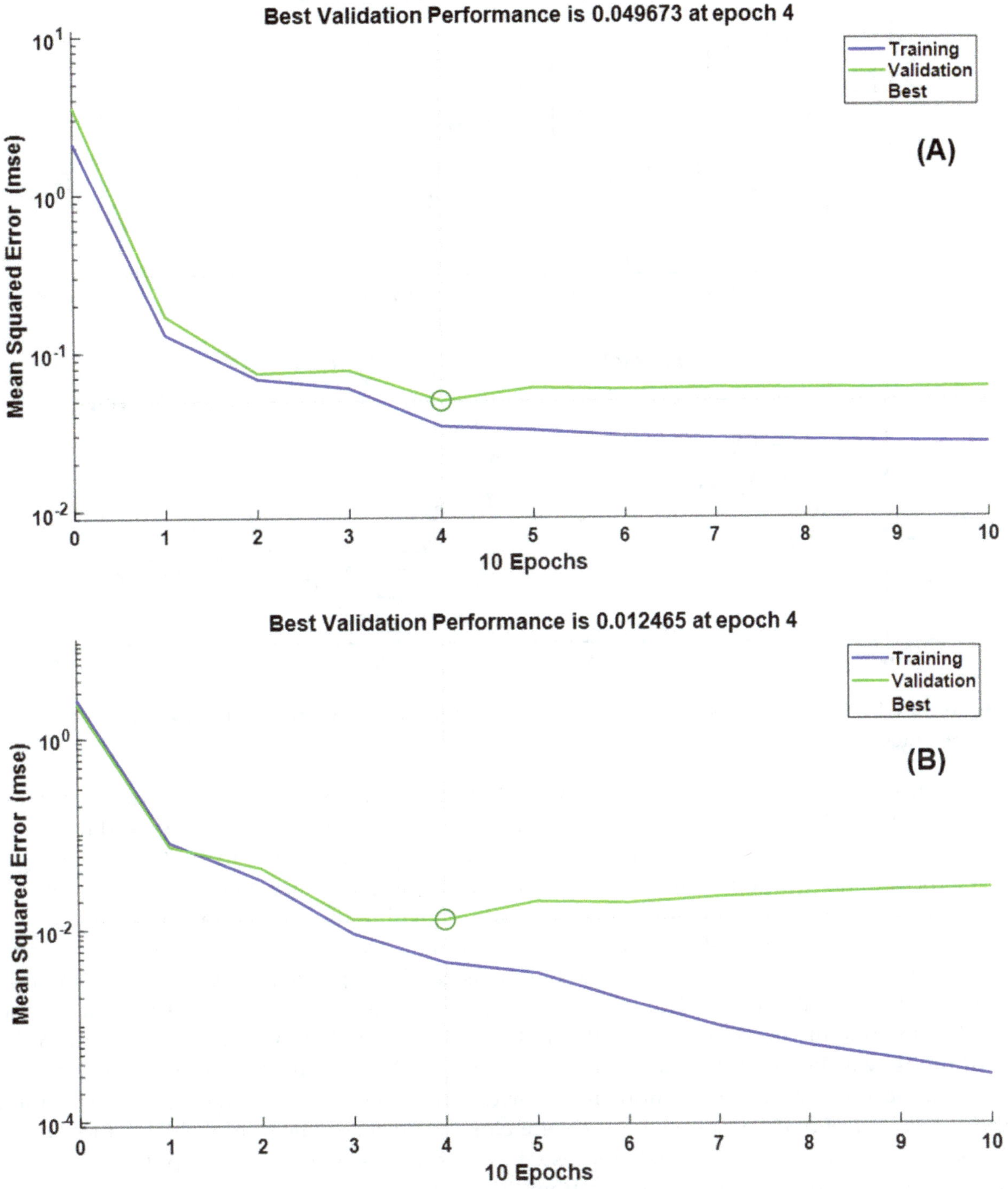

Figure 6. Variations in the mean squared error for different epochs in the training and validation periods for the (**A**) 12-month and (**B**) 24-month drought forecast models.

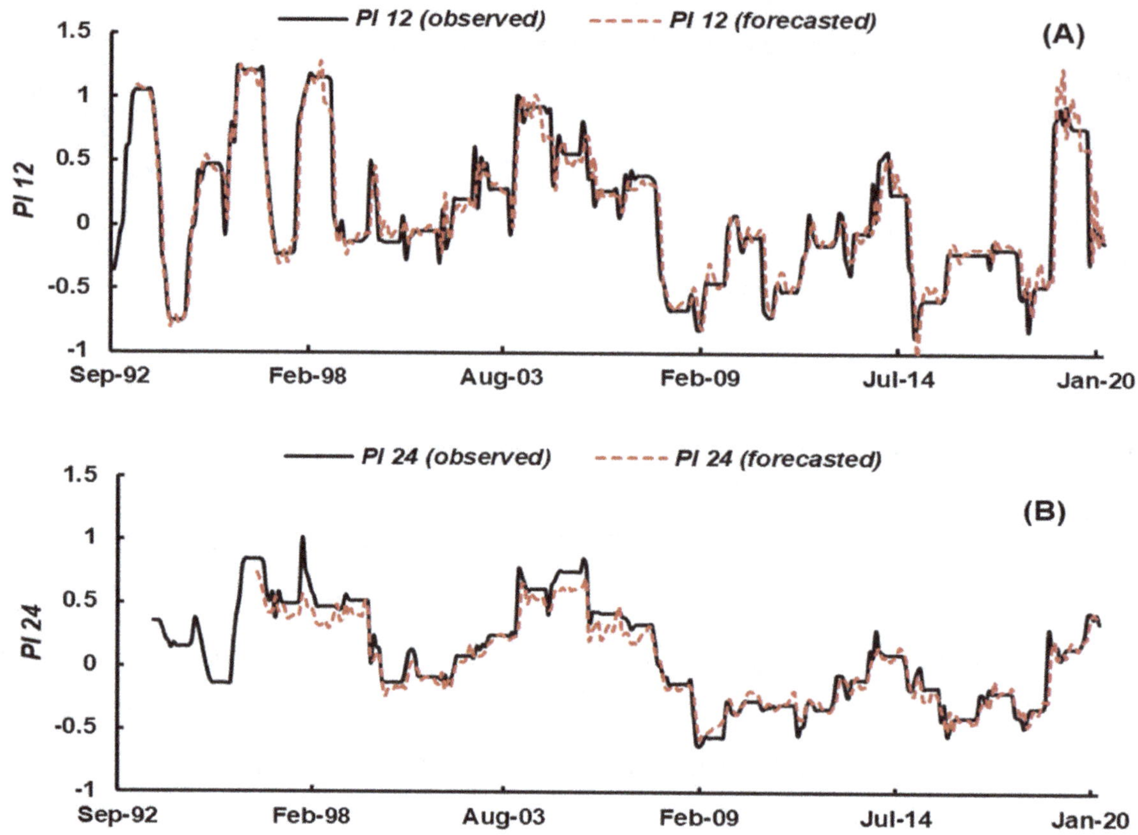

Figure 7. Observed and forecasted *PI* time series for the (**A**) 12-month and (**B**) 24-month drought forecast models.

Table 2 lists the evaluation metrics used for the NAR model's assessment. The computed statistical metrics indicated a better model performance in the training period than the validation period. This was logical as the models were optimized to match the target data during the training period. However, the reduction in the evaluation metric values during the validation period was considered minimal; therefore, the possibility of the model overfitting within the validation period can be rejected. Additionally, the NAR model efficiency when forecasting 24-month droughts was greater than the efficiency when forecasting 12-month droughts. This can be explained by considering the rainfall time series variability. For longer drought time scales (24 months in this case), the rainfall time series exhibited less variability in comparison with shorter time scales (12 months in this case). Accordingly, the NAR model was better at detecting the time series variability at the 24-month scale, resulting in better model performance. Furthermore, it is worth noting that the Pearson correlation coefficients and the coefficient of determination values were close to perfect match values. This indicated that the developed models could adequately represent the association between the input and output data. Hence, reliable predictions can be generated from the models. Additionally, the computed errors were found to be minimal, which objectively indicates the quality of the models.

Table 2. Statistical evaluation measures for the 12- and 24-month drought forecast models.

Evaluation Measure	12-Month		24-Month	
	Training	Validation	Training	Validation
NASH	0.893	0.761	0.931	0.878
Correlation	0.945	0.886	0.981	0.940
R^2	0.893	0.784	0.963	0.883
MAE	0.108	0.142	0.076	0.058

Figure 8 presents the forecasted versus observed precipitation indices at the 12- and 24-month time scales, which are plotted with the perfect match line ($y = x$ line) for comparison. The random distribution of points relative to the $y = x$ line shows that no bias in the predictions occurred, which signifies the model's reliability and that prediction biases were eliminated. As the 24-month drought index time series had less long-term variability, it also had a smoother scattering of points.

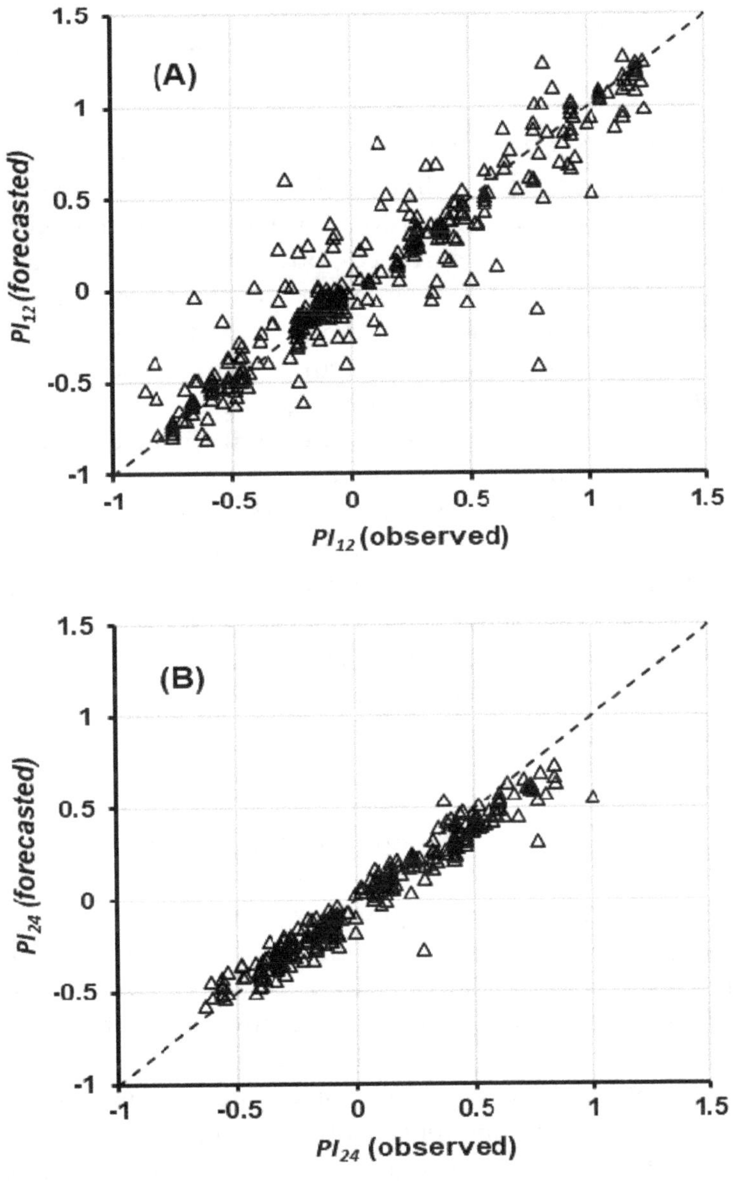

Figure 8. Forecasted vs. observed precipitation indices for the (**A**) 12-month and (**B**) 24-month scale models.

An essential feature of any drought forecasting model is its ability to robustly predict drought severities and durations. Equation (2) was used to compute the observed and forecasted drought severities at the 12- and 24-month time scales. Additionally, the drought durations were calculated for periods when the precipitation index was <0. Table 3 compares the computed and observed drought severities and durations. The NAR-based forecast models efficiently represented the drought duration and severity for the examined time scales, indicating that the NAR-based approach represents a powerful and reliable drought forecasting tool that can aid water managers when planning for anticipated droughts and possible water-shortage scenarios.

Table 3. Observed and forecasted drought severities and durations for 12- and 24-month drought time scales.

Drought Scale		Average Drought Severity	Average Drought Duration (Months)
PI_{12}	Observed	4.2	13.4
	Forecasted	4.4	15
PI_{24}	Observed	7.5	30
	Forecasted	9.2	36

3.2. Comparisons with Previous Studies

It essential to consider the NAR model's performance in the context of previous attempts of developing drought forecasting models. Almedeij [47] developed a stochastic periodic model to forecast drought occurrences in the study area using SPI. This model showed a considerable ability for forecasting long-term trends in drought occurrences, but a notable deficiency in detecting short-term droughts. The need for short-term forecasts is vital for water resource planners to develop resilient water-supply plans. A basic premise of the autoregressive model developed here is the capability to conduct short-term predictions. The results of the current study showed that the NAR model was superior to periodic models for short-term water management.

The current study's results can also be compared with previous studies that used artificial intelligence methods to forecast drought occurrences. A key feature of the current model is the relatively long lead time for the recursive neural network. The autocorrelation analysis shown in Figure 5 demonstrated significant autocorrelations (up to 9 and 35 months) in the 12- and 24-month drought time series, respectively, enabling the constructed model to provide early warnings of anticipated droughts. In comparison, Belayneh et al. [30] showed that autoregressive neural networks were efficient at forecasting the SPI drought index over 6 and 12 months in advance. The statistical assessment measures for the constructed NAR model (Table 2) were comparable with previous drought forecasting studies using ANN approaches (e.g., [28,30]).

3.3. Drought Forecasting in Hyper-Arid Climates

As the study area receives no precipitation from June to August and rainfall is exceedingly rare during May and September, the high frequency of no-rainfall months makes the probability distribution for the precipitation time series highly abnormal. Most existing drought indices require the normalization of precipitation data, making their use in arid climates problematic. In contrast, the drought forecast model developed here is based on the *PI*, which does not require fitting a probability distribution to the rainfall data. This makes it a more reliable index for drought analysis in arid climates.

As groundwater represents the only readily available water resource in the study area, developing efficient drought monitoring and forecasting tools relative to drought impacts on groundwater aquifers is highly important. A key element of the drought analysis process is the selection of an appropriate drought index time scale. The longer time scales (12 and 24 months) used in this study are more suitable for characterizing drought impacts on groundwater resources. Groundwater aquifers respond slowly to rainfall shortages [48]; therefore, shorter drought analysis scales (e.g., 1–6 months) would certainly fail to present the true drought status given Kuwait's long annual zero-rainfall period, making their usage for drought analysis meaningless.

3.4. Model Generalizability, Limitations, and Potential Improvements

The artificial-intelligence-based recursive model presented here was clearly efficient and effective at making reliable drought forecasts under the considered experimental conditions. However, in general, data-driven models applied to the analysis of hydrological processes should be done with caution. The model developed in the current study was site-specific; therefore, its potential application at other sites might require recalibration of network parameters, especially the feedback delay term, which is based on the autocorrelation analysis. Another limitation of the constructed model is its reliance on the periodic behavior of the drought time series. For example, the model does not account for other anthropogenic stresses on water resources that may accelerate the occurrence of drought events. Furthermore, the model was developed for a hyper-arid region; therefore, applying it to other climatic regimes requires consideration of other factors that could be more attributable to droughts in the target climatic region (e.g., irrigation of agricultural lands). Overall, though, the application of a data-driven approach was powerful and robust when forecasting drought events in the study area. In particular, managing water resources in such a harsh environment is crucial due to water scarcity; therefore, future research should be dedicated to exploring and enhancing data-driven approaches when monitoring and forecasting droughts in such water-limited regions.

4. Conclusions

This study investigated the application of an NAR model and the *PI* to forecasting drought events in a hyper-arid climate, using historical rainfall in Kuwait (KIA station) from 1992–2020 for the model construction and validation. The NAR methodology was both efficient and reliable at forecasting short-term drought events and clearly outperformed other periodic models developed for the same study area regarding short-term drought prediction. Moreover, the NAR model was able to efficiently represent long-term trends in observed drought events. A notable improvement in the drought forecasting ability was found for the 24-month drought forecast model in comparison with the 12-month model. This is due to less long-term variability in the 24-month drought index time series. The NAR model forecasted droughts in hyper-arid climates with similar efficiency to other ANN-based methods for drought forecasting in other climatic regions. Despite the robust performance of the presented NAR model, caution should be taken when applying it to other study sites, as the recalibration of network parameters might be needed.

Water resources in hyper-arid regions are severely limited, making the development of drought monitoring and forecasting tools vital for setting resilient water-supply plans during water-shortage periods. These tools are a cornerstone element when managing limited but readily available water resources in such regions. Future studies should work on extending these findings by further examining the validity of data-driven methods when forecasting drought events at local and regional scales. An especially appealing feature of using data-driven models for water management is the simplicity of computations using readily available meteorological data. Additionally, the use of these methods is appropriate in cases when the practical implications of predictions are more important than understanding the physical mechanisms of the hydrological processes involved.

Author Contributions: Conceptualization, A.A.A.; data curation, A.A.A. and M.S.A.; formal analysis, A.A.A.; investigation, A.A.A.; methodology, A.A.A.; project administration, A.A.A.; resources, A.A.A. and M.S.A.; software, A.A.A.; supervision, A.A.A.; validation, A.A.A.; visualization, A.A.A. and M.S.A.; writing—original draft, A.A.A.; writing—review and editing, A.A.A. and M.S.A. All authors have read and agreed to the published version of the manuscript.

References

1. Stott, P.A. How climate change affects extreme weather events. *Science* **2016**, *352*, 1517–1518. [CrossRef] [PubMed]
2. Yuan, X.-C.; Wei, Y.-M.; Wang, B.; Mi, Z. Risk management of extreme events under climate change. *J. Clean. Prod.* **2017**, *166*, 1169–1174. [CrossRef]
3. Ziolkowska, J.R. Socio-Economic Implications of Drought in the Agricultural Sector and the State Economy. *Economies* **2016**, *4*, 19. [CrossRef]

4. Kilimani, N.; Van Heerden, J.H.; Bohlmann, H.; Roos, L. Economy-wide impact of drought induced productivity losses. *Disaster Prev. Manag. Int. J.* **2018**, *27*, 636–648. [CrossRef]
5. Mosley, L.M. Drought impacts on the water quality of freshwater systems; review and integration. *Earth-Sci. Rev.* **2015**, *140*, 203–214. [CrossRef]
6. Field, C.B.; Barros, V.; Stocker, T.F.; Dahe, Q. *Managing the Risks of Extreme Events and Disasters to Advance Climate Change Adaptation: Special Report of the Intergovernmental Panel on Climate Change*; Cambridge University Press: Cambridge, UK, 2012; ISBN 1107025060.
7. Rummukainen, M. Changes in climate and weather extremes in the 21st century. *Wiley Interdiscip. Rev. Clim. Chang.* **2012**, *3*, 115–129. [CrossRef]
8. Yuan, X.-C.; Lall, U.; Mi, Z.; Wei, Y.-M.; Sun, X.; He, J. China's socioeconomic risk from extreme events in a changing climate: A hierarchical Bayesian model. *Clim. Chang.* **2016**, *139*, 169–181. [CrossRef]
9. Trenberth, K.E.; Dai, A.; Van Der Schrier, G.; Jones, P.D.; Barichivich, J.; Briffa, K.R.; Sheffield, J. Global warming and changes in drought. *Nat. Clim. Chang.* **2013**, *4*, 17–22. [CrossRef]
10. Seager, R.; Kushnir, Y.; Ting, M.; Cane, M.; Naik, N.; Miller, J.; Henderson, N. Would Advance Knowledge of 1930s SSTs Have Allowed Prediction of the Dust Bowl Drought? *J. Clim.* **2008**, *21*, 3261–3281. [CrossRef]
11. Nam, W.-H.; Choi, J.-Y.; Yoo, S.-H.; Jang, M.-W. A decision support system for agricultural drought management using risk assessment. *Paddy Water Environ.* **2012**, *10*, 197–207. [CrossRef]
12. Kunreuther, H.; Heal, G.; Allen, M.; Edenhofer, O.; Field, C.B.; Yohe, G. Risk management and climate change. *Nat. Clim. Chang.* **2013**, *3*, 447–450. [CrossRef]
13. Mishra, A.K.; Desai, V.R. Drought forecasting using stochastic models. *Stoch. Environ. Res. Risk Assess.* **2005**, *19*, 326–339. [CrossRef]
14. Sahu, R.K.; Mishra, S.K.; Eldho, T.I.; Jain, M. An advanced soil moisture accounting procedure for SCS curve number method. *Hydrol. Process.* **2007**, *21*, 2872–2881. [CrossRef]
15. Han, P.; Wang, P.X.; Zhang, S.Y.; Zhu, D.H. Drought forecasting based on the remote sensing data using ARIMA models. *Math. Comput. Model.* **2010**, *51*, 1398–1403. [CrossRef]
16. Han, P.; Wang, P.; Tian, M.; Zhang, S.; Liu, J.; Zhu, D. Application of the ARIMA Models in Drought Forecasting Using the Standardized Precipitation Index. In *Proceedings of the Integrity and Internal Control in Information Systems*; Springer Science and Business Media LLC: Berlin/Heidelberg, Germany, 2013; Volume 392, pp. 352–358.
17. Tian, M.; Wang, P.; Khan, J. Drought Forecasting with Vegetation Temperature Condition Index Using ARIMA Models in the Guanzhong Plain. *Remote Sens.* **2016**, *8*, 690. [CrossRef]
18. Mishra, A.K.; Desai, V. Drought forecasting using feed-forward recursive neural network. *Ecol. Model.* **2006**, *198*, 127–138. [CrossRef]
19. Morid, S.; Smakhtin, V.; Bagherzadeh, K. Drought forecasting using artificial neural networks and time series of drought indices. *Int. J. Clim.* **2007**, *27*, 2103–2111. [CrossRef]
20. Bacanli, U.G.; Firat, M.; Dikbas, F. Adaptive Neuro-Fuzzy Inference System for drought forecasting. *Stoch. Environ. Res. Risk Assess.* **2008**, *23*, 1143–1154. [CrossRef]
21. Barros, A.P.; Bowden, G.J. Toward long-lead operational forecasts of drought: An experimental study in the Murray-Darling River Basin. *J. Hydrol.* **2008**, *357*, 349–367. [CrossRef]
22. Cutore, P.; Di Mauro, G.; Cancelliere, A. Forecasting Palmer Index Using Neural Networks and Climatic Indexes. *J. Hydrol. Eng.* **2009**, *14*, 588–595. [CrossRef]
23. Karamouz, M.; Rasouli, K.; Nazif, S. Development of a Hybrid Index for Drought Prediction: Case Study. *J. Hydrol. Eng.* **2009**, *14*, 617–627. [CrossRef]
24. Marj, A.F.; Meijerink, A.M.J. Agricultural drought forecasting using satellite images, climate indices and artificial neural network. *Int. J. Remote Sens.* **2011**, *32*, 9707–9719. [CrossRef]
25. Mishra, S.S.; Nagarajan, R. Forecasting drought in Tel River Basin using feedforward recursive neural network. In Proceedings of the International Conference on Environmental, Biomedical and Biotechnology, Dubai, UAE, 4 August 2012; Volume 41, pp. 122–126.
26. Belayneh, A.; Adamowski, J.F.; Khalil, B. Short-term SPI drought forecasting in the Awash River Basin in Ethiopia using wavelet transforms and machine learning methods. *Sustain. Water Resour. Manag.* **2015**, *2*, 87–101. [CrossRef]
27. Vidyarthi, V.K.; Jain, A. Knowledge extraction from trained ANN drought classification model. *J. Hydrol.* **2020**, *585*, 124804. [CrossRef]

28. Khan, M.H.; Muhammad, N.S.; El-Shafie, A. Wavelet-ANN versus ANN-Based Model for Hydrometeorological Drought Forecasting. *Water* **2018**, *10*, 998. [CrossRef]
29. Qasem, S.N.; Samadianfard, S.; Kheshtgar, S.; Jarhan, S.; Kisi, O.; Shamshirband, S.; Chau, K.-W. Modeling monthly pan evaporation using wavelet support vector regression and wavelet artificial neural networks in arid and humid climates. *Eng. Appl. Comput. Fluid Mech.* **2019**, *13*, 177–187. [CrossRef]
30. Belayneh, A.; Adamowski, J.; Khalil, B.; Ozga-Zielinski, B. Long-term SPI drought forecasting in the Awash River Basin in Ethiopia using wavelet neural network and wavelet support vector regression models. *J. Hydrol.* **2014**, *508*, 418–429. [CrossRef]
31. Alsumaiei, A.A. Monitoring Hydrometeorological Droughts Using a Simplified Precipitation Index. *Climate* **2020**, *8*, 19. [CrossRef]
32. Al-Rashed, M.; Aliewi, A. Water resources sustainability in Kuwait against United Nations Sustainable Development Goals. In *Sustainability in the Gulf*; Informa UK Limited: Routledge, UK, 2017; pp. 37–56.
33. Alhumoud, J.M.; Al-Ruwaih, F.M.; Al-Dhafeeri, Z.M. Groundwater quality analysis of limestone aquifer of Al-Sulaibiya field, Kuwait. *Desalination* **2010**, *254*, 58–67. [CrossRef]
34. Mukhopadhyay, A.; Al-Awadi, E.; Quinn, M.; Akber, A.; Al-Senafy, M.; Rashid, T. Ground Water Contamination in Kuwait Resulting from the 1991 Gulf War: A Preliminary Assessment. *Ground Water Monit. Remediat.* **2008**, *28*, 81–93. [CrossRef]
35. Al-Sulaimi, J.; Viswanathan, M.N.; Székely, F. Effect of oil pollution on fresh groundwater in Kuwait. *Environ. Earth Sci.* **1993**, *22*, 246–256. [CrossRef]
36. Kozłowski, E.; Kowalska, B.; Kowalski, D.; Mazurkiewicz, D. Water demand forecasting by trend and harmonic analysis. *Arch. Civ. Mech. Eng.* **2018**, *18*, 140–148. [CrossRef]
37. Alhumoud, J.M.; Behbehani, H.S.; Abdullah, T.H. Wastewater Reuse Practices in Kuwait. *Environmentalist* **2003**, *23*, 117–126. [CrossRef]
38. Almedeij, J. Modeling Rainfall Variability over Urban Areas: A Case Study for Kuwait. *Sci. World J.* **2012**, *2012*, 1–8. [CrossRef]
39. McKee, T.B.; Doesken, N.J.; Kleist, J. The relationship of drought frequency and duration to time scales. In *Proceedings of the 8th Conference on Applied Climatology*; American Meteorological Society: Boston, MA, USA, 1993; Volume 17, pp. 179–183.
40. ASCE Task Committee on Application of Artificial Neural Networks in Hydrology. Artificial neural networks in hydrology. I: Preliminary concepts. *J. Hydrol. Eng.* **2000**, *5*, 115–123.
41. Nourani, V.; Mogaddam, A.A.; Nadiri, A.O. An ANN-based model for spatiotemporal groundwater level forecasting. *Hydrol. Process.* **2008**, *22*, 5054–5066. [CrossRef]
42. Kisi, O. Streamflow Forecasting Using Different Artificial Neural Network Algorithms. *J. Hydrol. Eng.* **2007**, *12*, 532–539. [CrossRef]
43. Adeloye, A.J.; De Munari, A. Artificial neural network based generalized storage–yield–reliability models using the Levenberg–Marquardt algorithm. *J. Hydrol.* **2006**, *326*, 215–230. [CrossRef]
44. Alsumaiei, A.A. A Nonlinear Autoregressive Modeling Approach for Forecasting Groundwater Level Fluctuation in Urban Aquifers. *Water* **2020**, *12*, 820. [CrossRef]
45. Alsumaiei, A.A. Utility of Artificial Neural Networks in Modeling Pan Evaporation in Hyper-Arid Climates. *Water* **2020**, *12*, 1508. [CrossRef]
46. Nash, J.E.; Sutcliffe, J. V River flow forecasting through conceptual models part I—A discussion of principles. *J. Hydrol.* **1970**, *10*, 282–290. [CrossRef]
47. Almedeij, J. Long-term periodic drought modeling. *Stoch. Environ. Res. Risk Assess.* **2015**, *30*, 901–910. [CrossRef]
48. AlRashidi, M.S.; Bailey, R.T. Estimating groundwater recharge for a freshwater lens in an arid region: Formative and stability assessment. *Hydrol. Process.* **2019**, *34*, 1063–1080. [CrossRef]

3

Innovative Variance Corrected Sen's Trend Test on Persistent Hydrometeorological Data

Wenpeng Wang [1,2,*], Yuelong Zhu [1], Bo Liu [3], Yuanfang Chen [3] and Xu Zhao [4]

1 College of Computer and Information, Hohai University, Nanjing 210098, China; ylzhu@hhu.edu.cn
2 State Key Laboratory of Hydrology-Water Resources and Hydraulic Engineering, Nanjing 210098, China
3 College of Hydrology and Water Resources, Hohai University, Nanjing 210098, China; bobol3705@163.com (B.L.); chenyuanfang@hhu.edu.cn (Y.C.)
4 College of Environment, Hohai University, Nanjing 210098, China; xuzhao@hhu.edu.cn
* Correspondence: wangwenpeng@hhu.edu.cn

Abstract: Trend detection in observations helps one to identify anthropogenic forces on natural hydrological and climatic systems. Hydrometeorological data are often persistent over time that deviates from the assumption of independence used by many statistical methods. A recently proposed Sen's trend test claimed to be free of this problem and thereby received widespread attention. However, both theoretical derivation and stochastic simulation of the current study implies that persistence inflates the trend significance, leading to false trends. To tackle this problem, we incorporate the feature of persistence into the variance of the trend test statistic, whereby an innovative variance-corrected Sen's trend test is developed. Two theoretical variances of the trend test statistic are newly derived to account for short-term and long-term persistent behavior. The original variance for independent data is also corrected because of its negative bias. A stepwise procedure, including steps to specify the underlying persistent behavior and to test trend with new statistic, is outlined for performing the new test on factual data. Variance-corrected Sen's trend test can effectively restore the inflated trend significance back to its nominal state. This study may call for the reassessment of published results of the original Sen's trend test on data with persistence.

Keywords: trend analysis; Sen's trend test; persistent behavior; Hurst phenomenon; random variability; variance correction

---

1. Introduction

Concerns are growing that hydrological systems have undergone substantial and unprecedented changes during the Anthropocene era. The IPCC's reports highlighted the potential impacts and associated risks of global warming on natural and human systems, among which freshwater-related responses and risks were elaborated [1]. In general, renewable water resources are projected to decline in many dry subtropical regions. Approximately 7% of the global population would suffer from a severe reduction in renewable water resources with a 1 °C rise in global mean temperature [2]. Global warming research has motivated the widespread application of statistical change analysis in observed time series of hydrometeorological variables [3–7]. Trend detection is one branch of change detection methods, focusing on gradual changes that have far-reaching effects on the effectiveness of current water management policies. The presence of significant trends would entail a movement toward attributing causative effects and quantitative assessment of future evolution. Both the knowledge of past and potential changes could be incorporated into designing adaptation strategies to minimize impact and manage risk.

A comprehensive trend detection study commonly involves graphical and statistical analysis on trends throughout the observed hydrometeorological series. Graphical trend analysis allows an exploratory inspection of the available data and suggests qualitative information about the trend characteristics. For example, the rescaled adjusted partial sums (RAPS) tools can visualize small but systematic trends over time that are often hidden by internal variability of the data [8,9]. Statistical trend analysis quantifies the significance of trends in order to confirm the existence (or nonexistence) of trends detected by visual inspection. Most commonly used statistical techniques are Mann–Kendall test, Spearman Rho test, and linear regression test.

Recently, an innovative trend detection method, introduced by Sen [10–12], has got graphical and statistical analyses together. The graphical part of the method, Sen's trend plot, was given the best technical note award by the American Society of Civil Engineers in 2014. Sen's trend plot offers two major merits. First, it allows one to visually identify sub-trends in low, medium, and high clusters. The information is of practical importance for engineers to avoid missing significant hidden trends in extreme events such as high floods or severe droughts. Second, Sen's trend plot can highlight trends even in short, skew distributed and persistent data. For these benefits, Sen's trend plot has been widely embraced by the hydrological community and it was broadly applied to investigate trends in several hydrometeorological observations, such as in studies on precipitation [13,14], streamflow [15,16], air temperature [17], pan evaporation [18], solar radiation [19], water quality [20], and drought indexes [21,22].

The statistical part of the method, Sen's trend test, was also claimed to be free of any assumptions, especially that of serial independence [11]. However, Sen's trend test was built based on original data directly, and hence it cannot naturally remove interference by data persistence. The effect of persistence on Sen's trend test still requires firm investigations via theoretical analysis and extensive simulation studies. This persistence, also called "serial dependence," has been recognized as an inherent component of many natural, economic, and even medical data [23]. A persistent stochastic process exhibits enhanced variability at large time scales, as evidenced by the tendency in time series for high or low values to cluster together in many groups. This behavior came to be known as Hurst phenomenon. It is a manifestation of frequent and irregular fluctuations of climatic and hydrological processes on several scales. The emergence of persistence may produce some intense local slopes, which might be falsely identified as deterministic trends but actually are only segments of large scale random fluctuations. This is the fundamental impetus for developing appropriate statistical tests to accurately discriminate between real deterministic trends and natural persistent processes.

Among several efforts to account for persistence in trend significance tests, two techniques, namely data whitening and variance correction of the test statistic, play the dominant roles. The former renders the persistent data into a serially independent counterpart, by either removing autocorrelation or adding white noise [24,25]. The latter adapts the original trend test to persistent data via explicitly incorporating the persistence into the variance of the test statistic [26,27]. The intriguing property of the variance correction approach is that the influence of persistence on the trend test is transparent to users with grounded theoretical expressions. Additionally, it avoids artificially dismantling original data that are subject to the mutual effect of trends and persistence. For these reasons, the variance correction approach has become an active option to acclimatize trend tests to real world observations exhibiting persistent behavior. However, no literature has attached the variance of Sen's trend test statistic to persistent data.

The main purpose of this study is to develop an innovative variance-corrected Sen's trend test in which exact expressions for the variance of the trend test statistic under the persistence assumption are given. Our study begins by analyzing the original expression of variance proposed by Sen [11]. Interestingly, the original variance is substantially biased even in the absence of persistence, because of an error in its derivation. The theoretical variances for both serially independent and dependent data are newly derived. Based on this finding, we suggest a stepwise procedure for the implementation of a new test using actual data.

Simulation experiments and factual data applications are then carried out to evaluate and discuss the effectiveness of variance correction on mitigating the side effects of persistence on trend detection.

2. Sen's Trend Detection Method

2.1. Sen's Trend Plot

Sen's method of trend detection proposes an innovative trend plot for the visual inspection of possible trends, including a quantitative assessment of trend slope and its statistical significance. In the preparation of the trend plot (Figure 1), the first step is to divide the given hydrometeorological time series into two equal halves and separately sort them in ascending order. The former half series is then placed on the horizontal axis and the latter half on the vertical axis of the Cartesian coordinate system. Finally, the scatter graph is rendered with the same scale on both axes and a 1:1 (45°) straight line is drawn. A monotonic upward (downward) trend clearly shows up when the scatter points wholly fall above (below) the 1:1 straight line. Nonmonotonic sub-trends are identified if the scatter points partially appear in the upper and lower triangular areas of the 1:1 straight line. If all the scatter points collect around the 1:1 straight line, the time series is nearly free of trend.

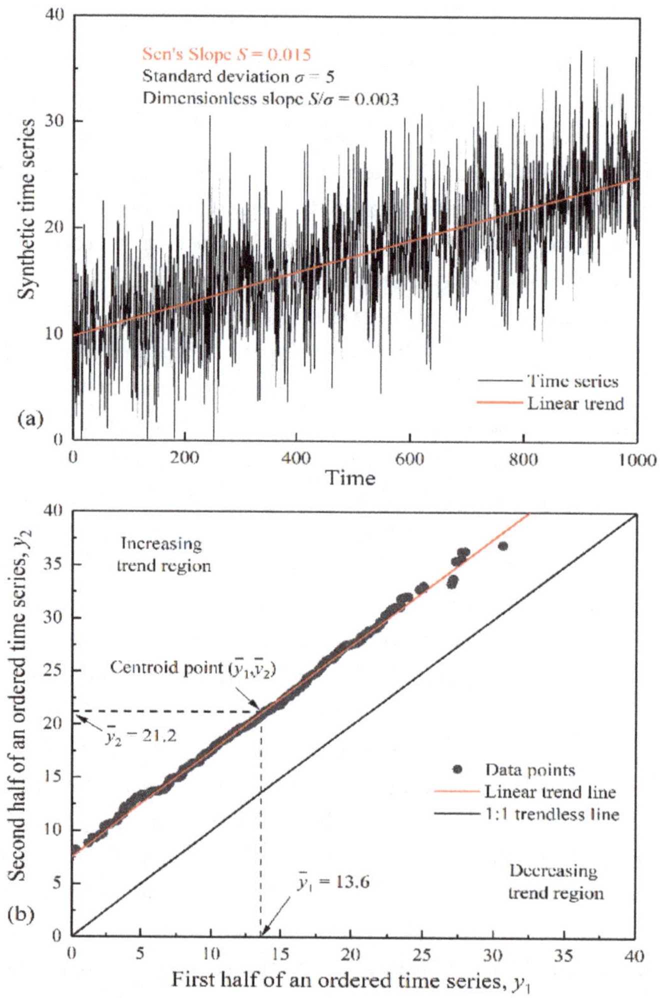

Figure 1. Illustration of linear trend identification with Sen's method: (**a**) Time series and trend; (**b**) Sen's trend plot.

A linear trend, as a special pattern of monotonic trend, is widely employed to approximate a gradual change of a hydrometeorological time series having a constant change rate. It appears along a straight line parallel to the 1:1 line in the trend plot. The vertical distance between this "linear trend line" and the 1:1 line indicates the magnitude of the trend slope. There are alternative ways to determine the slope of the linear trend. The simplest way is to find the arithmetic average of the two half sub-series, which is the centroid of the linear trend line. The slope estimate of the linear trend S is then formulated as:

$$S = \frac{2(\overline{y}_2 - \overline{y}_1)}{n} \tag{1}$$

where $\overline{y}_1$ and $\overline{y}_2$ are the means of the first and second half, respectively, and n is the total series length.

Figure 1 illustrates the application of the trend plot for a synthetic time series, which was generated according to a Normally distributed white noise process with mean $\mu = 10$, standard deviation $\sigma = 5$, and length $n = 1000$. A linear trend component with slope $S = 0.015$ was superimposed onto the simulated series. In the corresponding trend plot (Figure 1b), most data points fall on the linear trend line. Equation (1) yields $S = 0.0152$, which is very close to the embedded slope.

2.2. Sen's Trend Test

A trend significance test is constructed by considering the statistical confidence interval of the slope estimator S. If the trend slope estimator S is sufficiently far from zero, then we can deduce an existence of trends. Trend significance can be assessed by performing hypothesis tests on the slope estimator. The null hypothesis H_0 of the trend test is that the process underlying the observed series is trendless if S remains below a confidence limit at a pre-assigned significance level. Alternatively, H_0 is rejected and the detected trend is statistically significant. Under the null hypothesis, with the basic assumption that the data are independent and identically distributed, Sen [11] provided the mean and variance of the slope estimator as:

$$E(S) = \frac{2}{n}\left[E(\overline{y}_2) - E(\overline{y}_1)\right] = 0 \tag{2}$$

$$V_0(S) = \frac{8}{n^2}\sigma_{\overline{y}_1}\sigma_{\overline{y}_2}\left(1 - \rho_{\overline{y}_1\overline{y}_2}\right) = \frac{8}{n^2}\frac{\sigma^2}{n}\left(1 - \rho_{\overline{y}_1\overline{y}_2}\right) \tag{3}$$

where $E(\overline{y}_1)$ and $E(\overline{y}_2)$ are the expectations of the mean estimator of the first and second half-time series, respectively, $\sigma_{\overline{y}_1}$ and $\sigma_{\overline{y}_2}$ are the standard deviations, and $\rho_{\overline{y}_1\overline{y}_2}$ is the cross correlation, σ is the standard deviation of the entire time series.

Since all the odd-order moments of the slope estimator are equal to zero, the slope estimator approximately obeys the Normal distribution with a zero mean. For assessing the significance of the trend slope, the standardized S statistic $Z_S = S/\sqrt{V_0(S)}$ is compared with the standard Normal variates $Z_{\alpha/2}$ and $Z_{1-\alpha/2}$ at a desired significance level α, where $Z_{\alpha/2}$ and $Z_{1-\alpha/2}$ are the $\alpha/2$ and $1-\alpha/2$ quantiles of the standard Normal distribution $N(0,1)$. Significant upward (downward) trends are identified if $Z_S > Z_{1-\alpha/2}$ ($Z_S < Z_{\alpha/2}$).

Herein, we discuss the problems about the form of trend slope variance $V_0(S)$ derived by Sen [11]. In the case of the null hypothesis of no trend, the two half sub-series represent one identical population, which implies $E(\overline{y}_1) = E(\overline{y}_2)$ and hence $E(S) = 0$. Likewise, $\sigma_{\overline{y}_1} = \sigma_{\overline{y}_2} = \sqrt{2}\sigma/\sqrt{n}$ which is the standard deviation of the mean estimator of a sub-series with length $n/2$. However, Sen [11] took the form of $\sigma_{\overline{y}_1}$ and $\sigma_{\overline{y}_2}$ as $\sigma/\sqrt{n}$ by mistake. Additionally, this classical relationship $\sigma_{\overline{y}_1} = \sigma_{\overline{y}_2} = \sqrt{2}\sigma/\sqrt{n}$ is only valid for independent and identically distributed data. Sen [11] did not mention the use of assumption of independence explicitly but introduced it in the derivation of the trend test statistic unconsciously.

Another mistake is related to the fact that independent data are characterized by zero cross correlation among any two different time points, and therefore $\rho_{\overline{y}_1\overline{y}_2} = 0$. The computation of cross correlation between the two sorted half-time series, as suggested by Sen [11], is theoretically unnecessary. After correcting the two mistakes that have been explained, the variance of the slope estimator has the following expression:

$$V_0(S) = \frac{16\sigma^2}{n^3} \tag{4}$$

Figure 2 depicts the distribution histograms of the test statistic Z_S, using the original variance (Equation (3)) and the bias-corrected variance (Equation (4)), respectively, from 10,000 synthetic trendless independent data of length 100, with a Normal probability density function. The Normal distribution gives a satisfactory approximation to the distribution of Z_S. However, when using the original variance (Equation (3)), the distribution of Z_S is too wide to preserve an expectable rejection ratio that should be equal to the significance level, as illuminated by Figure 2a. Such a high rejection ratio will result in many false indications of significant trends and lead to a spurious statement that the Sen's trend test is much more powerful than other tests. After bias correction, the upper and lower tails of Z_S still align with the Normal curve and further provide a proper rejection ratio (see Figure 2b). In the following analysis, the usefulness of the original Sen's trend test was evaluated using the bias-corrected variance in Equation (4).

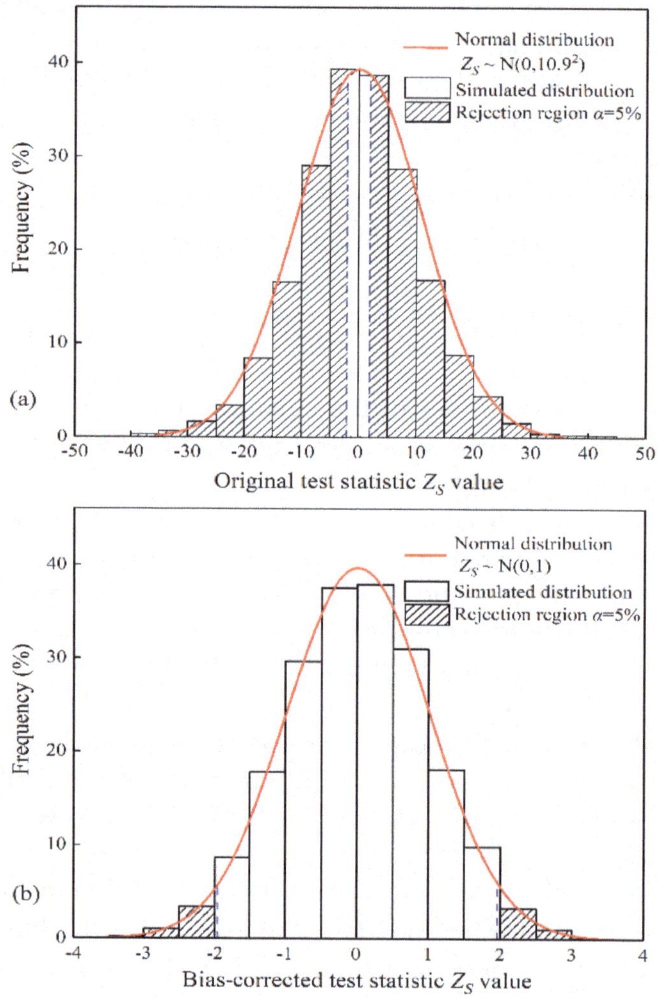

Figure 2. Comparison between the simulated distribution of the test statistic Z_S for stationary independent data of length $n = 100$ and the corresponding Normal approximation: (**a**) original test statistic (Sen, 2017a); (**b**) bias-corrected test statistic.

3. The Influence of Persistence on the Original Sen's Trend Test

3.1. Persistence in Hydrometeorological Data

The original Sen's trend test is tailored to independent data as mentioned above. Hydrometeorological data, on the other hand, are often time dependent and characterized by nonzero autocorrelation beyond the first lag. Two stochastic models that have been widely applied to describe natural processes were considered here to illustrate the effect of persistence on the original Sen's trend test. The first model is the lag-one autoregressive model, namely AR(1), which is characterized by an exponentially decaying autocorrelation function. The autocorrelation of an aggregated AR(1) process drops down to zero for large time scales; hence, AR(1) is commonly used to simulate short-term persistent processes. The autocorrelation function of an aggregated AR(1) process is given by (e.g., [28]):

$$\rho_j^{(k)} = \begin{cases} \frac{\rho(1-\rho^k)^2}{k(1-\rho^2)-2\rho(1-\rho^k)}; & |j| = 1 \\ \rho_1^{(k)}\rho^{k|j-1|}; & |j| > 1 \end{cases} \tag{5}$$

where, ρ is the lag-one autocorrelation coefficient at the basic time scale (e.g., one year), k is the time scale of interest (e.g., 5 or 10 years), and j denotes the time lag. Taking an annual series ($k = 1$) as an example, the autocorrelation becomes a quite simple and familiar expression:

$$\rho_j = \begin{cases} \rho; & |j| = 1 \\ \rho^{|j|}, & |j| > 1 \end{cases} \tag{6}$$

Likewise, the variances of an aggregated AR(1) process are also related to the time scale of aggregation (i.e., k). The variance of the process at scale k is given by:

$$\left(\sigma^2\right)^{(k)} = \sigma^2 \frac{k\left(1-\rho^2\right) - 2\rho\left(1-\rho^k\right)}{(1-\rho)^2} \tag{7}$$

where σ^2 is used to abbreviate the variance of the process at the basic time scale $k = 1$, i.e., ($\sigma^2 \equiv \left(\sigma^2\right)^{(1)}$). Also ρ denotes the lag-one correlation coefficient at $k = 1$.

For a large value of k, Koutsoyiannis [28] simplified the expression of Equation (7) as:

$$\left(\sigma^2\right)^{(k)} = k\frac{(1+\rho)}{(1-\rho)}\sigma^2 \tag{8}$$

The second model is Fractional Gaussian Noise (FGN), also known as the simple scaling stochastic process [28–32], which has a slowly-decaying autocorrelation function and is classified as a long-term persistent model. The FGN model is often more consistent with hydrometeorological reality than the classical AR(1) model, primarily because of its capability to reproduce natural random fluctuations on multiple time scales [33]. The autocorrelation and variance functions of an aggregated FGN process are given by:

$$\rho_j^{(k)} = \frac{1}{2}\left[|j+1|^{2H} - 2|j|^{2H} + |j-1|^{2H}\right] \tag{9}$$

$$\left(\sigma^2\right)^{(k)} = k^{2H}\sigma^2 \tag{10}$$

where H is the so-called "Hurst coefficient," ranging from 0 to 1.0. A stochastic process with H greater than, smaller than, or equal to 0.5, corresponds to the case of persistent, anti-persistent, or independent data.

A closer look at the two autocorrelation functions, Equations (5) and (9), reveals that the FGN autocorrelation holds a distinct property of scale invariance. For a specific time lag, the autocorrelation only depends on the Hurst coefficient and is constant for all time scales. This property implies a graphical means to differentiate between the AR(1) and FGN processes by plotting the lag-one autocorrelation coefficients of the aggregated processes versus time scales. This method of discrimination will be adopted as the first step in the practical application of the new Sen's trend test.

For ease of interpretation of the simulation results between these two persistent models in the rest of the paper, all simulated data have the same lag-one autocorrelation coefficient ρ at the basic time scale. The parameter H of the FGN model was computed in terms of ρ by setting $k = 1$ and $j = 1$ into Equation (9) and solving for H to obtain the following:

$$H = \frac{1}{2}\left[1 + \frac{\ln(1+\rho)}{\ln(2)}\right] \quad (11)$$

3.2. *Inflation of Trend Slope Variance and Type I Error*

The immediate reaction of the original trend test against persistence is to increase the variance of the slope estimator. The ratio of $V(S)/V_0(S)$ between the slope variance for persistent and independent data, referred to as the variance inflation factor [34], is widely used to indicate how much the variance has been inflated. The empirical values of $V(S)$ are estimated from synthetic persistent data here and the theoretical formula for calculating $V(S)$, denoted by $V_C(S)$, is derived in the next section. A total of 500 sets, each with 500 representations of the AR(1) and FGN time series, were generated, and subsequently $V(S)$ was estimated using these simulated time series. The theoretical values of $V_0(S)$ were computed using Equation (4). In this study, the synthetic FGN data were simulated using the function "simFGN0" in the R package "longmemo." The generated time series have Gaussian marginal distribution, which often approximates annual and over-annual hydrometeorological processes. Based on the present study, more comprehensive analyses using novel methods for the simulation of processes characterized by non-Gaussian marginal distributions, can assist in understanding the influence of marginal distribution on Sen's trend test [35,36].

Table 1 gives the variance inflation factor for AR(1) and FGN data with different series lengths n and lag-one autocorrelations ρ. The inflation factors are all larger than unity and generally grow in pace with the increases of n and ρ for both types of persistent data, as expected. However, the behaviors of the two persistent data types display some differences. For example, the inflation factor of the FGN data increases with ρ, but reaches a plateau and then decreases above a given value of n. For a given value of ρ, the FGN data are more likely to provide larger inflation factors than the AR(1) data, except at very high values of autocorrelation or in short series, where the situation is reversed.

Table 1. Variance inflation factor of Sen's slope $V(S)/V_0(S)$ calculated for different persistent data sets, given lag-one autocorrelation coefficient ρ and series length n.

n	ρ									
	0	0.1	0.2	0.3	0.4	0.5	0.6	0.7	0.8	0.9
AR(1) data										
30	1.00	1.21	1.44	1.74	2.11	2.61	3.24	4.15	5.19	5.69
50	1.00	1.21	1.46	1.79	2.20	2.77	3.54	4.73	6.63	9.21
100	1.00	1.21	1.48	1.82	2.26	2.89	3.77	5.18	7.80	13.61
150	1.00	1.22	1.49	1.83	2.29	2.92	3.83	5.36	8.16	15.45
200	1.00	1.22	1.49	1.84	2.31	2.94	3.89	5.44	8.38	16.31
FGN data										
30	1.00	1.31	1.64	1.95	2.24	2.43	2.52	2.39	1.99	1.23
50	1.00	1.41	1.87	2.36	2.85	3.28	3.55	3.53	3.07	1.97
100	1.00	1.54	2.24	3.06	4.02	4.91	5.69	6.03	5.55	3.74
150	1.00	1.63	2.51	3.57	4.90	6.25	7.47	8.21	7.77	5.45
200	1.00	1.70	2.71	3.98	5.62	7.37	9.09	10.20	9.93	7.09

As a result of slope variance inflation, the use of the original Sen's trend test on persistent data yields a much larger test statistic than for independent data, and thus indicates a significant trend when in fact no trend exists. This type of misidentification is the so-called Type I error of trend tests. The inflationary effect of data persistence on Type I error is clearly shown in Figure 3, based on 10,000 simulations. In the case of independent data when $\rho = 0$, Type I errors are equal to the chosen 5% significance level. Otherwise, Type I errors increase with ρ for AR(1) data, and increase with both n and ρ for FGN data.

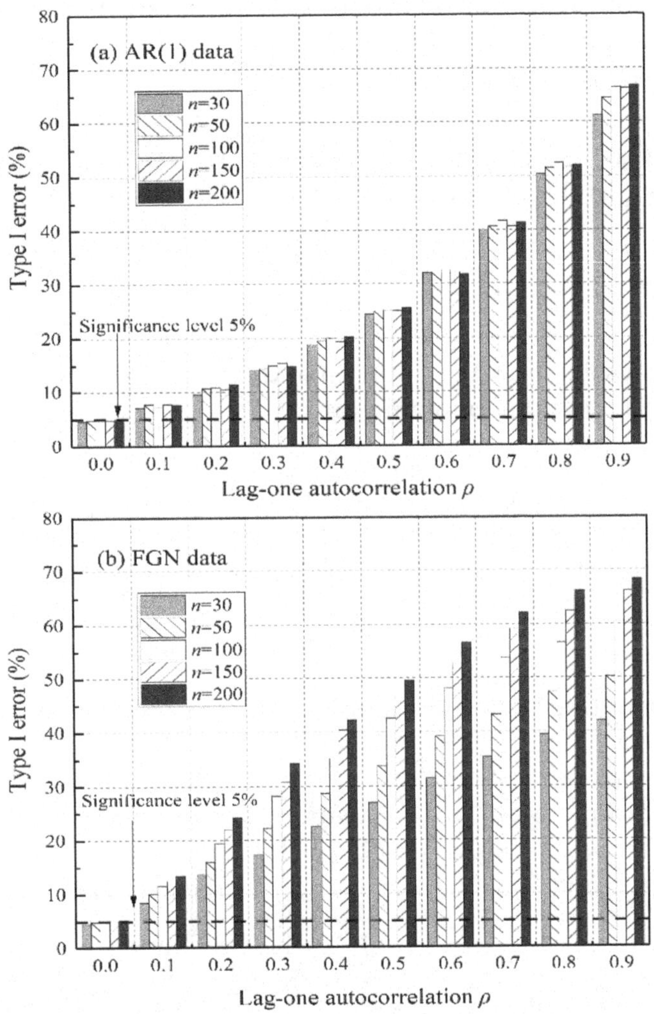

Figure 3. The effect of data persistence on Type I errors of the original Sen's trend test: (**a**) AR(1) data; (**b**) Fractional Gaussian Noise (FGN) data.

It is true that in hydrometeorological observations without considering the paleohydrological and paleoclimatic reconstructions, the level of persistence is not as high as in synthetic simulations, and the series lengths of annual observations are commonly less than 100. However, Type I errors are still inflated in many real observations. For example, the analysis of the global instrumental dataset (of length around 100 years) suggests that there are notable indications of long-term persistence in annual rainfall and runoff [37,38]. The global mean values of H are estimated close to 0.58 and 0.65 for annual rainfall and runoff, respectively. By setting and $H = 0.65$ in Equation (9), the equivalent values of ρ are 0.12 and 0.23. The use of the original Sen's trend test on FGN data of length $n = 100$, with $\rho = 0.12$ and $\rho = 0.23$, yields Type I errors of around 13% and 21% at the 5% significance level (2.6 and 4.2 times the nominal rate) from Figure 3b, which is unacceptable. As such, the adaptation of the original Sen's trend test to persistent data is necessary from both the theoretical and practical points of view.

4. Variance Corrected Sen's Trend Test

4.1. Theoretical Basis for Correcting Slope Variance

The original Sen's trend test is built on the basic assumption of serial independence. Under the alternative assumption of serial dependence, the original Sen's trend test misrepresents the significance of trends because of the inflated variance of the slope estimator. In this section, the slope variance expression is re-derived, accounting for short-term and long-term persistence by the use of variance functions of the aggregated AR(1) and FGN processes.

Revisiting the original expression of slope variance, Equation (3), the standard deviations $\sigma_{\overline{y}_1}$ and $\sigma_{\overline{y}_2}$ are no longer equal to $\sqrt{2}\sigma/\sqrt{n}$, and the cross correlation $\rho_{\overline{y}_1\overline{y}_2}$ is no longer equal to zero in persistent data. The cross correlation between the two mean estimators in the stochastic processes is given as follows:

$$\rho_{\overline{y}_1\overline{y}_2} = \frac{E(\overline{y}_1\overline{y}_2) - E(\overline{y}_1)E(\overline{y}_2)}{\sigma_{\overline{y}_1}\sigma_{\overline{y}_2}} \tag{12}$$

Substituting this expression into Equation (3) and considering stochastically that $E(\overline{y}_1) = E(\overline{y}_2)$ and $\sigma_{\overline{y}_1} = \sigma_{\overline{y}_2}$, Equation (3) takes the following form:

$$V_C(S) = \frac{8}{n^2}\left[\sigma_{\overline{y}_1}^2 + E^2(\overline{y}_1) - E(\overline{y}_1\overline{y}_2)\right] \tag{13}$$

After unfolding the expectation value $E(\overline{y}_1\overline{y}_2)$ using basic statistical and algebraic manipulations, one obtains:

$$V_C(S) = \frac{8}{n^2}\left[\sigma_{\overline{y}_1}^2 - \frac{4}{n^2}\sigma^2 \cdot R\right] \tag{14}$$

where R is a symbol denoting the summation of autocorrelation coefficients, given by $R = \sum_{i=1}^{n/2}\sum_{j=n/2+1}^{n}\rho_{(j-i)}$. The symbol $\rho_{(j-i)}$ is the $(j-i)th$ lag autocorrelation coefficient at the basic time scale. Its value is computed according to the autocorrelation function of the specified persistent model, i.e., Equation (6) for the AR(1) model or Equation (9) for the FGN model.

Notice that $\sigma_{\overline{y}_1}^2$ in Equation (14) is the variance of the mean estimator of a dependent time series with length $n/2$. It has the form of $\sigma_{\overline{y}_1}^2 = k^{-2}\left(\sigma^2\right)^{(k)}$, where the time scale k is equal to $n/2$. The specific form of the variances of the aggregated processes $\left(\sigma^2\right)^{(k)}$ depends on the choice of persistent model, i.e., Equation (8) if the data align with the AR(1) model or Equation (10) if aligned with the FGN model.

For the short-term persistent model of AR(1), the slope variance goes to its final form as:

$$V_C(S) = \frac{16}{n^3}\sigma^2\left[\frac{1+\rho}{1-\rho} - \frac{2}{n}\cdot R\right] \tag{15}$$

For the long-term persistent model of FGN, the final form of the slope variance goes to:

$$V_C(S) = \frac{16}{n^3}\sigma^2\left[\left(\frac{n}{2}\right)^{2H-1} - \frac{2}{n}\cdot R\right] \tag{16}$$

The expressions in the square brackets in Equations (15) and (16) are the theoretical formulas for correcting the original slope variance of the AR(1) and FGN processes, respectively. The values reduce to one when data are absolutely uncorrelated, in which case $H = 0.5$, $\rho = 0$, and $R = 0$. Equation (14) is a general expression for any legitimate theoretical persistent structure. If one has another type of persistent model with known autocorrelation and variance functions, it is straightforward to employ Equation (14) to estimate $V_C(S)$.

The distribution of the resulting standardized test statistic $Z_{SC} = S/\sqrt{V_C(S)}$, using the corrected slope variance, was empirically investigated. Figure 4 shows the upper tail of the simulated cumulative distribution of Z_{SC}, using 10,000 synthetic AR(1) and FGN data with length $n = 100$. Both of the parameters ρ and H were set to 0.9 to represent strong persistent cases. It is clear from Figure 4 that the tail of the standard Normal curve is in good agreement with the tails of Z_{SC} for persistent data. This agreement can be found in cases with lower levels of persistence. Thus, the use of the Normal distribution is still a reasonable approximation to assess the significance of trends in the presence of persistence.

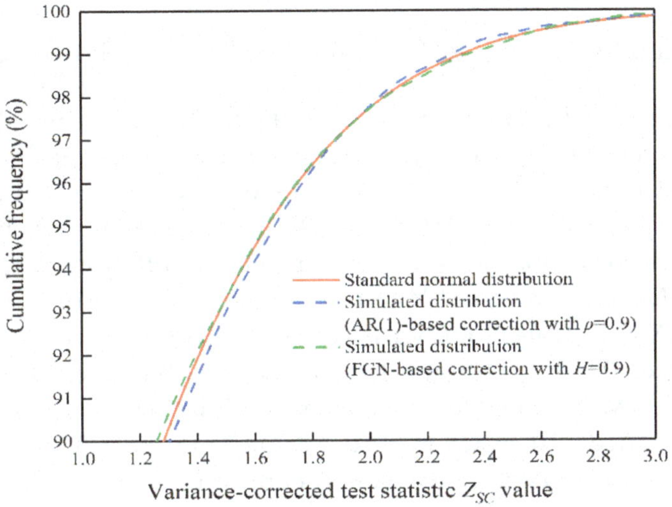

Figure 4. The upper tail of the simulated cumulative distribution of the test statistic Z_{SC} after variance correction for AR(1) and FGN data as compared to the standard Normal distribution.

4.2. Practical Procedure of the Method

Using the corrected slope variance described above, an innovative variance-corrected Sen's trend test can be performed on persistent data. The practical procedure of the method is outlined as follows:

Step 1: Persistent model specification. The empirical lag-one autocorrelation coefficients of the aggregated time series are plotted versus the time scales. If the lag-one autocorrelations are almost constant or decay slowly for large time scales, the time series is regarded as a representation of an FGN process. Otherwise, the lag-one autocorrelations will drop down to zero after a few time scales; in this case, the time series is classified as an AR(1) process.

Step 2: Persistent parameter estimation. For the AR(1) data, the empirical lag-one autocorrelation coefficient is estimated from the detrended series, and subsequently bias-corrected as recommended by Hamed [39]: $\rho = (n\hat{\rho} + 2)/(n - 4)$. For the FGN data, the Hurst coefficient H is estimated via the maximum likelihood method, which has been proven to be robust and to present low bias, as compared to several other methods [40].

Step 3: Slope variance correction. The corrected slope variance $V_C(S)$ is calculated using Equation (15) for AR(1) data or Equation (16) for FGN data, according to the results of the persistent model specification.

Step 4: Trend significance assessment. The variance-corrected test statistic Z_{SC} is compared with the quantiles of the standard Normal distribution at a desired significance level, and the trend significance is quantified.

5. Monte-Carlo Simulation

5.1. Simulation Design

To evaluate the applicability of the proposed method to persistent data, a set of Monte-Carlo simulation experiments were performed. The simulations applied a commonly assumed structure to describe hydrometeorological data, which is a combination of a stationary time series and a linear trend, written by:

$$x_t = A_t + St \tag{17}$$

where A_t is stationary data exhibiting short-term or long-term persistence, generated by the AR(1) or FGN model, and S is the slope of the linear trend with respect to time t.

Type I errors and the power of the test were estimated using the two-sided rejection ratio, given by:

$$\text{Rejection Ratio} = N_{rej}/N = \{\text{Type I error},\ S = 0; \text{Power},\ S \neq 0\} \tag{18}$$

where N is the total number of simulated series (10,000), and N_{rej} is the number of series that the null hypothesis of no trends is rejected by the test. The significance level is assigned at 5% in this study.

The simulations first generated a trendless series with various combinations of series length $n \in [30, 40, \ldots, 200]$ and lag-one autocorrelations $\rho \in [0, 0.1, \ldots, 0.9]$ to investigate how effectively the procedure of slope variance correction can alleviate Type I error inflation. After correction, the updated slope variance $V_C(S)$ replaced the original $V_0(S)$ as the denominator of the variance inflation factor, written as $V(S)/V_C(S)$. An effective slope variance correction needs to produce an inflation factor within ±20% from unity, i.e., 0.8–1.2, as suggested by Hamed [39]. According to simulations derived from the standard Normal distribution, which is the approximation of a trend test statistic distribution, the ±20% inflation in variance will result in actual Type I errors of 3.3–6.7% at the 5% level. This margin was determined because perfect correction cannot be achieved unless we can acquire the full knowledge of persistent data, including both the persistent model type and parameter, which is unrealistic in practice.

To highlight the effectiveness of slope variance correction in response to model misspecifications and parameter estimation errors, we discuss simulation results under three special cases. The first case assumes that the real persistent model and its parameter are known before correcting the slope variance. In the second case, the slope variance was corrected using the estimated value of a persistent parameter, where the real persistent model was well-specified. For the last case, attention focuses on the slope variance correction with the wrong choice of persistent model, even though its parameter is correctly identified.

The simulations also generated nonstationary series characterized by an upward linear trend to investigate the power of trend detection under varying levels of data persistence. The dimensionless slope of the linear trend S/σ ranges from 0 to 0.05. Power comparisons between Sen's trend test before and after slope variance correction, as well as the conventional Mann–Kendall test for uncorrelated data, were performed.

5.2. Slope Variance Correction and its Effectiveness on Mitigating Type I Error Inflation

The term "variance correction" means the conversion of the variance of the test statistic under the serial independence assumption into a variance that is appropriate to persistent data. Perfect correction can yield a variance inflation factor of unity, whereby Type I errors will be exactly equal to the significance level. Otherwise, large inflation factors (>1.2) result in unacceptably high Type I errors, namely "under-correction," while small inflation factors (<0.8) result in too few Type I errors, which is known as "over-correction."

Tables 2 and 3 provide the inflation factors after variance correction for AR(1) and FGN data, respectively. The inflation factors are greatly reduced for almost all combinations of n and ρ compared to the original values, as shown in Table 1. The reduction of inflation factors depends on the accuracy of the persistent model specification and parameter estimation. To be specific, nearly perfect correction can be achieved in an ideal case (Case I), where the persistent model and its parameter are exactly known beforehand. In a more practical case (Case II), where the persistent parameters ρ or H are estimated from data, the slope variance is under-corrected for high values of autocorrelation, and the situation occurs with larger autocorrelation as the series length n increases. Case III only produces a narrow band of effective variance correction because of persistent model misspecification. Correcting the slope variance of AR(1) data with an FGN model results in under-correction for high values of autocorrelation and over-correction for a long series. The situation reverses for FGN data, particularly when the slope variance is corrected using an AR(1) model.

Table 2. Variance inflation factor of Sen's slope $V(S)/V_C(S)$ after variance correction for AR(1) data. The underlined values denote that the slope variances are "over-corrected" (<0.8) or "under-corrected" (>1.2).

n	ρ									
	0	0.1	0.2	0.3	0.4	0.5	0.6	0.7	0.8	0.9
Case I: AR(1)-based correction with a known real value of ρ										
30	1.00	1.01	1.00	1.00	1.00	1.00	1.00	1.01	1.00	1.00
50	1.00	1.00	1.00	1.00	1.00	1.00	1.00	1.00	1.00	1.00
100	1.00	1.00	1.00	1.00	1.00	1.00	1.00	1.00	1.00	1.00
150	1.00	1.00	1.00	1.00	1.00	1.00	0.99	1.00	1.00	1.00
200	1.00	1.00	1.00	1.00	1.00	1.00	1.00	1.00	1.00	1.00
Case II: AR(1)-based correction with an estimated value of ρ										
30	0.91	0.91	0.89	0.88	0.89	0.90	0.97	1.22	1.76	2.89
50	0.95	0.94	0.92	0.92	0.90	0.90	0.91	0.95	1.15	2.17
100	0.97	0.96	0.96	0.96	0.95	0.95	0.93	0.91	0.89	1.19
150	0.99	0.98	0.98	0.97	0.97	0.96	0.95	0.94	0.91	0.97
200	0.99	0.99	0.99	0.98	0.98	0.96	0.97	0.96	0.93	0.91
Case III: FGN-based correction with a known real value of Hurst coefficient H										
30	1.00	0.92	0.88	0.89	0.95	1.07	1.29	1.74	2.61	4.63
50	1.00	0.86	0.78	0.76	0.77	0.84	1.00	1.34	2.16	4.67
100	1.00	0.79	0.66	0.59	0.56	0.59	0.66	0.86	1.41	3.64
150	1.00	0.75	0.60	0.51	0.47	0.47	0.51	0.66	1.05	2.84
200	1.00	0.72	0.55	0.46	0.41	0.40	0.43	0.53	0.84	2.29

Table 3. Variance inflation factor of Sen's slope $V(S)/V_C(S)$ after variance correction for FGN data.

n	ρ									
	0	0.1	0.2	0.3	0.4	0.5	0.6	0.7	0.8	0.9
Case I: FGN-based correction with a known real value of Hurst coefficient H										
30	1.00	1.00	1.00	1.00	1.00	1.00	1.00	1.00	1.00	1.00
50	1.00	1.00	1.00	1.00	1.00	1.00	1.00	1.00	1.00	1.00
100	1.00	1.00	1.00	1.00	1.00	1.00	1.00	1.01	1.01	1.00
150	1.00	1.00	1.00	1.00	1.00	1.00	1.00	1.00	1.00	1.00
200	1.00	1.00	1.00	1.00	1.00	1.00	1.00	1.00	1.00	1.00
Case II: FGN-based correction with an estimated value of H										
30	0.91	1.08	1.24	1.32	1.41	1.49	1.55	1.67	1.75	1.89
50	0.90	1.08	1.16	1.24	1.26	1.32	1.37	1.47	1.56	1.67
100	0.89	1.06	1.11	1.13	1.15	1.15	1.21	1.29	1.37	1.47
150	0.89	1.05	1.09	1.10	1.11	1.13	1.16	1.20	1.26	1.38
200	0.89	1.05	1.07	1.08	1.08	1.10	1.12	1.16	1.21	1.30
Case III: AR(1)-based correction with a known real value of ρ										
30	1.00	1.09	1.14	1.12	1.06	0.94	0.77	0.58	0.38	0.22
50	1.00	1.17	1.28	1.33	1.29	1.19	1.00	0.74	0.46	0.21
100	1.00	1.27	1.51	1.68	1.77	1.71	1.51	1.16	0.71	0.27
150	1.00	1.34	1.68	1.95	2.14	2.15	1.94	1.53	0.95	0.35
200	1.00	1.39	1.81	2.17	2.44	2.51	2.34	1.87	1.18	0.44

Figures 5 and 6 depict the results of Tables 2 and 3 using area plots, in which the range of effective slope variance correction is painted in light grey, and the over- and under- correction regions are painted in white and dark grey. In Case II (Figures 4a and 5a), there is a clear trend of increasing acceptable area as the series length increases. In Case III (Figures 4b and 5b), however, longer series do not produce wider acceptable areas. A long series tends to accentuate the discrepancy between the different models, as the data disobeys the hypothetical persistent model. As a result, the corrected test will inflate or deflate Type I errors beyond the acceptable margin of the significance level. To alleviate this weakness, the investigation of the persistent model must be performed initially, as suggested in the practical procedure. Some previous trend studies regarding data pre-whitening have also drawn attention to this serious issue [39,41]. Here, simulation results reinforce the standpoint that model specification is important for trend detection in real persistent hydrometeorological data, especially for long time series.

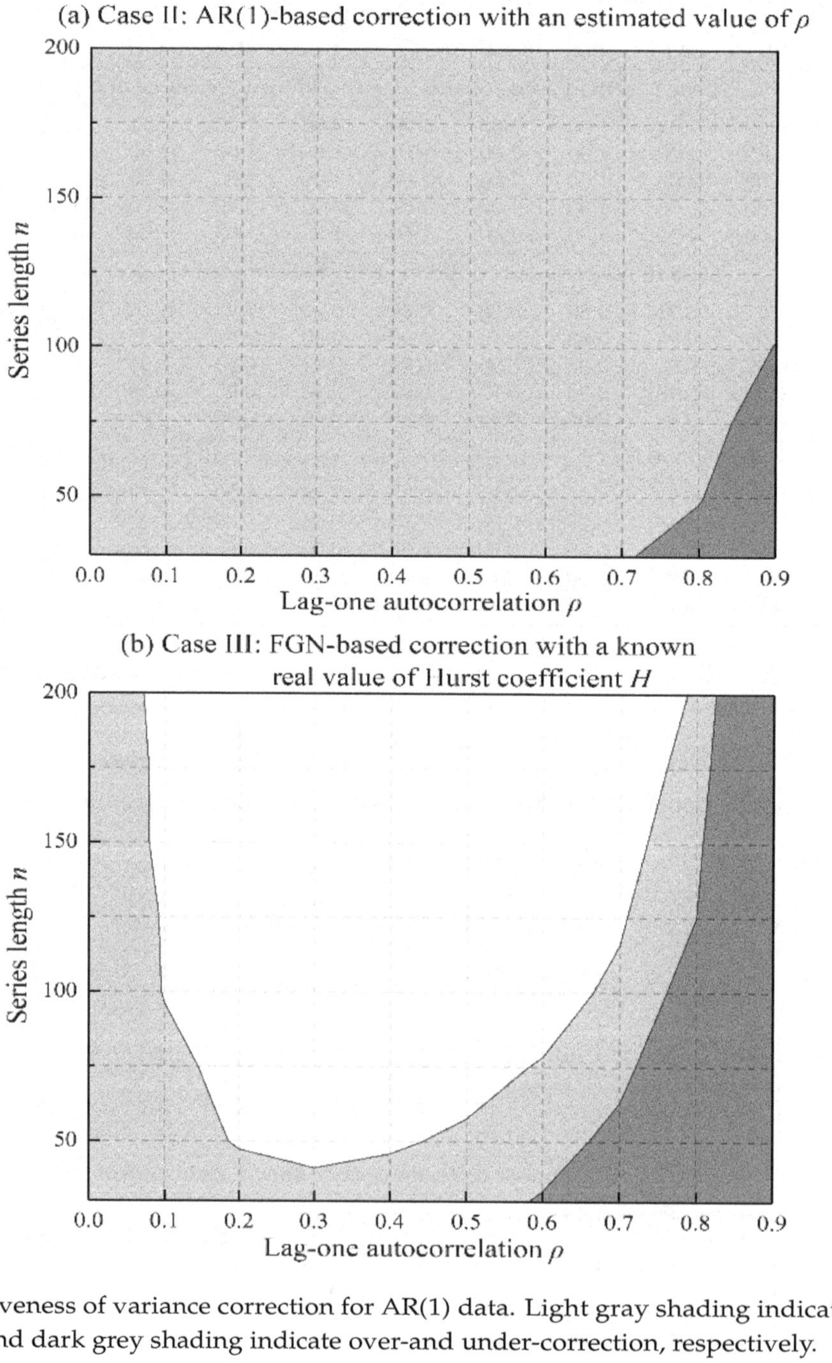

Figure 5. Effectiveness of variance correction for AR(1) data. Light gray shading indicates acceptable region. White and dark grey shading indicate over-and under-correction, respectively.

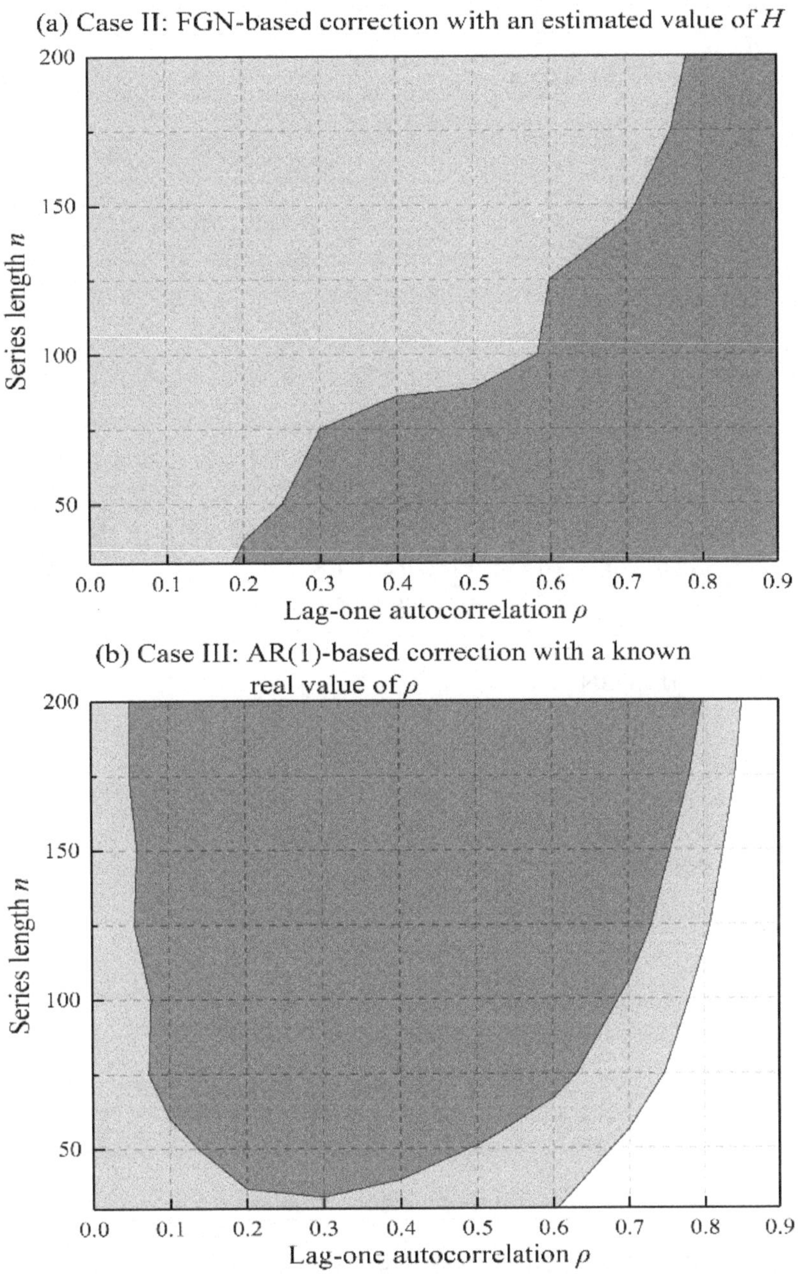

Figure 6. Effectiveness of variance correction for FGN data.

Figure 7 gives Type I errors associated with Sen's trend test using the correct persistent model. The inflated Type I errors are mostly restored to the acceptance range of the significance level, as opposed to those of the original test given by Figure 3. Considering a series with length $n = 100$ for instance, the new test apparently reduces Type I errors of the original test from the range (5–66%) to (5.2–6.7%) for AR(1) data, and from (5–61%) to (3.3–9.3%) for FGN data. Given that persistence in hydrometeorological records such as annual rainfall and runoff are usually moderate, $H \leq 0.80$ or equivalently $\rho \leq 0.52$, refer to global estimation results of H [37,38]. Thus, the new test requires at least around 50 observations of AR(1) data or 100 observations of FGN data to guarantee acceptable Type I errors. Numerous observations yield better results only if the underlying persistent structure is well specified.

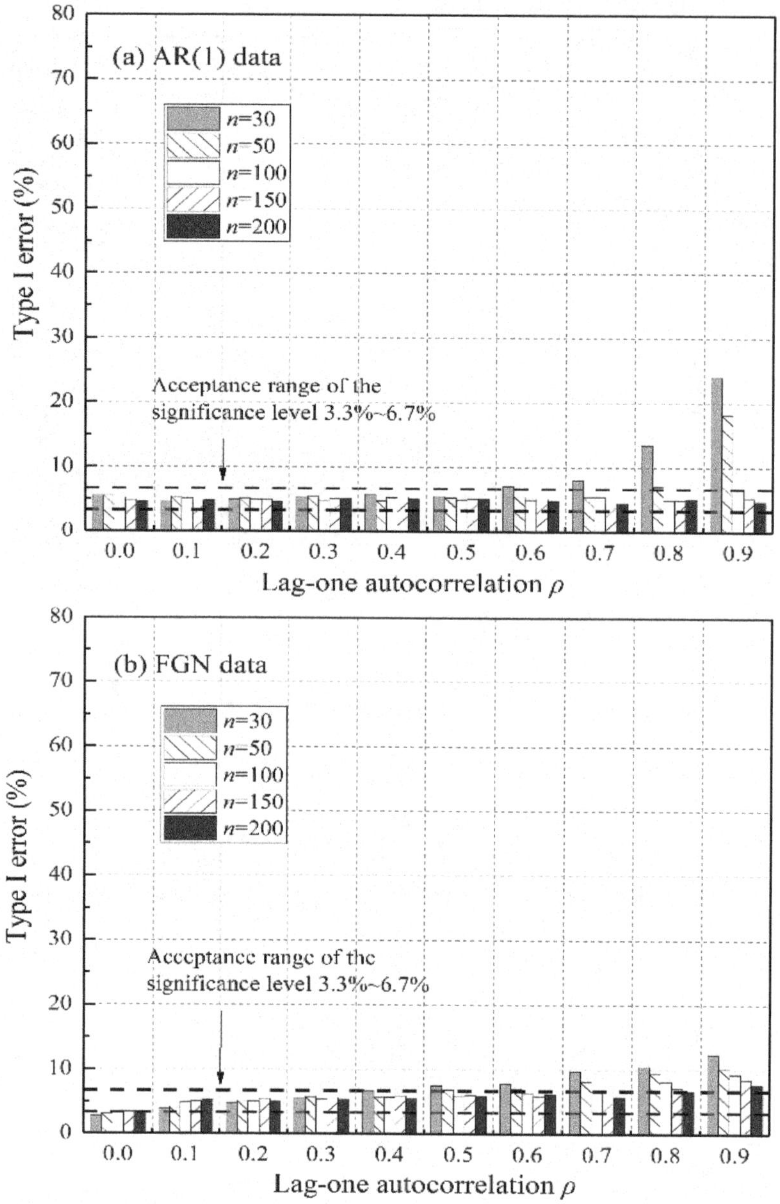

Figure 7. Type I errors of the variance-corrected Sen's trend test using a well-specified persistent model: (**a**) AR(1) data; (**b**) FGN data. The persistent parameters ρ and H are estimated from the data.

5.3. Power of Trend Detection

A test is powerful if it can detect weak trends and also maintain an acceptable Type I error. Herein, power curves are presented only in cases where the underlying persistent models are correctly specified, so that the inflated Type I errors can be restored to their normal states. The relationship among power, lag-one autocorrelation (reflecting strength of persistence), and the dimensionless slope of the linear trend for synthetic series with length 100 is depicted in Figure 8. A series length of 100 was chosen simply for illustrative purposes. An analogous pattern of power curves emerges for other series lengths.

The power of the new test is an increasing function of the dimensionless slope of the linear trend, irrespective of data persistence. In general, strong trends in data with weak variation are more likely to be detected than weak trends in highly variable data. In terms of the relationship with data persistence, the power is a monotonic decreasing function of the lag-one autocorrelation for AR(1) data. For FGN data, the power decreases to a minimum at around $\rho = 0.6$ and then rises again. Generally, the test is less powerful in detecting a trend in persistent data than in uncorrelated data. This weakness is the so-called power loss. In fact, power loss occurs because of the effect of persistence rather than the procedure of variance correction. Persistence inflates the variance of the trend slope and reduces the value of the test statistic numerically when the null hypothesis is false, thus weakening the ability of the test to detect a real trend from persistent data compared to the case when the same trend exists in uncorrelated data. The power losses depend on the extent of variance inflation. For FGN data, power restoration occurred when $\rho > 0.6$, since at that time the inflated variance gradually recovered spontaneously (see Table 1).

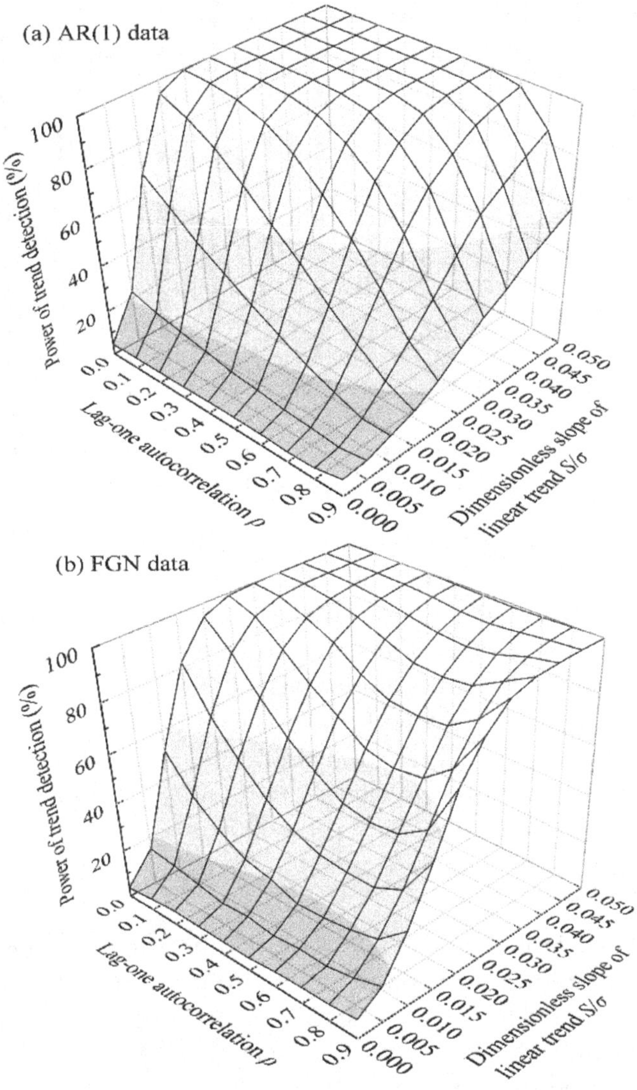

Figure 8. Power of the variance-corrected Sen's trend test with various combinations of data persistence and a dimensionless slope of linear trend (series length: $n = 100$) using a well-specified persistent model: (**a**) AR(1) data; (**b**) FGN data. The persistent parameters ρ and H are estimated from the data.

Without slope variance correction, the test may prevent power loss at the expense of producing a high Type I error rate, which is far beyond the reported significance level. However, this undermines the fundamental basis for the statistical trend test, which operates by controlling Type I errors within a known significance level, signifying users' confidence toward detected trends.

Sen's trend test with the help of slope variance correction is built for persistent data. When the data are independent of time, will this manipulation reduce the power? The answer is given in Figure 9. The power of the test based on AR(1) correction is comparable to that of the original.

However, the test based on FGN correction becomes less powerful, because of the sampling errors introduced by the Hurst coefficient estimation. This finding reasserts the importance of careful choice of a persistent model before performing variance correction. Revisiting the first step of the practical procedure, independent data can be classified as a special case of an AR(1) process. Thus, the adoption of AR(1)-based correction would not cause an apparent loss of power.

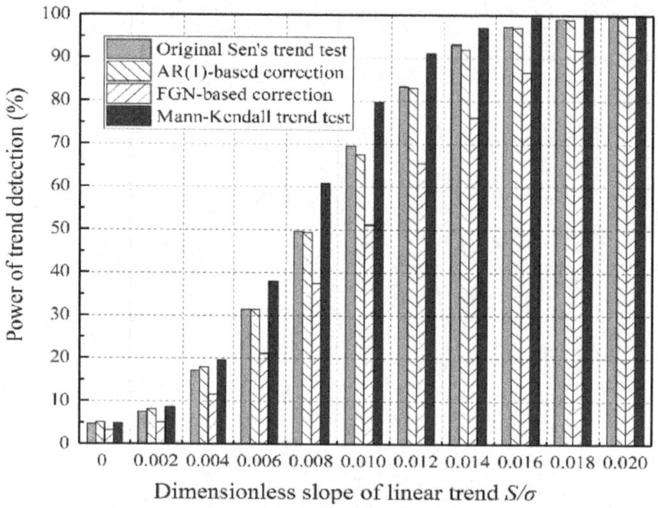

Figure 9. Power of the innovative variance-corrected Sen's trend test compared to that of the original test and the Mann–Kendall test, for the case of uncorrelated data (series length: $n = 100$).

Figure 9 also reveals that Sen's trend test is less powerful than the Mann–Kendall test for uncorrelated data with weak trends. However, the power of Sen's trend test achieves that of the Mann–Kendall test as the slope of the trend increases. This argues against the statement made by earlier comparison studies that Sen's trend test can identify many significant trends ignored by the Mann–Kendall test [19,42]. The discrepancy mainly arises from the incorrect slope variance, Equation (3), given by Sen [11]. In Section 2.2, it was demonstrated that this slope variance is substantially biased even for independent data. The expression should be replaced by a bias-corrected expression, Equation (4), so that the Type I error rate can return to its normal state. However, the power will be slightly lower than the Mann–Kendall test.

6. Application to Real-World Data

Hydrometeorological variables carries different persistence characteristics, for example independence, short-term, and long-term persistence. To demonstrate the applicability of the new method on series with different degree of persistence, three real-world records were selected for illustration, including rainy days, frost days, and streamflow. The annual series of rainy days, and frost days of China were obtained from the CRU CY 4.01 dataset (1901–2016), produced by the Climate Research Unit at the University of East Anglia [43]. The annual total flow (1824–2016) from Main River at Wuerzburg in Germany were kindly provided by the Global Runoff Data Centre in Koblenz, Germany (http://grdc.bafg.de).

The time series plot and its corresponding 1:1 trend plot are presented in Figure 10a,b. In the time series plots, increasing trends are recognizable in rainy days and annual total flow, and decreasing trends in frost days for the entire record. The 1:1 trend plots provide detailed information about sub-trends. Regarding the rainy days, the trend slopes at higher values are generally stronger than that for the lower value cases. The annual total flow series shares a similar behavior. The increasing trend is dominated by high values. In addition, the low cluster is almost trend-free, whereas an apparent reduction occurs at the extremely low values. In the series of frost days, downward trends are prevalent throughout high, medium, and low clusters and the trends are dominant in the low cluster.

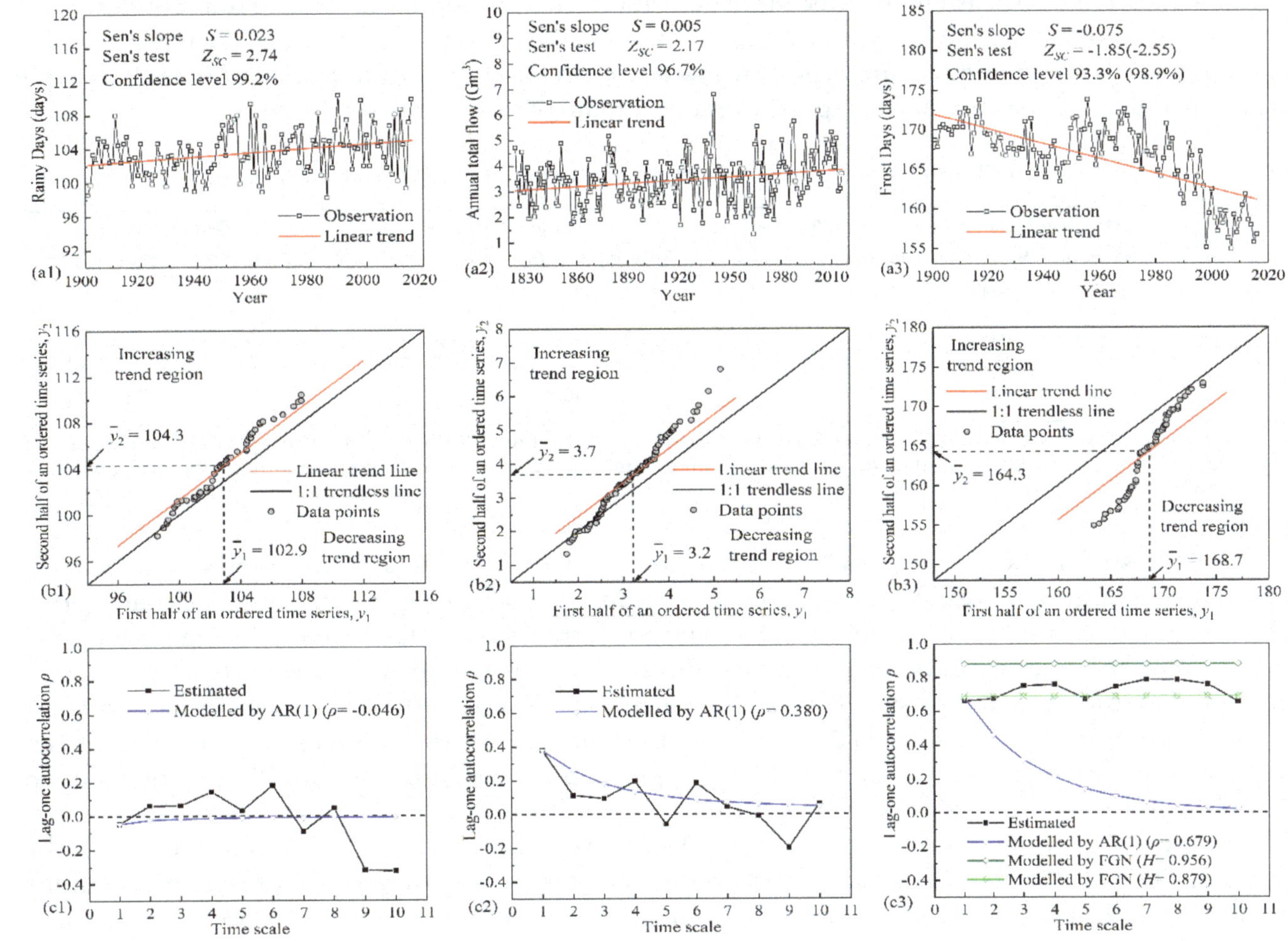

Figure 10. Time series of annual rainy days and frost days of China (**a1**,**a3**), and annual total flow from Main River at Wuerzburg in Germany (**a2**). Trend test results given in (**a1–a3**) are obtained from variance-corrected Sen's trend test. Sen's trend plots for the data sets in a1–3 (**b1–b3**). Lag-one autocorrelations of the aggregated series versus time scales as for a1–3 (**c1–c3**).

The persistence features of the three series were further examined by plotting the empirical lag-one autocorrelation of the aggregated series versus time scale, as depicted in Figure 10c. The trajectories of the theoretical autocorrelations of AR(1) and FGN models are also depicted in the same figure. The maximum scale of aggregation was set to ten, assuring that each aggregated series contains at least ten values. The rainy days exhibit low autocorrelations, which fluctuate around zero initially and become negative for large scales. The series is generally independent of time, and therefore we can regard the series as a representation of AR(1) data. The autocorrelations of annual total flow fluctuate around the value modelled by AR(1). Annual total flow is thus very likely to be AR(1) data. The long-term persistence of the frost days is clear in this example. Autocorrelation appears to be constant. Thus, the FGN model is a better approximation of the underlying process.

Table 4 gives the trend detection results when applying the original and variance-corrected Sen's trend test. The second and third columns show the basic statistical features of the series. The specified persistent model and its parameter are provided in the next two columns. In the frost days, the high value of the Hurst coefficient is in doubt because the existence of a real trend may result in overestimation. As such, the Hurst coefficient was also estimated from the detrended series (in parentheses) for comparison. The last four columns indicate the original and corrected slope variances and their corresponding trend test statistics. The new method only slightly alters the trend test statistic of rainy days, which is nearly independent of time. As for the persistent series, the new method effectively reduces the inflated statistic values. The confidence level (one minus significance level) of increasing annual total flow and decreasing frost days drops from 99% to 95%, when the effect of persistence is taken into account. The new method even rejects significant trend of frost days, when we corrected the slope variance using the higher estimate of the Hurst coefficient.

Table 4. Results of Sen's trend test before and after slope variance correction.

Variables	Statistical Features		Data Persistence		Slope Variance		Test Statistic [1]	
	Mean μ	Slope S/σ	Model	Parameter	Original $V_0(S)$	Corrected $V_C(S)$	Original Z_S	Corrected Z_{SC}
Rainy days	103 days	0.008	AR(1)	$\rho = -0.046$	7.9×10^{-5}	7.2×10^{-5}	2.62 ++	2.74 ++
Annual total flow	3.44 Gm^3	0.005	AR(1)	$\rho = 0.380$	2.1×10^{-6}	4.5×10^{-6}	3.22 ++	2.17 +
Frost days	166 days	−0.017	FGN	$H = 0.956$ (0.879)[b]	0.2×10^{-3}	0.2×10^{-2} (0.9×10^{-3})	−5.18 ++	−1.85 (−2.55 +)

[1] Significant trends at the 5% (\|Z\| > 1.96) level are indicated by "+", at the 1% (\|Z\| > 2.576) level by "++"; [2] Values in parentheses are the Hurst coefficients estimated from the detrended series and the corresponding slope variance and trend test statistic.

This result indicates a possibility that the statistical evidences of trends within many hydrometeorological processes are overestimated because of the ignorance of the intrinsic characteristic of persistence. Admitting persistence in the new Sen's trend test helps to avoid exaggerating trend significance and overuse of deterministic trends. In practical applications, properly accounting for deterministic t rends and persistent behavior are inherently linked with understanding and modelling hydrometeorological processes. On the one hand, the persistent process implies that there is nontrivial dependence between the present state and the previous states, and the handling of dependence in prediction model can be made through reasoning future state conditional on the known past information. Dependence improves one-step-ahead predictability in comparison to purely random process. On the other hand, the persistent process exhibits much wide range of natural variability at large time scales. Many statistical characteristics become time-related. For example, local means may vary in time and fluctuate irregularly around the global mean. This enhanced fluctuation reduce predictability of the process over long time horizon. The detected trend would initiate an investigation on the causative effects through deductive reasoning and data validation. The trend if at all valid can enrich our knowledge of the changes in external forces and in internal conditions of the hydrometeorological systems.

In the current application, increasing trend of annual rainy days was detected in China, which is a manifestation of precipitation regime changes. Decreasing frost days is an evidence of global warming over mainland China. But the statistical significance of trend markedly reduced when the long-term persistence is considered. Similar reductions in the significance of rising temperature are also evidenced by other studies [44,45]. We argue that the future evolution of temperature-like processes would be much more uncertain than we perceive. Annual total flow from Main River in Germany exhibits both the short-term persistence and the increasing trend. This feature reveals that the Main River was generally enriched, yet had a propensity of enduring consecutive years of natural excursions from the mean. This information has evident implication for the designing, operation, and management of hydro projects.

7. Conclusions

This study adapts the recently proposed Sen's trend test to persistent data via a variance correction approach. Two expressions for the variance of the trend test statistic were newly derived under the assumption of persistence. One variance is tailored to short-term persistent data, characterized by the AR(1) model. The other expression is designed for long-term persistent data and is represented by the FGN model. The computation of these new variances is convenient, as they only require one additional parameter, i.e., the lag-one autocorrelation coefficient or Hurst coefficient. We also fixed the original variance for independent data, thereby correcting its negative bias.

An innovative variance-corrected Sen's trend test was developed with the aid of newly derived variances. Simulation results verified the effectiveness of the new test on alleviating the side effect of persistence. By correcting the variance with the most approximate persistent model of the data, the rate of committing Type I errors remains within an acceptable range of the nominal significance level. The power of trend detection monotonically increases for higher trend slopes. The power losses depend on how much the persistence inflates the variance of the trend test statistic. The more seriously inflated the variance, the less powerful the test. A stepwise procedure was recommended for performing the new test on real hydrometeorological observations. Case studies explained in detail how to differentiate independent, short-term persistent, and long-term persistent data, so as to invoke the corresponding variance prudently.

The actual advantage of the Sen's trend detection method is to deduce the overall trend and hidden sub-trends in various categories simultaneously on a furnished 1:1 trend plot. The flexible categorization of observations provides detailed information about the trend characteristics for each cluster. However, in the context of trend significance analysis, Sen's trend test does not perform better than the conventional Mann–Kendall test. The purported novelty of Sen's trend test, i.e., it is free of any restrictive assumptions, is not supported by the current study. This finding renders some published trends based on the original Sen's trend test problematic and may therefore need to be reassessed. The proposed new Sen's trend test attaches its test statistic to the distinctive persistent feature of hydrometeorological data. Sen's trend plot with the help of variance-corrected Sen's trend test yields more reliable statistical evidence of the changes occurring and sheds light on the future evolution of hydrological and climatic processes.

Author Contributions: W.W. and Y.C. implemented the theoretical derivations and numerical simulation experiments. Case study were done by B.L. and verified by X.Z. The manuscript was written by W.W. and reviewed by Y.Z.

Acknowledgments: The authors are grateful to the University of East Anglia for providing CRU CY 4.01 dataset. We appreciate the Global Runoff Data Centre in Koblenz, Germany for providing flow data. We sincerely appreciate Guest editor Dr. Tsoukalas Ioannis and anonymous reviewers for reading the manuscript and giving many insightful recommendations about its content. We also thank LetPub (www.letpub.com) for its linguistic assistance during the preparation of this manuscript.

References

1. IPCC. Summary for Policymakers. In *Global Warming of 1.5 °C*; An IPCC special report on the impacts of global warming of 1.5 °C above pre-industrial levels and related global greenhouse gas emission pathways, in the context of strengthening the global response to the threat of climate change, sustainable development, and efforts to eradicate poverty; 2018; in press.
2. Jimenez Cisneros, B.E.; Oki, T.; Arnell, N.W.; Benito, G.; Cogley, J.G.; Doll, P.; Jiang, T.; Mwakalila, S.S. Freshwater Resources. In *Climate Change 2014: Impacts, Adaptation, and Vulnerability. Part A: Global and Sectoral Aspects*; Contribution of working group II to the Fifth Assessment Report of the Intergovernmental Panel on Climate Change; Cambridge University Press: Cambridge, UK; New York, NY, USA, 2014.
3. Durocher, M.; Requena, A.I.; Burn, D.H.; Pellerin, J. Analysis of trends in annual streamflow to the Arctic Ocean. *Hydrol. Process.* **2019**, *33*, 1143–1151. [CrossRef]
4. Arora, R.; Tockner, K.; Venohr, M. Changing river temperatures in northern Germany: Trends and drivers of change. *Hydrol. Process.* **2016**, *30*, 3084–3096. [CrossRef]
5. Gao, T.; Wang, H. Trends in precipitation extremes over the Yellow River basin in north China: Changing properties and causes. *Hydrol. Process.* **2017**, *31*, 2412–2428. [CrossRef]

6. Hodgkins, G.A.; Dudley, R.W.; Archfield, S.A.; Renard, B. Effects of climate, regulation, and urbanization on historical flood trends in the United States. *J. Hydrol.* **2019**, *573*, 697–709. [CrossRef]
7. Yeste, P.; Dorador, J.; Martin-Rosales, W.; Molero, E.; Esteban-Parra, M.J.; Rueda, F.J. Climate-driven trends in the streamflow records of a reference hydrologic network in southern Spain. *J. Hydrol.* **2018**, *566*, 55–72. [CrossRef]
8. Garbrecht, J.; Fernandez, G.P. Visualization of trends and fluctuations in climatic records. *Water Resour. Bull.* **1994**, *30*, 297–306. [CrossRef]
9. Basarin, B.; Lukic, T.; Pavic, D.; Wilby, R.L. Trends and multi-annual variability of water temperatures in the river Danube, Serbia. *Hydrol. Process.* **2016**, *30*, 3315–3329. [CrossRef]
10. Sen, Z. Innovative trend analysis methodology. *J. Hydrol. Eng.* **2012**, *17*, 1042–1046. [CrossRef]
11. Sen, Z. Innovative trend significance test and applications. *Theor. Appl. Climatol.* **2017**, *127*, 939–947. [CrossRef]
12. Sen, Z. Trend identification simulation and application. *J. Hydrol. Eng.* **2014**, *19*, 635–642. [CrossRef]
13. Wu, H.; Qian, H. Innovative trend analysis of annual and seasonal rainfall and extreme values in Shaanxi, China, since the 1950s. *Int. J. Climatol.* **2017**, *37*, 2582–2592. [CrossRef]
14. Caloiero, T.; Coscarelli, R.; Ferrari, E. Application of the innovative trend analysis method for the trend analysis of rainfall anomalies in southern Italy. *Water Resour. Manag.* **2018**, *32*, 4971–4983. [CrossRef]
15. Dabanli, I.; Sen, Z.; Yelegen, M.O.; Sisman, E.; Selek, B.; Guclu, Y.S. Trend assessment by the innovative-Sen method. *Water Resour. Manag.* **2016**, *30*, 5193–5203. [CrossRef]
16. Belihu, M.; Abate, B.; Tekleab, S.; Bewket, W. Hydro-meteorological trends in the Gidabo catchment of the Rift Valley Lakes basin of Ethiopia. *Phys. Chem. Earth* **2018**, *104*, 84–101. [CrossRef]
17. Cui, L.F.; Wang, L.C.; Lai, Z.P.; Tian, Q.; Liu, W.; Li, J. Innovative trend analysis of annual and seasonal air temperature and rainfall in the Yangtze River basin, China during 1960–2015. *J. Atmos. Sol. Terr. Phys.* **2017**, *164*, 48–59. [CrossRef]
18. Kisi, O. An innovative method for trend analysis of monthly pan evaporations. *J. Hydrol.* **2015**, *527*, 1123–1129. [CrossRef]
19. Zhou, Z.G.; Wang, L.C.; Lin, A.W.; Zhang, M.; Niu, Z.G. Innovative trend analysis of solar radiation in China during 1962–2015. *Renew. Energy* **2018**, *119*, 675–689. [CrossRef]
20. Kisi, O.; Ay, M. Comparison of Mann–Kendall and innovative trend method for water quality parameters of the Kizilirmak River, Turkey. *J. Hydrol.* **2014**, *513*, 362–375. [CrossRef]
21. Deng, S.L.; Chen, T.; Yang, N.; Qu, L.; Li, M.C.; Chen, D. Spatial and temporal distribution of rainfall and drought characteristics across the Pearl River basin. *Sci. Total Environ.* **2018**, *619*, 28–41. [CrossRef]
22. Tosunoglu, F.; Kisi, O. Trend analysis of maximum hydrologic drought variables using Mann–Kendall and Sen's innovative trend method. *River Res. Appl.* **2017**, *33*, 597–610. [CrossRef]
23. O'Connell, P.E.; Koutsoyiannis, D.; Lins, H.F.; Markonis, Y.; Montanari, A.; Cohn, T. The scientific legacy of Harold Edwin Hurst (1880–1978). *Hydrol. Sci. J.* **2016**, *61*, 1571–1590. [CrossRef]
24. Sen, Z.K. Hydrological trend analysis with innovative and over-whitening procedures. *Hydrol. Sci. J.* **2017**, *62*, 294–305. [CrossRef]
25. Burger, G. On trend detection. *Hydrol. Process.* **2017**, *31*, 4039–4042. [CrossRef]
26. Hamed, K.H. Trend detection in hydrologic data: The Mann–Kendall trend test under the scaling hypothesis. *J. Hydrol.* **2008**, *349*, 350–363. [CrossRef]
27. Wang, W.; Chen, Y.; Becker, S.; Liu, B. Linear trend detection in serially dependent hydrometeorological data based on a variance correction Spearman Rho method. *Water* **2015**, *7*, 7045–7065. [CrossRef]
28. Koutsoyiannis, D. The Hurst phenomenon and Fractional Gaussian Noise made easy. *Hydrol. Sci. J.* **2002**, *47*, 573–595. [CrossRef]
29. Hurst, H.E. Long-term storage capacity of reservoirs. *Trans. Am. Soc. Civ. Eng.* **1951**, *116*, 770–808.
30. Mandelbrot, B.B.; Wallis, J.R. Some long-run properties of geophysical records. *Water Resour. Res.* **1969**, *5*, 321–340. [CrossRef]
31. Koutsoyiannis, D. Climate change, the Hurst phenomenon, and hydrological statistics. *Hydrol. Sci. J.* **2003**, *48*, 3–24. [CrossRef]
32. Koutsoyiannis, D. Nonstationarity versus scaling in hydrology. *J. Hydrol.* **2006**, *324*, 239–254. [CrossRef]
33. Koutsoyiannis, D. Hydrology and change. *Hydrol. Sci. J.* **2013**, *58*, 1177–1197. [CrossRef]

34. Matalas, N.C.; Sankarasubramanian, A. Effect of persistence on trend detection via regression. *Water Resour. Res.* **2003**, *39*, 1342. [CrossRef]
35. Dimitriadis, P.; Koutsoyiannis, D. Stochastic synthesis approximating any process dependence and distribution. *Stoch. Environ. Res. Risk Assess.* **2018**, *32*, 1493–1515. [CrossRef]
36. Tsoukalas, I.; Makropoulos, C.; Koutsoyiannis, D. Simulation of stochastic processes exhibiting any-range dependence and arbitrary marginal distributions. *Water Resour. Res.* **2018**, *54*, 9484–9513. [CrossRef]
37. Iliopoulou, T.; Papalexiou, S.M.; Markonis, Y.; Koutsoyiannis, D. Revisiting long-range dependence in annual precipitation. *J. Hydrol.* **2018**, *556*, 891–900. [CrossRef]
38. Markonis, Y.; Moustakis, Y.; Nasika, C.; Sychova, P.; Dimitriadis, P.; Hanel, M.; Maca, P.; Papalexiou, M. Global estimation of long-term persistence in annual river runoff. *Adv. Water Resour.* **2018**, *113*, 1–12. [CrossRef]
39. Hamed, K.H. Enhancing the effectiveness of prewhitening in trend analysis of hydrologic data. *J. Hydrol.* **2009**, *368*, 143–155. [CrossRef]
40. Tyralis, H.; Koutsoyiannis, D. Simultaneous estimation of the parameters of the Hurst-Kolmogorov stochastic process. *Stoch. Environ. Res. Risk Assess.* **2011**, *25*, 21–33. [CrossRef]
41. Serinaldi, F.; Kilsby, C. The importance of prewhitening in change point analysis under persistence. *Stoch. Environ. Res. Risk Assess.* **2016**, *30*, 763–777. [CrossRef]
42. Li, J.; Wu, W.; Ye, X.; Jiang, H.; Gan, R.; Wu, H.; He, J.; Jiang, Y. Innovative trend analysis of main agriculture natural hazards in China during 1989–2014. *Nat. Hazards* **2018**, *95*, 677–720. [CrossRef]
43. Harris, I.; Jones, P.D.; Osborn, T.J.; Lister, D.H. Updated high-resolution grids of monthly climatic observations—The CRU TS3.10 dataset. *Int. J. Climatol.* **2014**, *34*, 623–642. [CrossRef]
44. Cohn, T.A.; Lins, H.F. Nature's style: Naturally trendy. *Geophys. Res. Lett.* **2005**, *32*, L23402. [CrossRef]
45. Ludescher, J.; Bunde, A.; Franzke, C.L.E.; Schellnhuber, H.J. Long-term persistence enhances uncertainty about anthropogenic warming of Antarctica. *Clim. Dyn.* **2016**, *46*, 263–271. [CrossRef]

4

Assimilation of Synthetic SWOT River Depths in a Regional Hydrometeorological Model

Vincent Häfliger [1,2,*], Eric Martin [1,3], Aaron Boone [1], Sophie Ricci [4] and Sylvain Biancamaria [5]

1 CNRM-GAME, UMR 3589, Météo-France, CNRS, 31057 Toulouse, France; eric.martin@irstea.fr (E.M.); aaron.a.boone@gmail.com (A.B.)
2 Meteorologisches Institut, Bonn Universität, 53113 Bonn, Germany
3 IRSTEA, UR RECOVER, 13100 Aix-en-Provence, France
4 CECI, CERFACS-CNRS, 42 avenue G. Coriolis, 31057 Toulouse, France; ricci@cerfacs.fr
5 CNRS, LEGOS, UMR 5566-CNRS-CNES-IRD-Université Toulouse III, 31057 Toulouse, France; sylvain.biancamaria@legos.obs-mip.fr
* Correspondence: vincent.haefliger@yahoo.fr

Abstract: The SWOT (Surface Water and Ocean Topography) mission, to be launched in 2021, will provide water surface elevations, slopes, and river width measurements for rivers wider than 100 m. In this study, synthetic SWOT data are assimilated in a regional hydrometeorological model in order to improve the dynamics of continental waters over the Garonne catchment, one of the major French catchments. The aim of this paper is to demonstrate that the sequential assimilation of SWOT-like river depths allows the correction of river bed roughness coefficients and thus simulated river depths. An extended Kalman filter is implemented and the data assimilation strategy was applied to four experiments of gradually increasing complexity regarding observation and model error over the 1995–2000 period. With respect to a "true" river state, assimilating river depths allows the proper retrieval of constant and spatially distributed roughness coefficients with a root mean square error of 1 $m^{1/3}$ s^{-1}, and the estimation of associated river depths. It was also shown that river depth differences can be assimilated, resulting in a higher root mean square error for roughness coefficients with respect to the true river state. Finally, the last experiment shows how one can take into account more realistic sources of SWOT error measurements, in particular the importance of the estimation of the tropospheric water content in the process.

Keywords: overland flow; satellite altimetry; hydrological modelling; data assimilation

---

1. Introduction

Surface water storage and fluxes in rivers, lakes, reservoirs, and wetlands are currently poorly observed at global and regional scales, even though they represent major components of the water cycle and impact human societies deeply [1], both in terms of water resources and catastrophic events, such as floods. In situ networks are heterogeneously distributed in space, and many river basins and most lakes–especially in the developing world and in sparsely populated regions–remain unmonitored [2]. The continental water cycle can be formulated in a simple mass balance equation, linking the total water storage variation in time (ΔS) with different water fluxes between the continental surface, the atmosphere, and the underground domain: These are precipitation (P), evapotranspiration (E), the infiltration in the subsurface (I), and the overland flow (Q). The mass balance equation simply assumes that the total water storage variation, ΔS, is equal to the precipitation, P, from which are subtracted the terms, E, Q, and I. Altimetry products from remote sensing are increasingly used for the monitoring of these hydrological cycle components [3–5]. Several altimetric satellites have been launched in the past to measure water surface elevations, among which ERS-1 (1991–2000), TOPEX/Poseidon (1992–2006), ERS-2 (1995–2003), Jason-1 (2001–2013), Envisat (2002–2012), Cryosat-2 (2010–now), SARAL (2013–now), Sentinel-3 (2016–now), and Jason-3 (2016–now). They provide repetitive water elevation measurements on a global scale, something which is particularly important for ungauged basins. They do, however, have many limitations, such as a longer revisit time (between 10 and 35 days [6]), and coarse spatial resolution. The instruments' footprints are several square km and they are only observed from their nadir (from the vertical of the satellite). Among them, the CryoSat-2 mission, launched in 2010, offers a very dense spatial sampling pattern with an across-track distance of only 7.5 km, but with a drifting ground track and a full repeat cycle of 369 days [7]. To overcome these limitations, a new satellite mission based on SAR (synthetic aperture radar) interferometry was proposed: This was called WATER (Water and Terrestrial Elevation Recovery), providing water elevation maps for two 50 km swaths [8]. In 2007, the National Research Council recommended this new satellite mission to NASA (The National Aeronautics and Space Administration), under the name SWOT (Surface and Ocean Topography), to measure both the ocean and land water surface topography. This new mission, conjointly developed by NASA, CNES (Centre National d'Etudes Spatiales), CSA/ASC (Canadian Space Agency/Agence Spatiale Canadienne), and UKSA (United-Kindom Space Agency), is planned for launch in 2021 and will observe the whole continental water-estuaries-ocean continuum. SWOT is designed to observe a large fraction of rivers and lakes globally and will provide observations of their seasonal cycles. SWOT will be the first altimetry mission to provide a quasi-global coverage of continental surfaces between 78° S and 78° N. Especially, it is designed to observe big and intermediate (or regional)-scale basins (i.e., 50,000–200,000 km^2 drainage area) in 21 days, the duration of a full orbital cycle [9]. Water level measurement errors are expected to be 10 cm aggregating pixels over a 1 km^2 water area (e.g., a 10-km reach length for a 100-m-wide river) [10]. This offers a new opportunity for the linking of open water surface elevations, land surface processes, and meteorology more closely on this scale. The SWOT mission will provide relevant information on the temporal evolution of the surface water storage. This information will allow a better understanding of the term, Q, both spatially and temporally, playing an important role in the mass balance equation.

DA (data assimilation) techniques are increasingly used by hydrologists to improve the performance of numerical models. The DA techniques are used to correct either the model parameters and/or the model state. The merits of remotely sensed DA have been shown in various studies. GRACE (Gravity Recovery And Climate Experiment) data were used on total water storage over the Mississipi river to improve the soil water content and different water fluxes in the basin [11]. In the field of wide swath altimetry, synthetic SWOT data were assimilated in the Arctic Ob River in Siberia to improve water elevation and discharge [12]: The combination of hydrologic and hydraulic models with the SWOT data allowed better description of temporal variations of open water surface elevations. At a large scale catchment, synthetic SWOT data were assimilated in the Niger basin to correct the roughness coefficient of the river bed, in order to better represent river discharge and flood plain dynamics [13]. These examples demonstrate the strong capability of spatialized satellite information to improve our knowledge of the continental hydrological cycle. Wide swath altimetry measurements made by the SWOT satellite mission will provide the potential for high-resolution characterization of open water surface elevations and will contribute to a fundamental understanding of the global water cycle by providing global measurements of terrestrial surface water storage changes and discharge.

In the specific context of SWOT pre-launch studies, most studies focused on big river basins to investigate the benefits of SWOT for hydrology modeling (e.g., [12,13]). The present paper aims to focus on a regional scale. The time evolution of physical processes is faster, and it is important to analyze whether or not the information given by SWOT will be able to improve our understanding of the hydrological cycle on this spatial and temporal scale. The ability of a regional hydrometeorological model to simulate river depths comparable with the future SWOT open surface water elevation observations was assessed [14]; the same model is used in the current study. The river depth is equal to the vertical distance between the river bed and the open surface water elevation of the river. The distributed ISBA/MODCOU (Interaction Soil Biosphere Atmosphere/MODèle COUplé) model was used over the Garonne catchment, taking into account a variable flow velocity in the river channel. This kinematic wave formulation is used in the present study in the form of the Manning-Strickler equations, derived from the full Saint-Venant equation system. It takes into consideration that there is no diffusion of the flood wave from upstream to downstream in the river. In the present study, synthetic SWOT-like data are assimilated into a regional scale hydrological model over the Garonne catchment in the South-West of France. The study is based on the previous numerical developments for the implementation of data assimilation achieved on the Niger basin (2,118,000 km^2) [13], but focuses on a smaller basin (Garonne, 56,000 km^2) located in another climatic region. The present study uses a different hydrological model, which has a much higher spatial resolution and a much more ramified river network than other studies (e.g., [13]).

This work is carried out in the framework of OSSE (observing system simulation experiment) with the ISBA/MODCOU hydrometeorological model to prepare future exploitation of the SWOT data. The paper is divided into five sections. Section 2 presents the Garonne catchment and the ISBA/MODCOU model that simulated river depths. The DA strategy is developed in Section 3: A short description of the SWOT mission is given, the OSSE framework is explained, and the filtering data assimilation algorithm is presented along with its implementation. Results are presented in Section 4 on four different experimental settings. Conclusions and perspectives concerning the DA strategy and link with the SWOT mission are finally given in Section 5.

2. Catchment and Hydrological Model Description

2.1. The Garonne River Catchment

The study is carried out over the Garonne River catchment (Figure 1). The Garonne catchment is one of the major catchments in France (56,000 km^2) with a river width ranging from 150 m in Toulouse to 300 m near Bordeaux. It is located in southwestern France and drains the northern slopes of the Pyrenees chain (along the French border with Spain). The Pyrenees and the Massif Central mountains border the basin to the south and the east, respectively. The main tributaries of the Garonne river are the Tarn and Lot rivers. The climate over the basin is under the influence of oceanic conditions over the western part of the domain, characterized by heavy rainfall events during winter and relatively warm weather during summer. There is a significant precipitation gradient from west to east, ranging from approximately 1200 mm year $^{-1}$ over the Atlantic coastal region to about 600 mm year^{-1} in the eastern part of the catchment. The upper Garonne and the Ariège rivers regime are characterized by spring snow melt in the Pyrenees [15], while summer flows are very low due to relatively dry summers.

2.2. The ISBA/MODCOU Hydrological Model

The ISBA land surface model [16] within the SURFEX (SURFface EXternalisée) platform is used to simulate the physical variables every 3 h in the upper soil, soil surface, and vegetation and to simulate water and energy exchanges within the soil–surface–atmosphere continuum [17]. Its parameters are derived from the ECOCLIMAP-II ecosystems and surface parameters database [18] at a 1-km resolution. ISBA is forced every 3 h by the SAFRAN (Système d'Analyse Fournissant des Renseignements Adaptés à la Nivologie) meteorological analysis [19]. ISBA uses a multilayer approach to solve the one-dimensional Fourier law and the mixed form of the Richards equation explicitly in order to calculate the time evolution of the soil energy and water budgets [20,21]. In terms of hydrology, the soil water balance accounts for infiltration, land surface evapotranspiration, and total runoff. The infiltration rate is provided by the difference between the through fall rate and the surface runoff. The throughfall rate is the sum of the rainfall not intercepted by the canopy, the dripping from the interception reservoir, and the snowmelt from the snowpack [22]. The total runoff is composed of the surface runoff, a lateral subsurface flow in the topsoil, and a free drainage condition at the bottom of the hydrological soil column.

Figure 1. The Garonne catchment with the names of main rivers (blue) and tributaries (black), and the names of main cities (dark black).

The hydrological and hydrogeological model platform, MODCOU, routes the continental surface water into the river. The surface runoff simulated by ISBA is transferred to the river by the ISO (ISOchrone transfer module) [23]. Water is then routed within the river by the parallel-computing-based RAPID module [24,25], with a spatial resolution of 1 or 2 km. MODCOU is the name given to the full platform with its two components, ISO and RAPID. The kinematic wave method is used for the river transfer routing RAPID, with the implementation of the Manning-Strickler equations. The equation linking the river depth and the roughness coefficient is the following (Equation (1)):

$$h(K_{str}) = \left(\frac{Q}{K_{str} \times W \times \sqrt{S_o}} \right)^{\frac{3}{5}} \quad (1)$$

where h is the river depth (m), K_{str} is the Strickler coefficient ($m^3\ s^{-1}$), Q is the discharge ($m^3\ s^{-1}$), W is the width (m), and S_o is the bed slope (-). The formulation is derived from the full Saint-Venant equation system, and supposes that the bed slope, S_o, of the river bed is entirely compensated by the friction slope, S_f. Thus, $S_o = S_f$ means that there is no diffusion, but a simple translation in the time of the flood wave along the river channel. The K_{str} coefficients are supposed to be constant over time and are spatially distributed for each grid cell of the model. The bed slopes, S_o, are calculated from the Digital Elevation Model SRTM 90 m [26] in the full river network of the Garonne catchment. The discharge regime is permanent and uniform in each grid cell of the model, where the values of the water storage, the discharge, and the river depth simulation are solved with a «Runge-Kutta order 4» method to prevent numerical bias caused by the nonlinearity of the Manning-Strickler formula [27]. This routing method is used in global hydrological models (e.g., in the TRIP model [27–29]) or in the regional ISBA/MODCOU model [14]. In the present paper, a time step of 300 s is chosen for the river routing. The geometry of the river channel is assumed to be rectangular. This approximation is based on the assumption that using rectangular or trapezoidal river channel cross-sections leads to close results in term of water depth [14]: in the Garonne river at Tonneins, by imposing discharge values ranging from 0 to 1500 $m^3\ s^{-1}$ and considering a low angle (30°) between the river banks and the vertical plane, the river depth difference between the rectangular and trapezoidal channel does not exceed 3 cm for high discharge values. By using a high angle (60°), the river depth difference is about 8 cm for a high discharge regime. This means that simulated river depths in RAPID are slightly overestimated for large flows in comparison with a high angle trapezoidal geometry.

3. Assimilation of Synthetic SWOT Data in ISBA/MODCOU: Experimental Design

3.1. The SWOT Mission

The observation of water elevations from space provides important information for many physical processes in hydrology. In the continental domain, the open water surface elevations of lakes and rivers can be observed by altimetry satellites. The SWOT mission (the launch of which is expected in 2021) is a wide swath altimetry mission able to measure the spatial and temporal evolution of the open water surface elevation of oceans, lakes and reservoirs, and rivers on the continental surface between 78° S and 78° N [1]. The nominal SWOT mission lifetime is 3 years. The principal SWOT payload is the SAR (Synthetic Aperture Radar) KaRIN (Ka-band Radar INterferometer), developed at the JPL (Jet Propulsion Laboratory). The instrument has two antennas separated by a baseline distance of 10 m (Figure 2a): By observing the same area from two different positions, the estimation of the elevation and the position of the target can be deducted. Each swath, from either side of the nadir area, is 50 km wide and they are separated by a distance of 20 km.

The SWOT's orbit will be at an 891-km altitude, have a 77.6° inclination, and function with a 21-day repeat period (the satellite flies over the same track every 21 days). Only 3.6% of the continental surfaces between 78° S and 78° N will be never observed by SWOT [1]. The observed variables over continents will be open water surface elevations maps, the surface slope, and the width of the river. The satellite will be designed to observe rivers wider than 100 m (requirement), with the goal being to observe rivers wider than 50 m [10]. In this study, water elevations correspond to the distance between the top of the surface water body and a reference surface, such as a geoid or an ellipsoid. The spatial resolution of the SWOT radar image in the range direction is 60 m for the near range of the swath and 10 m for the far range of the swath (Figure 2a). It is around 5 m in the azimuth direction: The range direction is perpendicular to the propagation axis of the satellite, and the azimuth direction is parallel to the propagation axis of the satellite (Figure 2b). At this resolution, however, the vertical accuracy of the instrument could be several meters. This accuracy tends to increase when water pixels are averaged together. The requirement for the measurement is to acquire a 10 cm vertical precision on water elevation, having averaged over 1 km^2 of the water area [10]. Measurement errors can be systematic or unsystematic. Unsystematic errors correspond to random errors (this, for example, is the case of errors induced by the instrument thermal noise). Systematic errors correspond to other sources of errors (such as errors induced by the atmosphere on the propagation of the signal emitted by SWOT). For many catchments in the world, only a few locations are sampled by in situ gauges. SWOT data will thus allow us to better understand the spatial and temporal evolution of open surface water elevations. SWOT will be the first wide-swath altimetry mission specifically designed to observe continental surfaces, providing maps of water elevations over regional-scale basins. SWOT will observe most areas between two and eight times during each 21-day repeat period, with the number of observations primarily depending on the distance from the equator [9]. The Garonne catchment will be observed from zero to four times per cycle, thanks to the satellite swaths, allowing several observations per cycle in most regions of the basin. However, due to these swaths overlapping, for locations with more than one observation per cycle, the temporal sampling between these observations will not be constant within a cycle. The number of observations inside one DA cycle, however, will be always the same: The distribution of the river reach observations in the catchment during one cycle will be shown in Section 3.4.1.

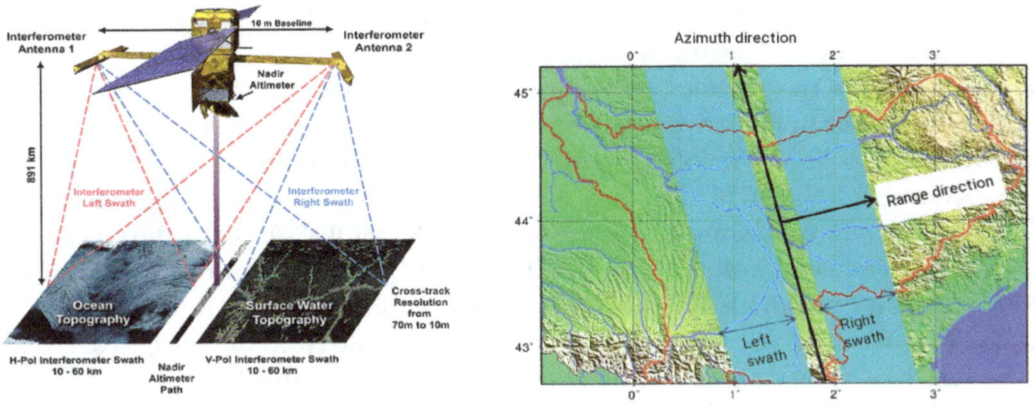

(a) (b)

Figure 2. (**a**) Illustration of SWOT The widths of the two rectangles represent two swaths of 50 km wide, observed by the two antennas of the satellite. They can observe either ocean or continental open water surfaces. (**b**) Illustration of the left and right swaths of SWOT with the range direction perpendicular to the satellite track (which is the azimuth direction).

3.2. Data Assimilation Algorithm and SWOT Observing System Simulation Experiment

The EKF (Extended Kalman Filter) was successfully employed in previous SWOT-related work over the Niger basin using ISBA/TRIP (Total Runoff Integrated Pathway) for model parameter correction, specifically friction coefficients [13]. This algorithm is well adapted for a weakly non-linear relation between the model outputs (noted y) and the DA control parameters to be corrected. EKF is used here over the Garonne basin to assimilate synthetic SWOT river depth observations in the RAPID module of the hydrological model, ISBA/MODCOU, to correct Strickler coefficients, K_{str} (noted x). Based on initial values of x, known as the background vector, x_b, the analyzed parameters, x_a, are calculated from Equation (2) [30]. We call "increment" the difference in value between the background, x_b, and the analysis, x_a.

$$x_a = x_b + (\mathbf{B}^{-1} + \mathbf{H}\mathbf{R}^{-1}\mathbf{H}^{T})^{-1} \times \mathbf{H}^{T}\mathbf{R}^{-1} \times (y_o - H(x_b)) \quad (2)$$

where **B** and **R** are, respectively, the model error matrix and the observation model error matrix, y_o represents the observation vector (river depth or river depth difference as a function of the experiment, see results). H is the observation operator that maps the control vector onto the observation space, it is the composition of the hydrometeorological model integration and the extraction of output values at observation time and space. **H** is the Jacobian matrix of H, calculated by using a finite difference scheme. It is a linear approximation of the observation operator, H [30].

The EKF method assumes that the system is linear. The following approximation (Equation (3)) should therefore be valid:

$$H(x + \Delta x) \approx H(x) + \frac{\partial H}{\partial x} \times \Delta x \quad (3)$$

To keep the system in the domain of the linearity, the term, Δx, should remain reasonably low, so that the value of the term, $H(x + \Delta x)$, is near the value of the term, $H(x) + \frac{\partial H}{\partial x} \times \Delta x$. The higher the Δx value, the higher the difference between the left and right term of the equation. The value of Δx should be similar to the maximal difference between x_a and x_b that we decided to impose on the model, and that we call the «limitation of increment».

By replacing the terms, H and x, in Equation (3) to the river depth, h, and its argument, K_{str}, we obtained the following equation (Equation (4)):

$$[h(K_{str} + \Delta K_{str})] - \left[h(K_{str}) + \frac{\partial h(K_{str})}{\partial K_{str}} \times \Delta K_{str}\right] = \Delta h \quad (4)$$

Δh should be equal to 0 in the case of a perfect linear model. In practice, it should remain sufficiently small to ensure good accuracy of the DA system.

OSSE are widely used to validate DA algorithms and strategies before working with real-data. They offer a convenient framework for the testing of various hypotheses on observation errors, observation spatial and temporal resolution, model error, the choice of DA algorithm, or the cycling of the analysis. Knowing a SWOT's orbit and pass plan over a 21-day cycle allows the extraction of SWOT-like data from a reference simulation output of ISBA/MODCOU over the Garonne catchment. The settings of the reference simulation (noted SIM_TRUE, blue box on the top, Figure 3) features unperturbed forcing and parameters resulting from the calibration of a RAPID model with a 1 or 2 km resolution [14]. An error of 10 cm standard deviation is added to the SIM_TRUE outputs to get some virtual SWOT-like observations averaged over a 1 km^2 area (blue box on the bottom, Figure 3). This simple hypothesis will be further investigated in Section 4.4. A perturbed set of forcing and parameters, x_b, is then used to run the background integration (noted SIM_PERT, red box, Figure 3) in which synthetic SWOT data are assimilated to retrieve the unperturbed forcing and parameters. The analysis branch (noted SIM_ANA, green box, Figure 3) for the control vector, x_a, allows then the calculation of the corrected parameters to provide corrected river depths and discharge. The DA analysis is carried out over a sliding time window covering several observation times beyond which a forecast can be issued. The tool used to set up our DA scheme is the Open-PALM (Parallel Assimilation with a Lot of Modularity) software [31,32] developed at CERFACS (Centre Européen de Recherche et Formation Avancée en Calcul Scientifique) and ONERA (Office National d'Etudes et de Recherches Aérospatiales). This software allows the coupling of codes of calculation between them, and in this way, allows the exchange of variables, such as vector or matrix, between these codes. OpenPALM provides a parallel environment based on high performance implementation of the Message Passing Interface standard: This interface is able to perform both data parallelism and task parallelism [33].

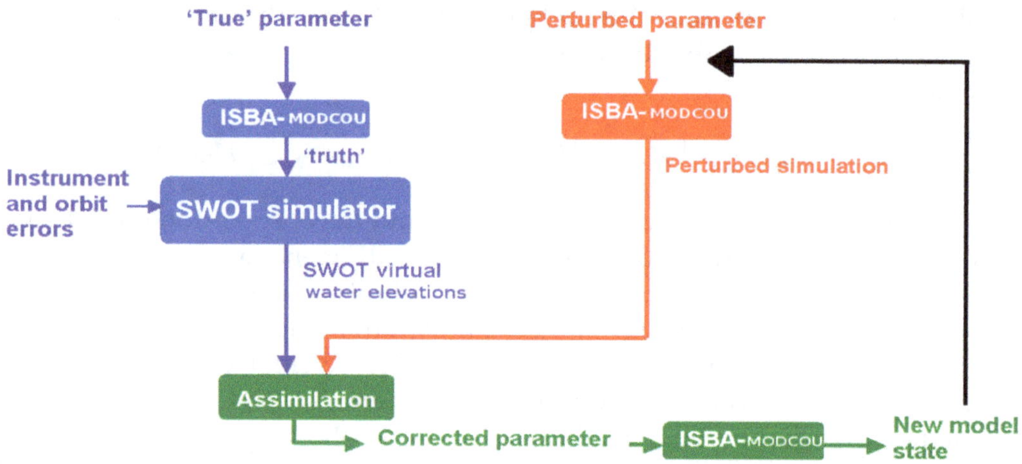

Figure 3. Schematic illustration of the twin experiment. The red boxes describe the perturbed parameters of the model (SIM_PERT). The blue boxes represent the truth SIM_TRUE (reference simulation) to which white noise is added to generate SWOT-like measurement error. The green boxes represent the corrected parameters of the model after a DA window: We obtained corrected parameters of the model (SIM_ANA) being used for the next DA window (black line), allowing the calculation of new corrected parameters at the next window's termination.

3.3. *Sensitivity of the River Depth to the Roughness Coefficient, K_{str}*

The sensitivity of the IBSA/MODCOU simulated river depth to a perturbation of the roughness coefficient, K_{str}, is illustrated in Figure 4, for one unique grid cell, with typical conditions of a large rectangular plain section of the lower Garonne river. This test case is mono-dimensional, without the need to analyze the consequences of the effect of the K_{str} perturbation on the upstream and downstream river depths. Discharge, Q, width, W, and bed slope, S_o, are equal to 500 m^3 s^{-1}, 150 m, and 0.0005 m m^{-1}, respectively. The K_{str} values vary between 20 and 40 $m^{1/3}$ s^{-1}. Figure 4 shows that for a difference of K_{str} equal to 5 $m^{1/3}$ s^{-1} (between 25 and 30 $m^{1/3}$ s^{-1}), the river depth variation is about 32 cm. For a difference of 1 $m^{1/3}$ s^{-1} (between 25 and 26 $m^{1/3}$ s^{-1}), the river depth variation is about 6.5 cm. It means that an error of one Strickler coefficient equal to $x_b - x_t = 1$ $m^{1/3}$ s^{-1} is related to an error of one river depth, $H(x_b) - H(x_t) = 6.5$ cm, x_t and $H(x_t)$ being the true Strickler coefficient and the true river depth. This result must be confirmed on the scale of the Garonne river under various hydrological conditions. In this present idealized test case considering one unique grid cell, the impact of a perturbation of K_{str} on the water level is not shown for downstream river grid cells. The results presented in Section 4 allow the study of how changes in the Strickler coefficient values impact the flood dynamics on the full river network of the Garonne catchment.

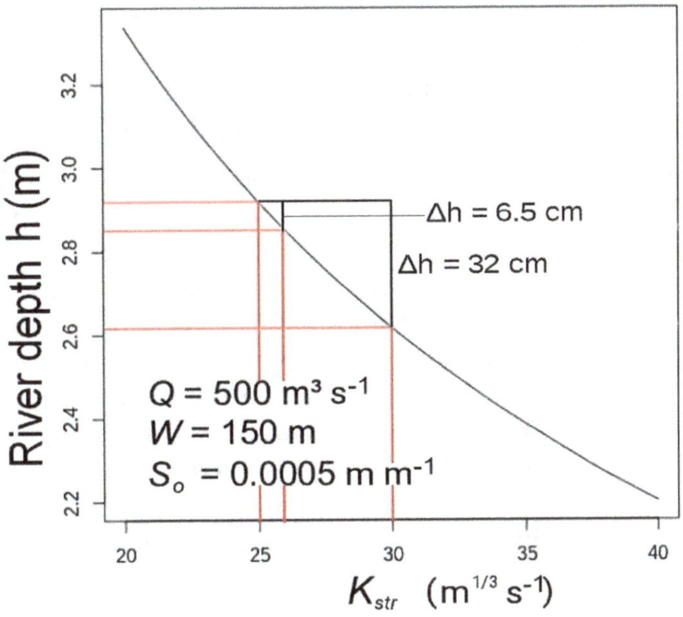

Figure 4. River depth, h, as a function of the roughness coefficient, K_{str}, for a large rectangular river channel, with geomorphological parameters and a discharge value typical of large plain rivers.

In order to verify that the model is close to linearity, we undertook an experiment on a river reach located in the downstream Garonne river: Tonneins (see Figure 1). This reach is representative of the river flows in the plain areas. The study was set up over a period of 6 weeks (between the 15th October and 30th November 1999) with a reference K_{str} value of 30 $m^{1/3}$ s^{-1}. The chosen period experienced strong climate variability with dry and wet periods, allowing the analysis of different ranges of discharge in the selected reach. We imposed four different perturbations to the reference K_{str} value and quantified the impact on the left balance term of Equation (4) (averaged value over the full period of study, by considering a daily time step). Table 1 shows that by comparing the four different perturbations, the linear approximation is only valid for small perturbations. In addition, it appears that the model is not symmetric. Hence, it is important to limit the perturbations to +5%, and to impose a maximum increment of 1.5 $m^{1/3}$ s^{-1}, in order to avoid any instability of the DA system. Table 1 shows that for a perturbation of +/−20%, the system is not at all linear.

Table 1. Impact of the relative and absolute perturbation of the K_{str} reference coefficient on the river depths, h (cm).

ΔK_{str}	Averaged Δh
−5% (−1.5 $m^{1/3}$ s^{-1})	+1.5 cm
+5% (+1.5 $m^{1/3}$ s^{-1})	+1.0 cm
−20% (−6 $m^{1/3}$ s^{-1})	+7.5 cm
+20% (+6 $m^{1/3}$ s^{-1})	+3.5 cm

3.4. Data Assimilation Experiment Setting

3.4.1. SWOT-Like Observations Generation and Their Processing

The methodology to generate SWOT-like observations has only been very briefly shown in Figure 3 (blue boxes). It is described in more detail in the following paragraphs.

SWOT data are generated using the simple methodology used in previous studies [12,13]. First, SWOT orbit ground tracks crossing the study domain are selected. For each orbit, polygons of SWOT swaths are created. These polygons are used to select all model pixels that lie in these swaths' polygons, for each observation time during a 21-day repeat period. Then, this pixel's selection is replicated for all repeat cycles (i.e., 21-day time spans) during the whole study time period. Then, for all observation times and all selected pixels, a noise on water depth is simply added to the reference simulation water depths (SIM_TRUE) to generate SWOT-like data. For the three first experiments, the instrument noise is a white noise with a 10 cm standard deviation, for all selected pixels. A less simplistic additive noise on water depth is computed for the last experiment (see Section 4.4).

With this methodology, all pixels within the river network that lies in the swath polygons could be potentially observed. Yet, SWOT is conceived to observe rivers wider than 100 m (requirement) and could even observe rivers wider than 50 m (goal). To take this limitation into account, only river portions that were wider than 50 m were considered in this study. Furthermore, the 10 cm accuracy on water elevation is only met after aggregating SWOT data over 1 km^2 [10]. That is why Strickler coefficients, K_{str}, and river depths, h, from the reference simulation are aggregated over 10 km reaches (5 to 10 RAPID grid cells) to meet this requirement, before adding the noise on water depths. The aggregation is operated by averaging the values in the considered grid cells, with a weighting proportional to their length (1 or 2 km). The area of each reach varies from 0.5 to 2 km^2, knowing that the computed width varies from 50 m to 200 m. A total of 165 reaches (Figure 5a) were constituted. Over the study domain in our experimental set-up, reaches were observed between one and four times per repeat period (Figure 5b).

Many sources of errors (orbit errors, SAR interferometry processing errors, impact of the surrounding topography and vegetation, delays induced by the crossing of the atmosphere by the electromagnetic wave) are therefore not considered, as the corresponding errors are not Gaussian.

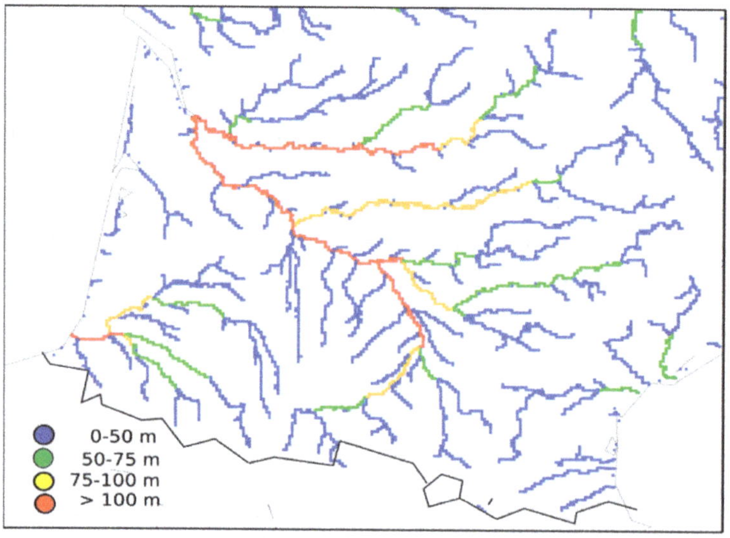

(a)

Figure 5. *Cont.*

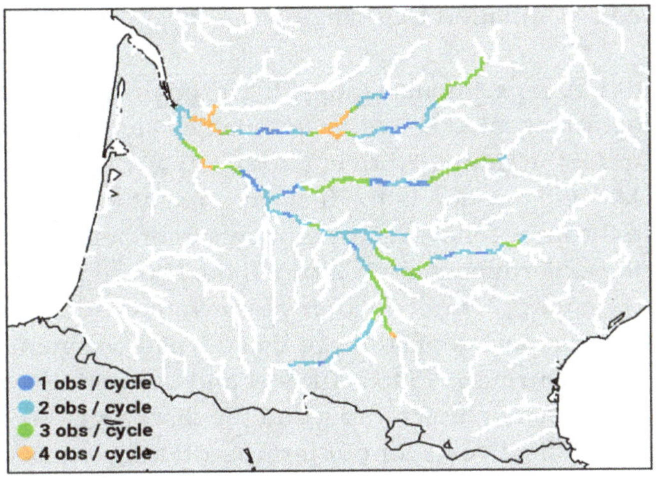

(b)

Figure 5. (**a**) River widths computed in the DA platform over the Garonne catchment, considering three river width classes where virtual SWOT observations are assimilated; (**b**) SWOT observation number for a 21-day cycle for every 10-km reaches of the Garonne catchment.

3.4.2. Description of the Variables Used in the DA Platform

In the DA platform, windows with a duration of 2 to 42 days were tested, as a function of the conditions imposed in the chosen DA experiment (see Section 4). The background vector, x_b, contains the 165 Strickler coefficient values for each river reach. The vector, $H(x_b)$, contains the 165 river depths simulated by the model for each river reach. The observation vector, y_o, contains the p SWOT observations by a DA window. If two examples are taken with five river reaches: p = 10 when these five reaches are observed two times during a DA window, or more complicated: p = 20 when the first reach is observed two times, the second reach is observed three times, and the third, fourth, and fifth reaches are observed five times during the window. It presents a sum of 10 SWOT observations in the first example, and 20 observations in the second. The p value is then a function of the number of SWOT passes across the catchment during one DA window, and of the number of rivers reaches observed during one pass of the satellite. The Jacobian matrix, **H**, contains the sensitivity of the 165 simulated river depths with a 5% perturbation of the p Strickler coefficients located in the p observed reaches. The value of +5% was chosen due to Sub-Section 3.3 (see also Table 1), having a minor impact on the Δh value calculated in Equation (4), for a typical large plain river. The observation error matrix, **R**, and the background error matrix, **B**, are diagonal: **R** contains the errors of the p river depth observations, associated with p diagonal terms. **B** contains the errors of the 165 background Strickler coefficients, associated with 165 diagonal terms. According to Sub-Section 3.3, in order to avoid problems of non-linearity, we decided to limit the limitation of increment to 1 $m^{1/3}$ s^{-1} in each experiment. The different variables of the Extended Kalman Filter equation (Equation (2)) and their dimension are described in Table 2.

Table 2. Synthetic description of the variables used in the assimilation equations (symbol, description, and dimension). The p value corresponds to the number of SWOT observations during a DA window.

Symbol	Variable	Dimension
x_b	A priori vector of the Strickler coefficients	165
x_t	Truth vector of the Strickler coefficients	165
x_a	Analysis vector of the Strickler coefficients	165
y_o	Observation vector of the synthetic SWOT river depths during a DA window	p
$H(x_b)$	Model equivalent of the y_o vector, with the river depths simulated by RAPID during a DA window	p
R	Observation error matrix of river depths	p × p
B	Model error matrix of Strickler coefficients	165 × 165
H	Jacobian matrix: sensitivity of $H(x_b)$ to a perturbation of the Strickler coefficient in the model	165 × p

3.4.3. Description of the Data Assimilation Experiments

We set-up four different DA experiments (Table 3): The goal was to analyze whether the a priori values, x_b, of the Strickler coefficient could converge to a stable value and tend to the truth, x_t, through the assimilation windows, and if the simulated river depth, $H(x_b)$, could be well estimated and tend to the true depth, $H(x_t)$. The DA configuration in the first experiment was the same as in previous works related to SWOT DA [13]. In the three following experiments, more realistic experiments were then set up in the DA platform. In the experiments, n° 1, 2 and 4, the chosen period of study was 1995–1998, and 1995–2001 in the experiment, n° 3, three further years were needed to get a full convergence of the system. We will show that the period of study of this third experiment is longer, due to a longer convergence time of the system. During 1995 to 2001, wet and dry periods were observed. The strong climate variability of these years had a direct impact on the discharge: We therefore conclude that it was a good idea to have selected these years for our test case. It allowed us to analyze a wide range of discharge signals. In the following paragraphs, a short description of all four experiments is proposed and synthesized in Table 3.

Table 3. Synthetic description of the four SWOT DA experiments.

Criterion	Experiment 1: River Depth Assimilation	Experiment 2: River Depth Assimilation by Perturbing ISBA Outputs	Experiment 3: Difference of River Depth Assimilation	Experiment 4: River Depth Assimilation Considering More Realistic SWOT Errors
Assimilation type	River depths	River depths; ISBA runoff and drainage perturbed ±10%	river depth differences	River depths
Window duration	48 h	48 h	42 days	48 h
Beginning of the experiment	1 August 1995	1 August 1995	1 August 1995	1 August 1995
End of the experiment	31 July 1998	31 July 1998	15 April 2001	31 July 1998
First *a priori* value x_b	Each value of Strickler coefficient are equal to 25 $m^{1/3}\ s^{-1}$	"Truth" of the 165 Strickler coefficient values perturbed with a centered Gaussian noise, σ_{xb}, equal to 5 $m^{1/3}\ s^{-1}$	"Truth" of the 165 Strickler coefficient values perturbed with a centered Gaussian noise, σ_{xb}, equal to 5 $m^{1/3}\ s^{-1}$	"Truth" of the 165 Strickler coefficient values perturbed with a centered Gaussian noise, σ_{xb}, equal to 5 $m^{1/3}\ s^{-1}$
Definition of the matrix **B**	Diagonal, each value $\sigma_B{}^2$ is equal to the variance of all the x_b values around the truth. Minimum imposed value = $(1.5\ m^{1/3}\ s^{-1})^2$	Diagonal, each value $\sigma_B{}^2$ is equal to the variance of all the x_b values around the truth. Minimum imposed value = $(1.5\ m^{1/3}\ s^{-1})^2$	Diagonal, each value $\sigma_B{}^2$ is equal to the variance of all the x_b values around the truth. Minimum imposed value = $(2.12\ m^{1/3}\ s^{-1})^2$	Diagonal, each value $\sigma_B{}^2$ is equal to the variance of all the x_b values around the truth. Minimum imposed value = $(1.5\ m^{1/3}\ s^{-1})^2$
Definition of the matrix **R**	Diagonal, each value $\sigma_R{}^2$ is equal to $(10\ cm)^2$	Diagonal, each value $\sigma_R{}^2$ is equal to $(10\ cm)^2$	Diagonal, each value $\sigma_R{}^2$ is equal to $(10\ cm)^2$	Diagonal, each value $\sigma_R{}^2$ varies in the time and the space

In the first experiment, we assimilated river depths over 48 h duration windows, across a period of three years. The choice of this duration was based on one main argument: The duration corresponds to the time taken by the water to flow from the upstream to the downstream section of the Garonne river observed by SWOT, i.e., from Saint-Gaudens to Bordeaux (see Figure 1). This means that by using a window duration of 48 h, a roughness coefficient perturbation at Saint-Gaudens impacts all river reaches located downstream (until Bordeaux) during this window duration. It was decided that the experiment will commence with a background vector, x_b, initialized with values equal to 25 $m^{1/3}\ s^{-1}$. This value corresponded to the averaged reference Strickler coefficient, x_t, in the catchment. Each diagonal term, $\sigma_B{}^2$, of the model error matrix, **B**, contained values equal to the variance of all the x_b parameters around the truth, x_t, with a minimum value of $(1.5\ m^{1/3}\ s^{-1})^2$. Each diagonal term, $\sigma_R{}^2$, of the observation error matrix, **R**, contained values equal to $(10\ cm)^2$.

In the second experiment, we assimilated data over 48 h duration windows by perturbing the amount of runoff and drainage produced by ISBA (increasing or decreasing the production with a bias of $\pm$10%), and tested the impact on the convergence through the assimilation windows. It is important to represent well the atmospheric forcing: In DA, an error of quantification or repartition of precipitation will have a direct impact on the convergence of the parameters to correct, and thus on the quality of the river flow simulation. In this way, it is important to quantify the impact of a forcing error in ISBA on the convergence of the Strickler coefficients. It was decided that the experiment should start with a background vector, x_b, initialized with values equal to the reference Strickler parameters on which a Gaussian centered noise, σ_{xb}, of 5 $m^{1/3}$ s^{-1} was added. Each diagonal term, $\sigma_B{}^2$, of the model error matrix, **B**, contained values equal to the variance of all the x_b parameters around the truth, x_t. A minimum value of (1.5 $m^{1/3}$ $s^{-1})^2$ was imposed in the DA platform. Note that σ_{xb} was always equal to σ_B when $\sigma_B \geq 1.5$ $m^{1/3}$ s^{-1}, and $\sigma_{xb} \leq \sigma_B$ when $\sigma_B = 1.5$ $m^{1/3}$ s^{-1}. Each diagonal term, $\sigma_R{}^2$, of the observation error matrix, **R**, contained values equal to (10 cm)^2.

In the third experiment, river depth differences were assimilated over a 42-day duration window in the model, considering that the satellite would not observe river depths, but, instead, open water surface elevations. A solution for the use of water surface elevation in our system was to assimilate the water surface elevation difference, δh, between two consecutive observations, which was equal to the river depth difference. In order to assimilate δh in the model, two or more river depth observations in one DA window were required. One can imagine two different scenarios:

1. In a DA window, there are zero or one river depth observations: It is impossible to assimilate a δh term; and
2. In a DA window, there are n observations (n $\geq$ 2): It is possible to assimilate n $-$ 1 δh terms.

To respect the case, n° 2, for all reaches of the Garonne basin, we decided to extend the assimilation window to 42 days, to be sure that at least two observations were present for each reach in the assimilation window. It was decided to start the experiment with a background vector, x_b, initialized with values equal to the reference Strickler parameters on which a Gaussian centered noise, σ_{xb}, of 5 $m^{1/3}$ s^{-1} was added. Each diagonal term, $\sigma_B{}^2$, of the model error matrix, **B**, contained values equal to the variance of all the x_b parameters around the truth, x_t. A minimum value of (2.12 $m^{1/3}$ $s^{-1})^2$ was imposed in the DA platform. Each diagonal term, $\sigma_R{}^2$, of the observation error matrix, **R**, contained values equal to (14.1 cm)^2.

In the fourth experiment, we increased the realism of the SWOT error measurement, by implementing time-and-space variable errors of observation: Each diagonal term, $\sigma_R{}^2$, of the observation error matrix, **R**, contained values varying for every reach at each DA window. Data were assimilated over a 48 h duration window. The decision was made to start the experiment with a background vector, x_b, initialized with values equal to the reference Strickler parameters on which a Gaussian centered noise, σ_{xb}, of 5 $m^{1/3}$ s^{-1} was added. Each diagonal term, $\sigma_B{}^2$, of the model error matrix, **B**, contained values equal to the variance of all the x_b parameters around the truth, x_t. A minimum value of (1.5 $m^{1/3}$ $s^{-1})^2$ was imposed in the DA platform.

Note that the common criterion for the comparison of these four different DA experiments is the quality of the convergence of the Strickler coefficients, K_{str}, to the truth, x_t. The convergence time of the K_{str} is irrelevant for the comparison between the four experiments, because the duration of the DA windows, the first guess, x_b, of the K_{str} at the beginning of the experiment, and the attribution of the diagonal terms in the R and B matrices vary in function to the experiment.

4. Results

4.1. Assimilation of River Depths (Experiment 1)

We decided in this first DA experiment to impose a minimum value of the diagonal terms, $\sigma_B{}^2$, in the **B** matrix to (1.5 $m^{1/3}$ $s^{-1})^2$: When the values are too low (<(1 $m^{1/3}$ $s^{-1})^2$), the convergence of the terms ($x_b - x_t$) to the truth, x_t, is complicated, because the analysis is closer to the background than the observation (for the EKF, there are fewer uncertainties in the background than in the observation). This limitation criterion thus improves the convergence of the terms ($x_b - x_t$) to the truth, x_t. Concerning the value of the increment ($x_a - x_b$), when no limitation is imposed, the values of the terms ($x_b - x_t$) of the DA system are very high in the first windows (between -40 to $+70$ $m^{1/3}$ s^{-1}), and the convergence time is about 800 days. When a limitation of increment of 5 $m^{1/3}$ s^{-1} is imposed, the values of the terms ($x_b - x_t$) of the DA system during the early windows are between -40 and +30$m^{1/3}$ s^{-1}until 100 days of assimilation, and the convergence to the truth occurs after 240 days. In the current experiment illustrated in Figure 6, the majority of the K_{str} values converge to the truth after 700 days of assimilation, considering a limitation of increment fixed to 1 $m^{1/3}$ s^{-1}. We consider that the Strickler coefficient convergence to the truth, x_t, is reached when the standard deviation, σ_{xb}, between all ($x_b - x_t$) values is equal or lower than 1 $m^{1/3}$ s^{-1}.

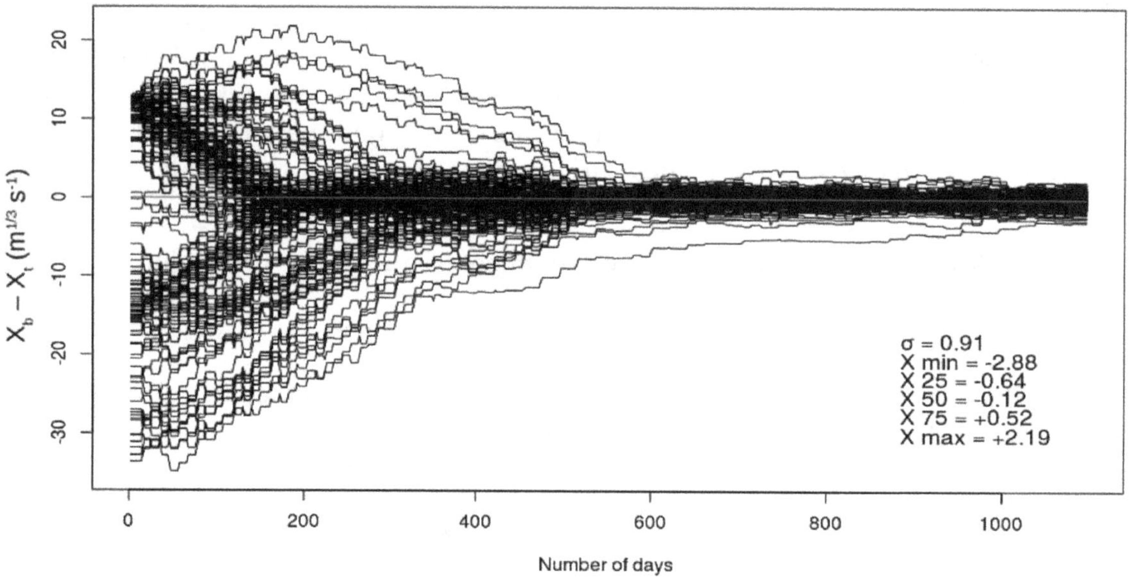

Figure 6. Temporal evolution (1 August 1995 to 31 July 1998) of the difference between the a priori control vector (x_b) and the true control vector (x_t) of the Strickler coefficients in the DA system. The 165 black curves represent the value ($x_b - x_t$) at all 165 reaches defined over the Garonne catchment. The red horizontal line is equivalent to a zero difference between x_b and x_t. After 3 years (i.e., 548 48 h-assimilation windows), we represent the standard deviation, σ_{xb}, between the 165 values ($x_b - x_t$) in the basin. The X terms represent, from top to bottom, the minimum value, the 1st quartile, the 2nd quartile, the median, the 3rd quartile, and the maximum value of ($x_b - x_t$) at the end of the experiment (day n° 1096).

In Figure 7, we show the time evolution of the simulated and the true river depth for the reaches of Lamagistère (a) and Bergerac (b) located, respectively, in the downstream Garonne and Dordogne river (see Figure 1). We have chosen these two reaches, firstly because they are representative of plain rivers where SWOT will provide observations, and secondly because their initial Strickler coefficient in the DA system is far from the truth, x_t.

At Lamagistère, the initial value was 25 $m^{1/3}$ s^{-1}, while the true value was 38.7 $m^{1/3}$ s^{-1}. The simulated river depth, $H(x_b)$, before SWOT data assimilation was then higher than the true river depth $H(x_t)$, with a positive difference of about 20 cm in normal flow, and more than 80 cm during flood periods. The average true river depth, $H(x_t)$, over the full period of study (365 days) was equal to 2.11 m, with peak values up until 7.5 m. Furthermore, at the beginning of the experiment, the river flow velocity at Lamagistère and upstream was too low (due to low x_b values compared to x_t): The temporal delay between the river depth time series before assimilation and the true time series was negative (with about one day of delay).

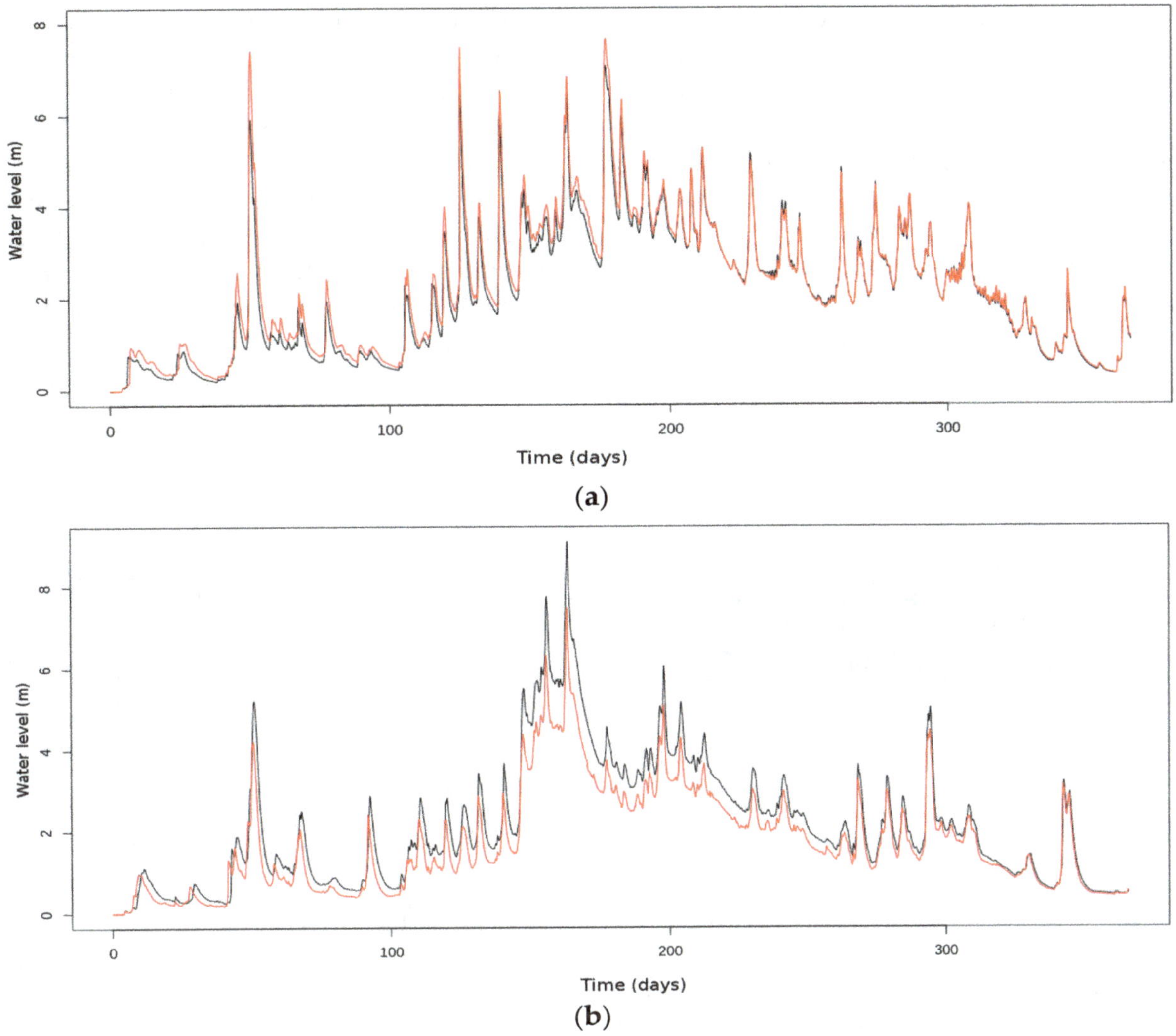

Figure 7. Temporal evolution of the river depth at Lamagistère (**a**) and Bergerac (**b**), over the period of 1 August 1995 to 31 July 1996. Black curves correspond to the truth and red curves are the result from the assimilation of the river depth.

At Bergerac, the case was the opposite: We started the experiment with a Strickler coefficient value higher than the truth, x_t (positive difference of 7.5 $m^{1/3}\ s^{-1}$), which was equal to 17.5 $m^{1/3}\ s^{-1}$: The relative difference between the first x_b value and x_t was equal to +42.9%. The simulated river depth, $H(x_b)$, before SWOT data assimilation was then lower than the true river depth, $H(x_t)$, with a negative difference of 20–30 cm in normal flow, and more than 1 m during flood periods. The average true river depth, $H(x_t)$, over the full period of study (365 days) was equal to 2.10 m, with peak values up to 9 m. Furthermore, at the beginning of the experiment, the river flow velocity at Bergerac and upstream was too high (due to high x_b values compared to x_t): The delay between the river depth time series before assimilation and the true time series was positive (with about one day of advance).

For Lamagistère and Bergerac after 365 days of SWOT data assimilation, both the phasing and signals between $H(x_b)$ and $H(x_t)$ were improved: The time difference between the arrival of one $H(x_b)$ peak and one $H(x_t)$ peak was less than 3 h, and the difference between $H(x_b)$ and $H(x_t)$ was inferior to 10 cm for every discharge regime.

4.2. *Assimilation of River Depths, in case of Atmospheric Forcing Biases (Experiment 2)*

In this experiment, we propose two different scenarios. In the first (a), the water produced by ISBA (runoff + drainage) is decreased by 10%. In the second scenario (b), the water produced by ISBA is increased by 10%. The main goal of the experiment is to analyze what the impact of these errors on the Strickler coefficient convergence is. The effect of the over-production of water in the river is the following: The DA system will tend to increase the K_{str} values to decrease the river depth. The effect of an under-production of water in the river is the opposite: The DA system will decrease the K_{str} values to increase the river depth. The illustration in Figure 8 shows that after about 1000 days of assimilation, the Strickler coefficient values converge to a stable state, but differ from the true x_t with an average bias of $-1.56\ m^{1/3}\ s^{-1}$ in scenario (a), and an average bias of $+1.83\ m^{1/3}\ s^{-1}$ in scenario (b). For a water production bias of +10% or −10%, we showed that the impact on the terms ($x_b - x_t$) at the end of the DA experiment was inferior to $2\ m^{1/3}\ s^{-1}$, having a minor impact on the flow velocity (phasing delay of 3 h at Tonneins between the "true" hydrograph and the "analyzed" hydrograph).

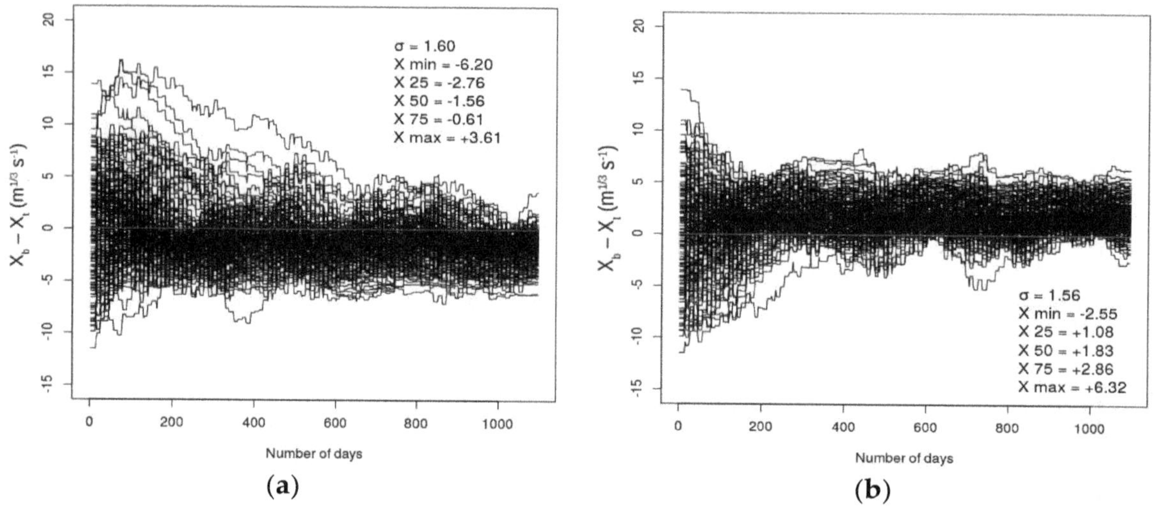

Figure 8. Temporal evolution (1 August 1995 to 31 July 1998) of the difference between the a priori control vector (x_b) and the true control vector (x_t) of the Strickler coefficients in the DA system. Scenario (**a**) represents a −10% perturbation of water produced by ISBA, and scenario (**b**) represents a +10% perturbation. The 165 black curves represent the value ($x_b - x_t$) of all 165 reaches defined over the Garonne catchment. The red horizontal line is equivalent to a zero difference between x_b and x_t. After 3 years (i.e., 548 48 h-assimilation windows), we represent the standard deviation, σ_{xb}, between the 165 values ($x_b - x_t$) in the basin. The X terms represent from top to bottom the minimum value, the 1st quartile, the 2nd quartile, the median, the 3rd quartile, and the maximum value of ($x_b - x_t$) at the end of the experiment (day n° 1096).

4.3. Assimilation or River Depth Differences (Experiment 3)

In previous experiments, only observed river depths have been considered. To obtain river depths from SWOT surface elevations, it is required to know a priori the river bathymetry, or at least to get some estimates of the river bed elevation. To overcome this stringent assumption, assimilation of river elevation differences (equal to river depths differences) is tested in this experiment.

The variance of the observation error for two consecutive observations, h_t and h_{t+1}, follows Equation (5):

$$Var(h_{t+1} - h_t) = Var(h_t) + Var(h_{t+1}) - 2cov(h_t, h_{t+1}) \tag{5}$$

Assuming that two consecutive observation errors are not correlated and that the SWOT error is 10 cm (standard deviation), $cov(h_t, h_{t+1}) = 0$ and the observation error $(Var(h_{t+1} - h_t))^{1/2}$ thus becomes $\sigma_R = 14.1$ cm.

To keep the same ratio between the diagonal terms, σ_R and σ_B (see Equation (6)), of the two matrix, **R** and **B** (as the experiments detailed in the Sections 4.1 and 4.2), the minimum value of σ_B is now fixed to $2.12\ m^{1/3}\ s^{-1}$. This value is calculated from the diagonal terms, σ_B, equal to $1.5\ m^{1/3}\ s^{-1}$

in the **B** matrix used in the two previous experiments. We show in Figure 9 the temporal evolution of the terms ($x_b - x_t$) for all reaches in the basin.

$$\frac{\sigma_{R(exp1)}}{\sigma_{B(exp1)}} = \frac{\sigma_{R(exp3)}}{\sigma_{B(exp3)}}, \sigma_{B(exp3)} = \sigma_{R(exp3)} \times \frac{\sigma_{B(exp1)}}{\sigma_{R(exp1)}} \quad (6)$$

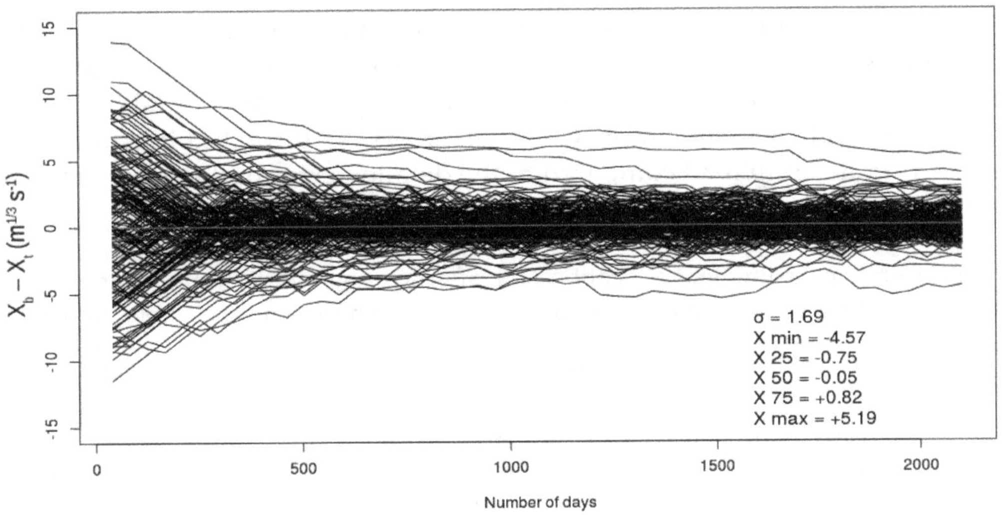

Figure 9. Temporal evolution (1 August 1995 to 15 April 2001) of the difference between the a priori control vector (x_b) and the true control vector (x_t) of the Strickler coefficients in the DA system. The 165 black curves represent the value ($x_b - x_t$) at all 165 reaches defined over the Garonne catchment. The red horizontal line is equivalent to a zero difference between x_b and x_t. After 5 years (i.e., 50 42 day-assimilation windows), we represent the standard deviation, σ_{xb}, between the 165 values ($x_b - x_t$) in the basin. The X terms represent from top to bottom the minimum value, the 1st quartile, the 2nd quartile, the median, the 3rd quartile, and the maximum value of ($x_b - x_t$) at the end of the experiment (day n° 2100).

After 15 DA windows (2 years or 750 days), the majority of the Strickler coefficients in the DA system converged to the truth, x_t. The convergence of four reaches occur more slowly: One plausible hypothesis is that the observation error is on average larger than for the other reaches, driven by the random Gaussian noise, σ_R, which can be lower or higher than 14.1 cm across the full study period. There are only 50 DA windows in this experiment, and the impact of σ_R on the convergence of x_b is different for one given reach after these 50 windows, when the experiment is remade. After 50 windows or 2100 days of assimilation, the standard deviation, σ_{xb}, of all ($x_b - x_t$) terms is about 1.70 $m^{1/3}$ s^{-1}: This value is bigger than the σ_{xb} value presented in the first experiment (Section 4.1), because the diagonal terms, $\sigma_R{}^2$ and $\sigma_B{}^2$, imposed in the observation error matrix, **R**, and the model error matrix, **B**, are more significant.

4.4. *Assimilation of River Depths Considering More Realistic SWOT Errors (Experiment 4)*

In this experiment, the diagonal terms, σ_R, in the observation error matrix, **R**, vary in space and time (at each DA window). The total error (taking into account the variable and constant errors averaged in time and space) should be equal to 10 cm, and is composed of systematic and non-systematic errors [10]. The non-systematic errors are random and Gaussian, and the systematic errors are constant (see Section 3.1). We are able to approximate by a very simple method the different sources of instrumental error contributing to the changing of the SWOT error measurement values in space and time. Four factors with an impact on the diagonal term values, σ_R, of the observation error matrix, **R**, are introduced, allowing σ_R to vary at each time step and for every river reach. The first three factors (non-systematic source of errors) are: The surface of the reach, the SWOT look angle, and the radar water surface roughness. The calculation of these simple instrumental-like errors is based on previous works [34]. The last factor having an impact on the σ_R of the **R** matrix is the wet-troposphere error (systematic source of errors). This error is estimated in this study by the intra-day variability of water content in the troposphere. There are other sources of errors, more difficult to quantify, but they should have a lower impact on the total error budget at the reach scale: They are the ionosphere signal, the dry troposphere signal, the orbit radial component, and the KaRIn random and systematic errors after cross-over corrections. Table 4 summarizes the SWOT error budget [35], after averaging a 1 km^2 area [10]. It is important to note that all the equations used to calculate these different sources of instrumental errors are greatly simplified, the more realistic and complex full radar equations are not used in our study. The main goal of this fourth experiment is to take into account errors of measurement more realistic than simple Gaussian errors of 10 cm, but it is important to keep in mind that these errors are simplified compared to the detailed error budget calculated by the SWOT simulator used by the JPL and the CNES (see [36]).

Table 4. SWOT error measurement allocation [35], after averaging over a 1 km^2 area [10]. The left column represents the error component, and the right column represents the values of these errors.

Hydrology Error Component	Height Error (cm)
Ionosphere signal (constant)	0.80
Dry troposphere signal (constant)	0.70
Wet troposphere signal (variable)	4.00
Orbit radial component (constant)	1.62
KaRIn Random and systematic errors after cross-over correction (constant)	7.74
Surface of the reach + look angle + wind speed (variable)	4.40
Total allocation	9.95
Unallocated margin (constant)	1.04
Total error	10.0

4.4.1. SWOT Instrumental Error along the Swath

The SWOT error measurement is linked to the surface of the observed reach: It is a non-systematic source of error. When several raw SWOT radar images are averaged over a bigger area, the error decreases with the squared root of the number of raw SWOT radar images [10]. The surface of the reach is proportional to the width, *W*. When the river width decreases, the error increases because the observed surface by the satellite is lower. Furthermore, the measurement errors increase with the look angle of the satellite because the surface of the raw SWOT radar image is not the same: 70 m $\times$ 5 m for the far range image, and 10 m $\times$ 5 m for the near range image (see Section 3.1). These measurement errors then vary along the swath. Based on previous works developping a simple algorithm for calculating the sensitivity of the measurement error as a function of the river width and the look angle of the satellite [34], the SWOT look angle will vary from 0.6° to 4.3°, corresponding, respectively, to the inner and outer board of each swath.

The roughness of the open water surface will determine the amount of energy reflected to the satellite and will thus impact SWOT's error budget. In our case study, the roughness is simulated very simply, considering that it depends only on the wind velocity, and was previously validated over the ocean domain only (see [34]). Even if the important complexity of these sources of errors is not considered well, the goal for our study is to consider the main processes and to understand how the measurement errors vary along the swath. In Figure 10, we show the sensitivity of the error we considered in our study to the wind speed near the water surface, considering a river reach 10 km long and 100 m wide. Over the full period of 1 August 1995–31 July 1998, 75% of the wind speed values are between 0 and 3 m s^{-1}. Wind speed values higher than 8 m s^{-1} are rare, representing about 1% of all wind speed values over the study period. The wind speeds correspond to SAFRAN reanalysis data at 10 m (3-h time step) averaged over a 10 min period.

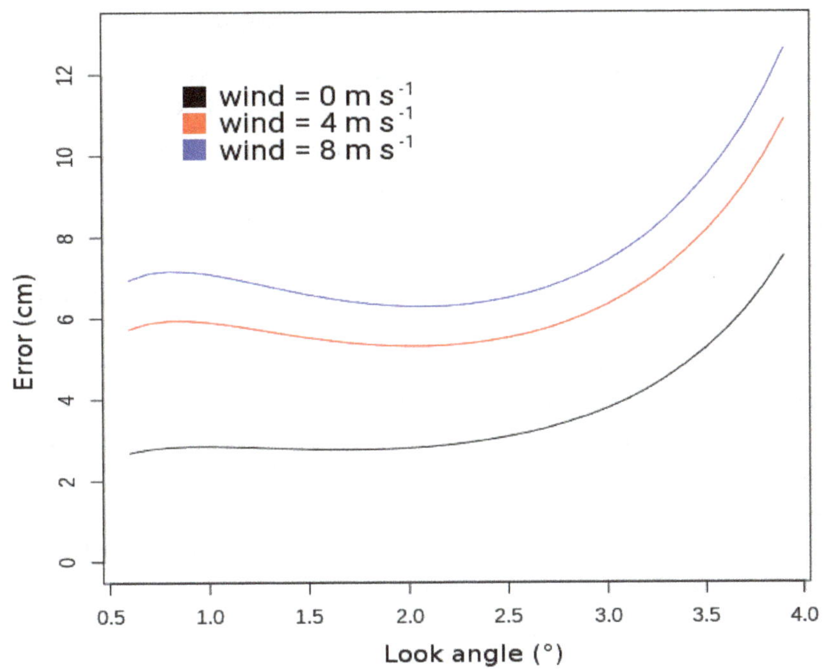

Figure 10. SWOT error measurement as a function of the look angle for a wind speed of 0 m s^{-1}, 4 m s^{-1}, or 8 m s^{-1}.

4.4.2. The Intra-Day Variability of Water Content in the Troposphere: Impact on the SWOT Error of Measurement

According to previous works [35], the estimation of the wet troposphere error is about 4 cm (Table 4). Water content changes in the troposphere have a direct impact on the velocity of electromagnetic waves. Numerical atmospheric models are able to produce a correction of the wet troposphere error. The water content simulated by models in the atmosphere, however, also shows an error. Several studies have looked at the impact of errors in meteorological model outputs or in in situ observations concerning the representation of water content in the troposphere [37,38]: These errors of the atmospheric state should be considered to determine the error on the measured open water surface elevation. We would like to base our study on previous works showing that it is more difficult to quantify the water content in a transition meteorological period than during a stable period [39]: We have defined a transition period equivalent to an interval of time with an important variability of the troposphere state, and a stable period equivalent to an interval of time with a low variability of the troposphere state. The variable chosen to describe this temporal variability is the water content, and the chosen interval of time is 24 h: This time value is well representative of the meteorological variability, especially in summer with important changes of the troposphere state during a full 24-h period (important contrasts between day and night). The SAFRAN reanalysis data of water content (g kg^{-1}) at 10 m (3-h time step) are used to calculate intra-day variances *Vi* (g $kg^{-1})^2$ across the study period 1 August 1995–31 July 1998. Considering the 165 river reaches of the DA platform over the full study period, a simple relationship between the intra-day variance of water content, *Vi*, and the SWOT error measurement, σ_R, is set up: The greater *Vi*, the greater the measurement error. We would like to build a linear relationship between *Vi* and σ_R. Once the relationship between *Vi* and σ_R is established, one can attribute a SWOT error measurement error, σ_R, at each time step for every river reach of the catchment. The slope of the function linking *Vi* and σ_R is equal to 6.67 cm/(g $kg^{-1})^2$. This relationship is based on the assumption that the spatial and temporal averaged wet troposphere error is equal to 4.0 cm (see Table 4), and that the full spatio-temporal SAFRAN reanalysis data distribution of water content is well-known in each river grid cell of the catchment. Eighty percent of the *Vi* values range between 0 and 1 (g $kg^{-1})^2$. Values higher than 3 (g $kg^{-1})^2$ are rare, representing about 1% of all *Vi* values over the full study period.

4.4.3. SWOT DA Experiment Using Realistic Errors of Measurement

Considering the sensitivity of the SWOT measurement error to the two described errors in Sections 4.4.1 and 4.4.2, it is possible to conduct this last experiment with slightly more realistic errors. The observation error matrix, **R**, is built considering the two sources of errors described in Sections 4.4.1 and 4.4.2: Each term, σ_R, varies at each time step of the experiment for every river reach, and the spatio-temporal average of σ_R over the full period of the study and domain is equal to 10 cm. After 365 days of assimilation, the majority of the parameters converges to the truth, x_t, with a standard deviation of ~1 $m^{1/3}$ s^{-1}. During the DA windows n° 259 and 448, and 518 and 896 days of assimilation, the standard deviation, σ_{xb}, between all ($x_b - x_t$) values decreases, respectively, to 0.76 and 0.79 $m^{1/3}$ s^{-1} (Figure 11). To understand these low values, we analyzed the temporal evolution of the *Vi* variable over the full study period: It showed that during the winter months the values of *Vi* are equal to 0.5 (g $kg^{-1})^2$ on average (the days 518 and 896 of this experiment are in the winter season), and 1.5 (g $kg^{-1})^2$ on average during the summer months. The low values of *Vi* during the winter season lead to low SWOT error measurements, σ_R, and then to a better convergence of the K_{str} values to the truth, x_t. The convergence time of this DA experiment is around 600 days when the x_b values of each reaches convergence at the truth, x_t.

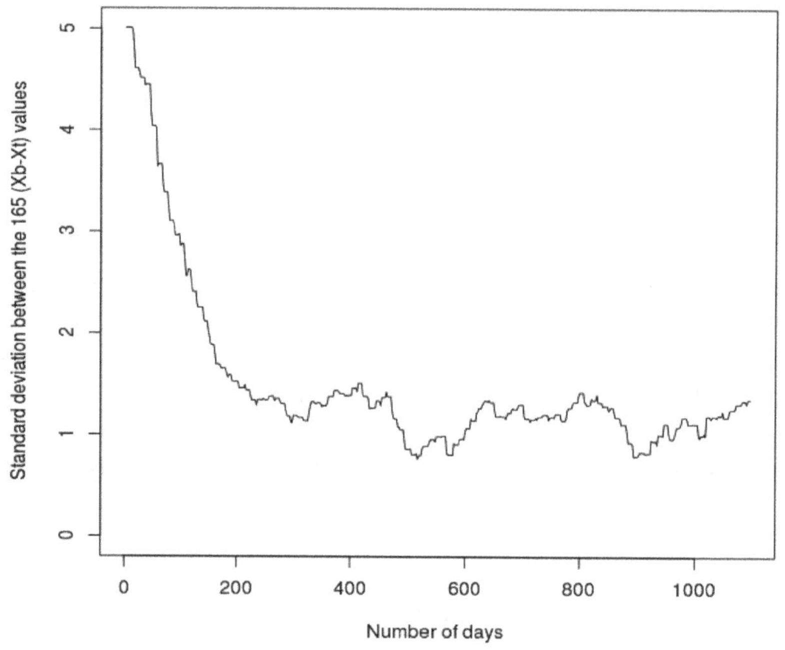

Figure 11. Temporal evolution of the standard deviation, σ, between the 165 ($x_b - x_t$) terms, over the 1 August 1995–31 July 1998 period.

5. Conclusion and Perspectives

This study is a contribution to the building of a DA scheme over the Garonne basin within the framework of the SWOT mission. More precisely, the objective of this study was to evaluate the ability of a regional hydrometeorological model to assimilate synthetic SWOT river depths in the Garonne catchment, and to correct the Strickler roughness coefficient, K_{str}, of the river bed.

In every experiment, we showed the ability of the DA system to calibrate the K_{str} parameter around the known truth. In the first experiment, we showed that the parameters converge to the truth after ~2 years, with an average error of ± 1 $m^{1/3}$ s^{-1}, and with an associated average error on river depths of ± 5 cm. The choice of the maximum increment value (the difference between the analysis, x_a, and the a priori value of the Strickler coefficient, x_b) has an impact on the time of convergence. In the second experiment, we showed the importance of the atmospheric forcing and its impact on the convergence of the K_{str} parameters. An error of $\pm 10\%$ water produced by ISBA led to an error of

$\pm 1.5\ m^{1/3}\ s^{-1}$, but had a minor impact on the river flows. The impact of greater errors on the discharge phasing should be taken in consideration. The third experiment illustrated the interest of assimilating the difference of river depths instead of absolute open surface water elevations that are equivalent to river depths in our study. The bathymetry elevation is unknown in RAPID, and the satellite will not provide measurements of river depths. The duration of a DA window should be equal to 42 days, in order to be sure that every reach is observed at least two times during one DA window. The quality of the convergence ($\pm 1.69\ m^{1/3}\ s^{-1}$) was slightly lower in comparison with the first experiment, but the time of convergence was equivalent (~2 years). In the last experiment, we proposed the introduction of more realistic SWOT error measurements varying in time and space, which depended on the surface of the observed reach, the look angle of the satellite, the surface water roughness, and the intra-day variability of the water content in the troposphere. The impact of the last factor seems to play an important role: The convergence was on average better during the winter months than during the summer months, because of a low temporal variability of the troposphere water content.

Several points concerning data assimilation in the context of the SWOT mission could be explored. The four main ones are as follows:

5.1. The Choice of the Data Assimilation Method and the Experimental Design

The main goal of each presented experiment in Section 4 was to calibrate the Strickler coefficients to converge them to the truth, in the context of twin experiments. The choice of the EKF requires the use of the hydrological river routing model, RAPID, in the domain of the linearity. It has its limitations: The perturbation of the parameters we want to correct (roughness coefficient) should be small, and they require limited increment values (a maximum difference between the first guess, x_b, and the analysis, x_a, calculated by the DA system). Other methods, such as the Ensemble Kalman Filter (EnKF) [40], could be tested to see whether it is possible to reduce the convergence time of the parameters, or to decrease the standard deviation, σ_{xb}, around the truth, x_t, at the end of one given DA experiment. In our study, we assumed that the model is linear: We should use it in the domain of linearity by imposing small perturbations to the parameter to correct. The EnKF can be used in the domain of non-linearity and could be tested. It has gained popularity because of its simple conceptual formulation and relatively easy implementation, requiring no derivation of a tangent linear operator or adjoint equations, and no integrations backward in time [40]. Concerning the description of the different variables and data assimilation experiments described in Section 3.4, another way in which to build the experimental design could be proposed: One important one is the testing of what the impact is on different lengths of river reaches on the convergence of the Strickler coefficients in all four DA experiments, the SWOT error measurement being directly linked to the length of the reaches. Another point is the attribution of the different parameters used in the DA platform: Choice of the DA window duration, threshold of the increment ($x_a - x_b$) limitation, and attribution of the model error matrix. The choice of the attribution of the first a priori value, x_b, at the beginning of each experiment also plays an important role, having a direct impact on the time of convergence of the Strickler coefficients. The study period can present a more or less spatial and temporal variability of precipitation reaching the surface, impacting directly the flood dynamics in the river network, and thus the increment term ($x_a - x_b$), which depends on the discharge regime. All these factors shortly described could have different impacts on the convergence of the system.

5.2. Choice of the Observation to Assimilate and the Parameter to Correct

In the study, we decided to assimilate synthetic SWOT river depths in order to calibrate the roughness coefficient of the river bed in the Garonne catchment. The possibility to assimilate river widths or water surface bed slopes should be investigated. In our work, we considered that the river channel geometry is rectangular and that the width is thus constant in time in each river grid cell. In reality, the amplitudes of the river width between summer (low discharge) and winter (high discharge) can be several tens of meters. The implementation of a trapezoidal geometry should be investigated, in order to take into account the temporal variability of the river width. Furthermore, instead of a parameter calibration, one could propose the correction of the state of prognostic variables in RAPID (initial water volumes in the river reach) or in ISBA (soil water content), or the correction of the description of the atmospheric forcing SAFRAN. The assimilation of discharge over France in ISBA/MODCOU was proposed in order to correct the initial value of the soil moisture [41]. This kind of study could also be realized with synthetic river depths instead of discharge. The correction of prognostic variables instead of parameters could lead to significant advances for operational hydrology, such as the monitoring of water resources and better representation of floods on a regional scale.

5.3. Assimilation of River Depths Considering More Realistic Meteorological and SWOT Errors

In the last proposed DA experiment (Section 4.4), we decided to set up a more realistic scheme, leading to a better representation of the SWOT error measurements. Several points can be considered. The dependence of the error to the reach surface is well established [10], but other factors could have an important impact on the measurement error, like the presence of topography and/or vegetation around the reach, or the orientation of the reach compared to the azimuthal direction of the satellite: The area of the SWOT raw radar image is not the same as a function of the propagation direction of the satellite, and it will be therefore possible to aggregate more or fewer raw radar images over a 1 km^2 water area; only the non-systematic errors are impacted by this last factor. They should be taken into account in order to better quantify the measurement error. However, to do so, a more complex SWOT simulator is required: The high-resolution SWOT hydrology simulator (https://swot.jpl.nasa.gov/) will allow a better quantification of the error than in the present study. This simulator is currently used at fine spatial and temporal scales in comparison with those used in this study. It would be interesting to test whether it works on a regional or global scale or, based on the results from this simulator, whether it will be possible to parametrize SWOT errors in a more realistic way than what was achieved in this study.

Concerning the wet tropospheric signal, the chosen criterion was the intra-day variability, *Vi*, of the water content at 10 m. We showed that this variability is greater in summer than in winter, and proposed that the associated measurement error is proportional to *Vi*. We assumed in this way that it is more difficult to represent variable states than stable states of the troposphere in the time. This hypothesis needs more investigation by considering carefully the synoptic weather situation. For example, important changes of water content due to a warm or cold front (case A) are more difficult to represent than the diurnal temporal evolution of the water content in a homogeneous synoptic situation (case B), because there may be uncertainties about the arrival time of the front: A *Vi* value of 1 (g kg^{-1})^2 could therefore be more difficult to estimate in case A than the same *Vi* value during case B.

Finally, it was shown that meteorological input errors impact the determination of the Strickler coefficients. Further efforts are needed to evaluate the impact of meteorological errors in the hydrological model, by, e.g., replacing data from the high quality SAFRAN analysis by ouput from meteorological models.

5.4. Extension of SWOT DA in Other Catchments of the World

The main perspective of this study was to extend the assimilation of SWOT observations not only across the Garonne basin, but all over the world where observations are available. The initial decision to use SRTM 90 m [26] as a DEM (Digital Elevation Model) will make this operation possible. Finally, the use of real satellite data (e.g., altimetry nadir JASON-2/-3, ENVISAT or SARAL, Sentinel-3) could be tested over big river basins, such as the Amazon or the Congo basin, and will allow inter-comparisons between actual experiments and our SWOT twin experiments.

Finally, it would be valuable to build an OSSE to compare benefits of assimilating current nadir altimetry mission(s), for example, Jason-3, Sentinel-3A/B, SARAL, and/or CryoSat-2, to the benefits of assimilating SWOT data. It was shown that across 12 in situ stations along the Po river (Italy), the average root mean squared error of CryoSat-2 compared with hourly water level observations was 38 cm [7], this value being comparable to the SWOT error measurements used in this study. Furthermore, they showed the strong interest to use satellite products to cover the gaps due to the spacing of in situ measurements. Our study should be completed by providing more quantitative information on the specific added value of SWOT compared to current missions.

Author Contributions: Hydrological modelling: V.H., E.M. and A.B., Assimilation technique: S.R., A.B. and S.B., Experiment design: E.M., A.B., and S.R., Realistic SWOT errors: S.B. and V.H., Original draft preparation: V.H., Review and editing: E.M., S.B., S.R. and A.B., Funding acquisition: A.B. and E.M.

Acknowledgments: The authors wish to thank Vanessa Pedinotti and Charlotte Emery for scientific discussion, methods and assimilation code transfer, Thierry Morel for continuous support with respect to the Open-Palm software, and two anonymous reviewers who helped to improve the manuscript.

References

1. Biancamaria, S.; Lettenmaier, D.P.; Pavelsky, T.M. The SWOT mission and Its capabilities for land hydrology. *Surv. Geophys.* **2016**, *37*, 307–337. [CrossRef]
2. Biancamaria, S.; Andreadis, K.M.; Durand, M.; Clark, E.A.; Rodríguez, E.; Mognard, N.M.; Alsdorf, D.E.; Lettenmaier, D.P.; Oudin, Y. Preliminary characterization of SWOT hydrology error budget and global capabilities. *IEEE J. Sel. Top. Appl. Earth Obs. Remote Sens.* **2010**, *3*, 6–19. [CrossRef]
3. Bjerklie, D.M.; Lawrence Dingman, S.; Vorosmarty, C.J.; Bolster, C.H.; Congalton, R.G. Evaluating the potential for measuring river discharge from space. *J. Hydrol.* **2003**, *278*, 17–38. [CrossRef]
4. Crétaux, J.F.; Jelinski, W.; Calmant, S.; Kouraev, A.; Vuglinski, V.; Berge-Nguyen, M.; Gennero, M.C.; Nino, F.; Del Rio, R.A.; Cazenave, A.; et al. SOLS: A lake database to monitor in the Near Real Time water level and storage variations from remote sensing data. *Adv. Space Res.* **2011**, *47*, 1497–1507. [CrossRef]
5. Seyoum, W.M. Characterizing water storage trends and regional climate influence using GRACE observation and satellite altimetry data in the Upper Blue Nile River Basin. *J. Hydrol.* **2018**, *556*, 274–284. [CrossRef]
6. Santos da Silva, J.; Calmant, S.; Seyler, F.; Corrêa Rotunno Filho, O.; Cochonneau, G.; João Mansur, W. Water levels in the Amazon basin derived from the ERS2 and ENVISAT radar altimetry missions. *Remote Sens. Environ.* **2010**, *114*, 2160–2181. [CrossRef]
7. Schneider, R.; Tarpanelli, A.; Nielsen, K.; Madsen, H.; Bauer-Gottwein, P. Evaluation of multi-mode CryoSat-2 altimetry data over the Po River against in situ data and a hydrodynamic model. *Adv. Water Resour.* **2018**, *112*, 17–26. [CrossRef]
8. Alsdorf, D.E.; Rodríguez, E.; Lettenmaier, D.P. Measuring surface water from space. *Rev. Geophys.* **2007**, *45*. [CrossRef]
9. Pavelsky, P.; Durand, M.; Andreadis, K.M.; Beighley, R.E.; Paiva, R.C.D.; Allen, G.H.; Miller, Z.F. Assessing the potential global extent of SWOT river discharge observations. *J. Hydrol.* **2014**, *519*, 1519–1525. [CrossRef]
10. Rodríguez, E.; Surface Water and Ocean Topography (SWOT). Science Requirements Document, JPL Docment D-61923. 2016.
11. Zaitchik, B.F.; Rodell, M.; Reichle, R.H. Assimilation of GRACE Terrestrial Water Storage Data into a Land Surface Model: Results for the Mississippi River Basin. *J. Hydrometeorol.* **2008**, *9*, 535–548. [CrossRef]
12. Biancamaria, S.; Durand, M.; Andreadis, K.M.; Bates, P.D.; Boone, A.; Mognard, N.M.; Rodríguez, E.; Alsdorf, D.E.; Lettenmaier, D.P.; Clark, E.A. Assimilation of virtual wide swath altimetry to improve Arctic river modeling. *Remote Sens. Environ.* **2011**, *115*, 373–381. [CrossRef]
13. Pedinotti, V.; Ricci, S.; Biancamaria, S.; Mognard, N. Assimilation of satellite data to optimize large scale hydrological model parameters: A case study for the SWOT mission. *Hydrol. Earth Syst. Sci.* **2014**, *18*, 4485–4507. [CrossRef]
14. Häfliger, V.; Martin, E.; Boone, A.; Habets, F.; David, C.H.; Garambois, P.A.; Roux, H.; Ricci, S.; Berthon, L.; Thévenin, A. Evaluation of Regional-Scale River Depth Simulations Using Various Routing Schemes within a Hydrometeorological Modeling Framework for the Preparation of the SWOT Mission. *J. Hydrometeorol.* **2015**, *16*, 1821–1842. [CrossRef]

15. Caballero, Y.; Voirin-Morel, S.; Habets, F.; Noilhan, J.; Le Moigne, P.; Lehenaff, A.; Boone, A. Hydrological sensitivity of the Adour-Garonne river basin to climate change. *Water Resour. Res.* **2007**, *43*. [CrossRef]
16. Noilhan, J.; Planton, S. A simple parameterization of land surface processes for meteorological models. *Mon. Weather Rev.* **1989**, *117*, 536–549. [CrossRef]
17. Masson, V.; Le Moigne, P.; Martin, E.; Faroux, S.; Alias, A.; Alkama, R.; Belamari, S.; Barbu, A.; Boone, A.; Bouyssel, F.; et al. The SURFEX v7.2 land and ocean surface platform for coupled or offline simulation of earth surface variables and fluxes. *Geosci. Model Dev.* **2013**, *6*, 929–960. [CrossRef]
18. Faroux, S.; Kaptué Tchuenté, A.T.; Roujean, J.-L.; Masson, V.; Martin, E.; Le Moigne, P. ECOCLIMAP-II/Europe: A twofold database of ecosystems and surface parameters at 1 km resolution based on satellite information for use in land surface. meteorological and climate models. *Geosci. Model Dev.* **2013**, *6*, 563–582. [CrossRef]
19. Quintana-Seguí, P.; Le Moigne, P.; Durand, Y.; Martin, E.; Habets, F. Analysis of near-surface atmospheric variables: Validation of the SAFRAN analysis over France. *Appl. Meteorol. Clim. J.* **2008**, *47*, 92–107. [CrossRef]
20. Boone, A.; Masson, V.; Meyers, T.; Noilhan, J. The influence of the inclusion of soil freezing on simulations by a soil–vegetation–atmosphere transfer scheme. *Appl. Meteorol. J.* **2000**, *39*, 1544–1569. [CrossRef]
21. Decharme, B.; Boone, A.; Delire, C.; Noilhan, J. Local evaluation of the interaction between Soil Biosphere Atmosphere soil multilayer diffusion scheme using four pedotransfer functions. *Geophys. Res. J.* **2011**, 116. [CrossRef]
22. Decharme, B.; Martin, E.; Faroux, S. Reconciling soil thermal and hydrological lower boundary conditions in land surface models. *Geophys. Res. Atmos. J.* **2013**, *118*, 7819–7834. [CrossRef]
23. Ledoux, E.; Girard, G.; Villeneuve, J.P. Proposition d'un modèle couplé pour la simulation conjointe des écoulements de surface et des écoulements souterrains sur un bassin hydrologique. *La Houille Blanche* **1984**, *1–2*, 101–110. [CrossRef]
24. David, C.H.; Habets, F.; Maidment, D.R.; Yang, Z.-L. RAPID applied to the SIM France model. *Hydrol. Proess.* **2011**, *25*, 3412–3425. [CrossRef]
25. David, C.H.; Maidment, D.R.; Niu, G.Y.; Yang, Z.L.; Habets, F.; Eijkhout, V. River network routing on the NHDPlus dataset. *J. Hydrometeorol.* **2011**, *12*, 913–934. [CrossRef]
26. Farr, T.G.; Rosen, P.A.; Caro, E.; Crippen, R.; Duren, R.; Hensley, S.; Kobrick, M.; Paller, M.; Rodríguez, E.; Roth, L.; et al. The Shuttle Radar Topography Mission. *Rev. Geophys.* **2007**, *45*. [CrossRef]
27. Decharme, B.; Alkama, R.; Douville, H.; Becker, M.; Cazenave, A. Global Evaluation of the ISBA−TRIP Continental Hydrological System. Part II: Uncertainties in River Routing Simulation Related to Flow Velocity and Groundwater Storage. *Hydrometeorol. J.* **2010**, *11*, 601–617. [CrossRef]
28. Alkama, R. Global evaluation of the ISBA-TRIP continental hydrological system. Part I: Comparison to GRACE terrestrial water storage estimates and in situ river discharges. *J. Hydrometeorol.* **2010**, *11*, 583–600. [CrossRef]
29. Pedinotti, V.; Boone, A.; Decharme, B.; Crétaux, J.F.; Mognard, N.; Panthou, G.; Papa, F.; Tanimou, B.A. Evaluation of the ISBA-TRIP continental hydrological system over the Niger basin using in situ and satellite derived datastes. *Hydrol. Earth Syst. Sci.* **2012**, *16*, 1745–1773. [CrossRef]
30. Bouttier, F.; Courtier, P. *Data Assimilation Concepts and Methods*; ECMWF Lecture Note: Reading, UK, 1999.
31. Buis, S.; Piacentini, A.; Déclat, D. PALM: A Computational framework for assembling high performance computing applications. *Concurr. Comput. Pract. Exp.* **2006**, *18*, 247–262. [CrossRef]
32. Fouilloux, A.; Piacentini, A. The PALM Project: MPMD Paradigm for an Oceanic Data Assimilation Software. *Comput. Sci.* **1999**, *1685*, 1423–1430.
33. Barthélémy, S.; Ricci, S.; Rochoux, M.C.; Le Pape, E.; Thual, O. Ensemble-based data assimilation for operational flood forecasting—On the merits of state estimation for 1D hydrodynamic forecasting through the example of the "Adour Marine" river. *J. Hydrol.* **2017**, *552*, 210–224. [CrossRef]
34. Enjolras, V.; Vincent, P.; Souyris, J.-C.; Rodríguez, E.; Phalippou, L.; Cazenave, A. Performances study of interferometric radar altimeters: From the instrument to the global mission definition. *Sensors* **2006**, *6*, 164–192. [CrossRef]
35. Brown, S.; Obligis, E. SWOT Wet Tropospheric Correction Working Group Report. In Proceedings of the 3rd SWOT Science Defintion Team Meeting, Washington, DC, USA, 14–16 January 2014.
36. Fernandez, D.E. SWOT Project, Mission Performance and Error Budget, JPL Document D-79084. 2017.

37. Cimini, D.; Pierdicca, N.; Pichelli, E.; Ferretti, R.; Mattioli, V.; Bonafoni, S.; Montopoli, M.; Perissin, D. On the accuracy of integrated water vapor observations and the potential for mitigating electromagnetic path delay error in InSAR. *Atmos. Meas. Tech.* **2012**, *5*, 1015–1030. [CrossRef]
38. Ning, T.; Elgered, G.; Willen, U.; Johansson, J.M. Evaluation of the atmospheric water vapor content in a regional climate model using ground-based GPS measurements. *J. Geophys. Res. Atmos.* **2013**, *118*, 329–339. [CrossRef]
39. Flentje, H.; Dörnbrack, A.; Fix, A.; Ehret, G.; Holm, E. Evaluation of ECMWF water vapour fields by airborne differential absorption lidar measurements: A case study between Brazil and Europe. *Atmos. Chem. Phys.* **2007**, *7*, 5033–5042. [CrossRef]
40. Evensen, G. The ensemble Kalman filter: Theoritical formulation and practical implementation. *Ocean Dyn.* **2003**, *53*, 343–367. [CrossRef]
41. Thirel, G.; Martin, E.; Mahfouf, J.-F.; Massart, S.; Ricci, S.; Habets, F. A past discharge assimilation system for ensemble streamflow forecasts over France—Part 1: Description and validation of the assimilation system. *Hydrol. Earth Syst. Sci.* **2010**, *14*, 1623–1637. [CrossRef]

5

Discharge Measurements of Snowmelt Flood by Space-Time Image Velocimetry during the Night using Far-Infrared Camera

Ichiro Fujita

Department of Civil Engineering, Graduate School of Engineering, Kobe University, Kobe 657-8501, Japan; ifujita@kobe-u.ac.jp

Academic Editor: Roberto Ranzi

Abstract: The space time image velocimetry (STIV) technique is presented and shown to be a useful tool for extracting river flow information non-intrusively simply by taking surface video images. This technique is applied to measure surface velocity distributions on the Uono River on Honshu Island, Japan. At the site, various measurement methods such as a radio-wave velocity meter, an acoustic Doppler current profiler (ADCP) or imaging techniques were implemented. The performance of STIV was examined in various aspects such as a night measurement using a far-infrared-ray (FIR) camera and a comparison to ADCP data for checking measurement accuracy. All the results showed that STIV is capable of providing reliable data for surface velocity and water discharge that agree fairly well with ADCP data. In particular, it was demonstrated that measurements during the night can be conducted without any difficulty using an FIR camera and the STIV technique. In particular, using the FIR camera, the STIV technique can capture water surface features better than conventional cameras even at low resolution. Furthermore, it was demonstrated that measurements during the night can be conducted without any difficulty.

Keywords: river flow measurement; surface flow; image-based technique; Space Time Image Velocimetry (STIV); far-infrared camera; Large Scale Image Velocimetry (LSPIV)

1. Introduction

Intensive and record-breaking rainfall caused severe flood disasters in the world and in Japan in recent years partially due to the global warming [1]. For example, in July 2011, Niigata and Fukushima areas suffered from torrential rain with levee breaches and inundation disasters at many locations due to the maximum flood in recorded history. In September 2011, typhoon No. 12 directly hit Nara and Wakayama Prefectures causing landslide and sediment-related disasters in a wide area. Furthermore, in July 2012, a total of eight first-class river systems in the northern part of Kyushu Island were severely damaged twice by a torrential rain brought by a seasonal rain front. In 2015, the Kinugawa River, a first-class river in Kanto area, suffered from a levee breach, causing a huge inundated area while destroying nearby houses. In the Japanese river management system, 109 rivers are designated as first-class rivers, for which the Japanese government is responsible for the management of major river reach. The other smaller rivers of more than four thousand are classified as second-class rivers, for which the local government takes care of the river reach.

It should also be mentioned that huge flood disasters are taking place every year in the world; e.g., Australia, Brazil, China and Thailand were struck by a record-breaking flooding in the past decades [2]. A specific feature of the recent flood is characterized by the fact that heavy rain with an hourly rainfall intensity of about 100 mm continues for several hours, and this causes a sudden increase of river discharge, which results in an inundation in the worst case. In order to be prepared for such disasters, it is of vital importance to collect and analyze hydrological data such as rainfall intensity, water level and flow discharge. The acquisition of such data is important for the purpose of understanding rainfall run-off processes correctly while tuning various parameters included in a run-off model [3,4]. Among the hydrological data, peak discharge is the most important information because it can determine the flow capacity of a river section in response to a heavy rainfall. However, contrary to the floods in continental large-scale rivers, which last several days or weeks, typical floods in Japan have a duration of several hours; therefore, it becomes difficult to correctly measure the flow at the right timing especially when the peak flow occurs during the night time. In the past, discharge measurements in Japan have been conducted mainly by using a float [5] with its length dependent on the water depth. In Japan, float is made of cardboard pipe in which a certain amount of sand is contained to control its specific gravity close to unity when it floats. According to the Japanese regulation, floats with different lengths are used depending on the water depth; e.g., two-meter float is used when the water depth is greater than 2.6 m and less than 5.2 m.

However, there exist several critical defects in this method because a measurement site of a bridge can be submerged in a huge flood, as actually occurred in the 2011 Niigata Flood. In this case, the measurement itself became impossible and one missed a chance to measure the peak flow. Furthermore, the road to the measurement site was closed due to inundation. In the case of the flood of Kyushu in 2012, the water level rose to the height of the river bank and overflowing flooding actually occurred. Particularly in the dark night time, it is difficult to apply the float method as the measurement might become dangerous in such a condition. In addition, even if measurement is possible, the float may seriously meander under the influence of large-scale river turbulence, which may greatly reduce measurement accuracy.

In the light of such a present status of discharge measurement, the so-called non-intrusive measurement techniques have been paid attention to in recent years, such as a radio-wave velocity meter (RVM) [6,7] or imaging techniques [8–12]. The RVM is usually installed at a bridge to measure streamwise surface velocity continuously at one point by using the effect of the Doppler shift of the surface flow [6,7]. However, the instrument has to be shifted back and forth along the bridge when trying to measure a cross-sectional velocity distribution or a large number of the instruments have to be installed along a bridge with a high cost. An alternative method is the use of surface-flow video images taken from a river bank. The idea behind the use of video images is that surface flow features composed of surface ripples or floating objects are assumed to be advected with the surface velocity as long as the wind effect is negligible. Among the image-based techniques, such as a large-scale image velocimetry (LSPIV) [13–15] or the space-time image velocimetry (STIV) [16–20], STIV seems to be more robust under deteriorated image condition and utilized in the present research. The non-intrusive methods are very useful and effective as they can measure the flow velocity information safely from a remote place. However, the image obtained by a conventional video camera has a shortcoming incapable of recording favorable images during the night. Since peak flows frequently occur in a dark condition in Japan, the existing river monitoring systems' installed conventional cameras are difficult to use for flow measurements. In order to overcome this weak point, we propose a method to use a far-infrared ray (FIR) camera in river flow measurements. It is worth mentioning that traditional invasive instruments such as propeller ammeters, electromagnetic flowmeters or ADCP cannot be applied during high-speed floods. On the other hand, STIV, for example, can measure velocities over 5 m/s [16]. In this paper, we examined the traceability of the surface texture composed of surface ripples without tracers to the surface flow, and discussed the measurement accuracy of river flow rate by using the image obtained by the FIR camera. Furthermore, robustness of the STIV technique to LSPIV in measuring streamwise velocity components when using deteriorated images of low resolution will be shown.

2. Outline of Space-Time Image Velocimetry

2.1. Image Rectification by Camera Calibration and Search Lines

STIV is an image analysis technique to measure streamwise water surface velocity distributions from video images usually taken from a riverbank. This technique assumes that features that appeared on the water surface would follow the surface velocity when the wind effect is negligible. The idea to use surface feature advection is the same as the radio-wave velocity meter or LSPIV. The aspect of STIV different from the other techniques is that STIV pays attention to the time- and space-averaged velocity along a line segment set in the streamwise direction. The line segment is treated as a search line for velocity estimation.

In the first step of STIV, establishing a relation between the screen coordinates (x, y) and the physical coordinates (X, Y, Z), by using the following collinearity equation is required:

$$x = -c\left[\frac{a_{11}(X-X_0)+a_{12}(Y-Y_0)+a_{13}(Z-Z_0)}{a_{31}(X-X_0)+a_{32}(Y-Y_0)+a_{33}(Z-Z_0)}\right]+x_0 \tag{1}$$

$$y = -c\left[\frac{a_{21}(X-X_0)+a_{22}(Y-Y_0)+a_{23}(Z-Z_0)}{a_{31}(X-X_0)+a_{32}(Y-Y_0)+a_{33}(Z-Z_0)}\right]+y_0 \tag{2}$$

where c is the focal distance, a_{ij} with $1 \le i \le 3$ and $1 \le j \le 3$ are mapping coefficients, (X_0, Y_0, Z_0) is the physical coordinates of the camera, and (x_0, y_0) is the center of the screen image. The coefficients can be determined via a camera calibration procedure by using at least six ground control points. The image rectification can be performed by substituting the water level data H to Z in the above equations. An example of the image rectification for an image obtained by using a normal high definition video camera is shown in Figure 1, together with the search lines set parallel to the streamwise direction. The length of the search line is 17.6 m and the transverse spacing is 5.6 m in this case. The set of search lines covers almost the whole width of the river of 140 m. The direction of search lines is determined by visual inspection after generating a rectified image, i.e., the lines are drawn parallel to the riverbank.

(a) (b)

Figure 1. Example of an ortho-rectified image (**b**) for an image (**a**) obtained by using a normal high definition video camera, each with search lines.

2.2. Space Time Image

After the set of search lines, space time images (STIs) are generated for each search line by taking brightness distribution in the search line as the horizontal axis and its evolution with time as the vertical axis. Several examples of STI in a transverse direction is shown in Figure 2. The horizontal length is 17.6 m and the downward vertical axis is ten seconds. Since the frame rate of image sampling is 30 frames per second, the vertical length is composed of 300 images. The images are enhanced for a better recognition of surface features by applying the histogram equalization technique. It is clearly seen that an inclined pattern is generated in each STI, indicating that overall surface features are advected along the search line with the flow at almost constant speed, although there are some other noisy patterns created by surface wave motion moving in the direction different from the main flow. The average gradient of the pattern corresponds to the local time- and space-averaged mean velocity along the search line.

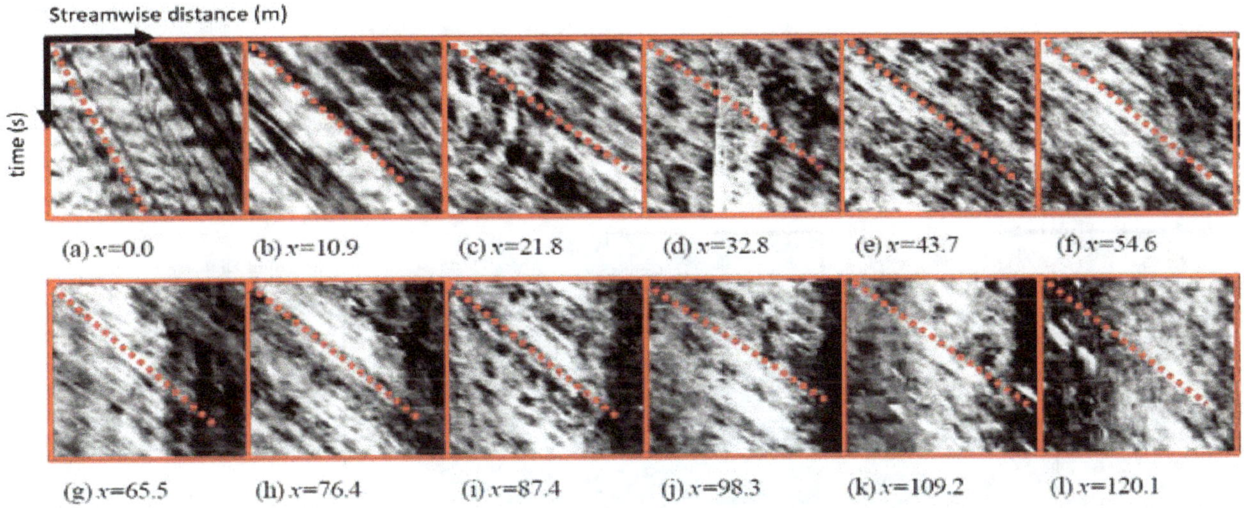

Figure 2. Space time images obtained by normal high definition camera at several transverse location (unit in meter): horizontal scale is 17.6 m and downward vertical scale is ten seconds. Image are enhanced for a better recognition. The dotted line is the main orientation angle of the pattern.

2.3. Measurement of Orientation Angle of the Pattern

In the original STIV, the average orientation angle f of a STI is obtained by dividing the STI into small segments and calculating the local gradient for each segment as indicated in Figure 3. The local orientation angle ϕ can be calculated using the following relations [17]:

$$\tan 2\phi = \frac{2J_{xt}}{J_{tt} - J_{xx}} \tag{3}$$

where

$$J_{xx} = \int_A \frac{\partial g}{\partial x}\frac{\partial g}{\partial x} dxdt \tag{4}$$

$$J_{xt} = \int_A \frac{\partial g}{\partial x}\frac{\partial g}{\partial t} dxdt \tag{5}$$

$$J_{tt} = \int_A \frac{\partial g}{\partial t}\frac{\partial g}{\partial t} dxdt \tag{6}$$

$g(x,t)$ is the grey level intensity in STI and A is the area of the small segment indicated in Figure 3a. Typically, A is a square with a side length of fifteen pixels. Figure 3b provides a distribution of the coherency defined by

$$C_{oh} = \frac{\sqrt{(J_{tt} - J_{xx})^2 + 4J^2{}_{xt}}}{J_{xx} + J_{tt}} \tag{7}$$

which is a measure of the image pattern coherence and takes a value between zero and one; for an ideal local orientation, its value becomes one, and, for an isotropic gray image, it becomes zero. Therefore, it is possible to calculate the mean orientation angle by preferably obtaining clearer orientation information by taking coherency as a weighting function:

$$\overline{\phi} = \int \phi C_{oh}(\phi) d\phi \Big/ \int C_{oh}(\phi) d\phi \tag{8}$$

Figure 3c is a weighted histogram of the orientation angle, showing a distribution with a peak. Since the length and time scales of the STI are given, the mean velocity along the line segment can be calculated by the following relation:

$$U_s = \frac{S_x}{S_t} \tan \overline{\phi} \tag{9}$$

where S_x is the unit length scale along the search line (m/pixel) and S_t is the unit time scale of the time axis.

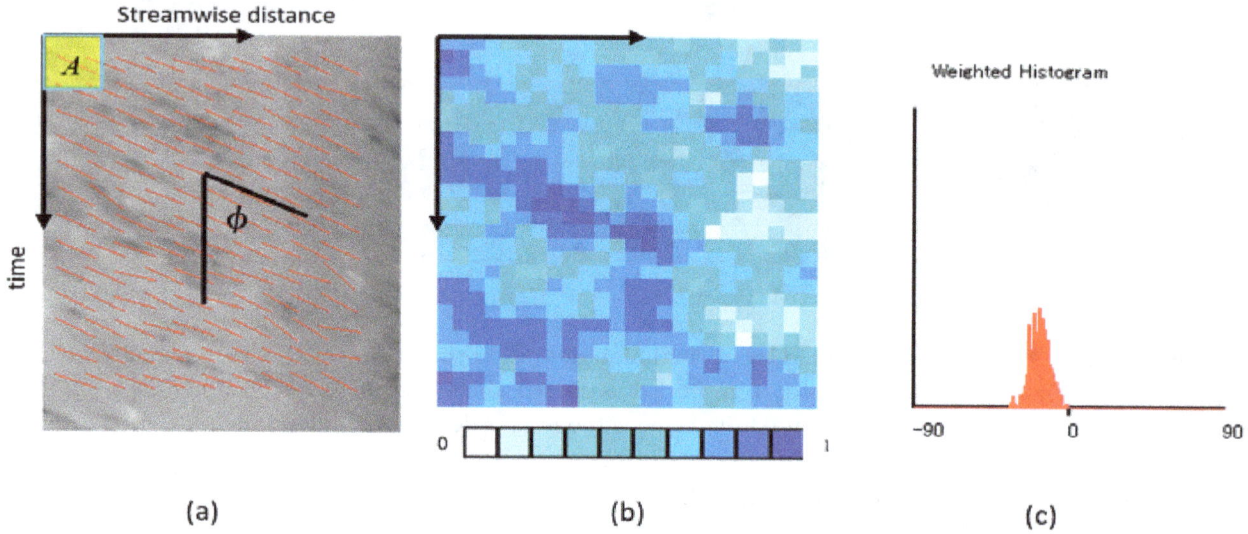

Figure 3. Application of gradient tensor method: (**a**) local angle in segment (ϕ); (**b**) coherency distribution; (**c**) histogram of orientation angles.

3. Outline of the Field Measurement

3.1. Study Area

Field investigations were conducted from 19 April to 22 April in 2012 at the Negoya Bridge of the Uono River, which is a major tributary of the Shinano River, the longest river in Japan. The site is located at Horinouchi in Uonuma City of Niigata Prefecture in the middle of Japan. This location was chosen because daily snowmelt flood occurs frequently in early spring, and we could actually observe the increase of water level of about 40 cm due to the snowmelt during our measurements every day as shown in Figure 4. The water level sensor of a pressure type was temporally installed near the panel R04 (Figure 5). On the measurement day, the water level rose up to the full width of the lower channel with a width of about 150 m, while the bank to bank width is about 220 m. The measurements were mainly conducted at the cross-section downstream of the bridge every hour in order to compare the performance of various measurement methods, i.e., STIV with normal cameras, STIV with a FIR camera, RVM, and an ADCP. Floats were also supplied from the bridge to compare their velocity with the surface velocity measured by the above methods. Figure 5 is the measurement site showing the location of cameras and mark point panels to be used for a camera calibration in STIV measurement. The panels were carefully set so that at least six of them are visible for each camera screen. The measurements were conducted jointly with staffs from International Centre for Water Hazard and Risk Management (ICHARM) under the auspices of UNESCO and several consulting companies.

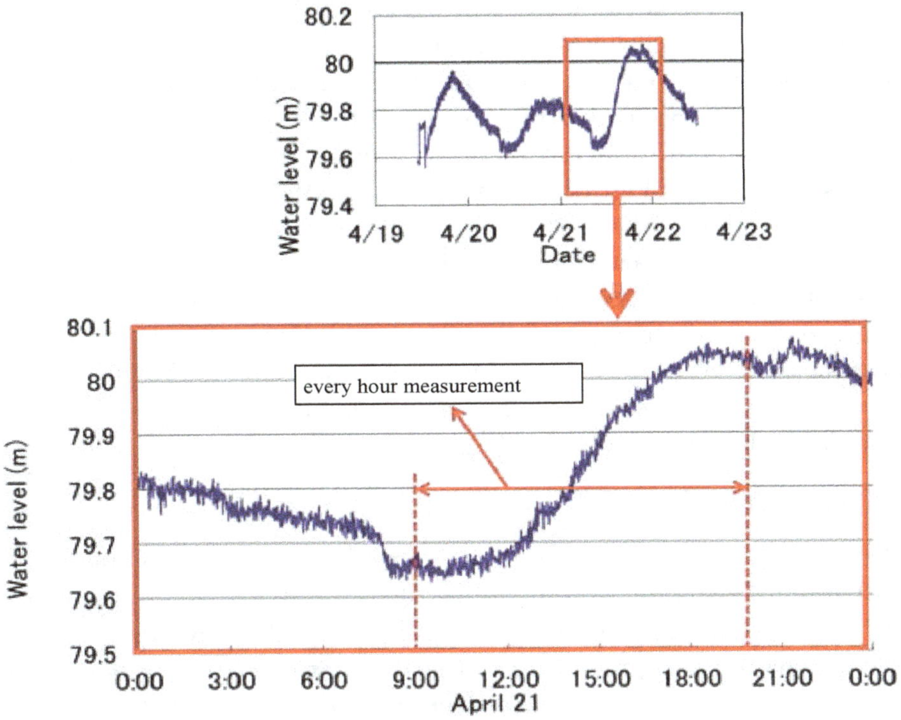

Figure 4. Water level hydrograph at the measurement site.

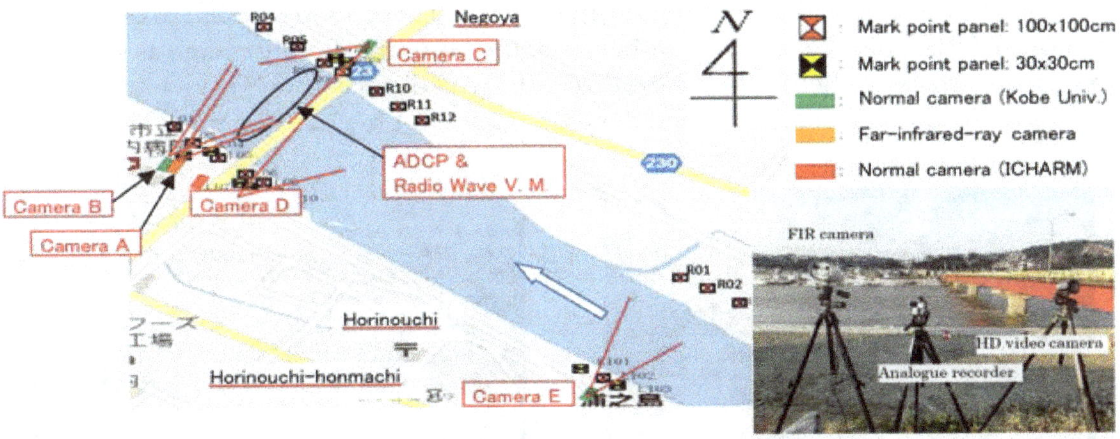

Figure 5. Measurement site description at the Negoya Bridge on the Uono River. L: Panels on the left bank; R: Panels on the right bank; ADCP: acoustic Doppler current profiler; V.M.: velocity meter; FIR: far-infrared-ray; ICHARM: International Centre of Excellence for Water Hazard and Risk Management.

3.2. *Image Acquiring Methods*

In the present observation, several cameras took images of the water surface from various angles. The cameras are an FIR camera (Camera A in Figure 4) and four normal high-definition video cameras (cameras B to E). Cameras A, B and C viewed the same water surface zone downstream of the bridge where ADCP and radio-wave measurements were executed. Camera D and Camera E took images just upstream and far upstream of the bridge, respectively. Figure 5 shows how the measurement were conducted at the site. It shows that the image-based measurement technique is quite simple, i.e., just take surface images from a riverbank, taking care about the size of the field of view. The important point is to set at least six ground reference points (GRPs) within the field of view for the ortho-rectification of images.

The measurements were repeated every hour for about five minutes from 10:00 a.m. to 8:00 p.m. The FIR camera (FLIR Systems, SR-334, Wilsonville, OR, USA), has a resolution of 320 by 240 pixels and its image sampling frequency is 7.5 fps, while normal high definition (HD) cameras have a resolution of 1920 by 1080 pixels with 30 fps. The video images of the FIR camera were recorded as an NTSC (National Television System Committee) format on an analogue video recorder with a size of 640 by 480 pixels at 30 fps as indicated in Figure 4. Therefore, the recorded images were somewhat blurred with a staggering frame sequence. The relationship between the screen and physical coordinates were determined from the location of mark point panels for each coordinate. Typical panels are made of plywood with a size of 1 m by 1 m. As the resolution of the FIR camera is not high, and, at the same time, it was difficult to detect the panels on the other side of the river about 150 m away from the camera, pocket heating pads were attached on the panel for improving the visibility of panel locations. Since the air temperature is very low (about 5 °C), heated spots were clearly visible in FIR camera screen even from a distant location.

3.3. Comparison of Captured Images

In order to demonstrate the interest of a FIR camera for river flow measurement, images captured in the morning and at night are compared in Figure 6 together with the images captured by a normal camera. It is obvious from the upper images in Figure 6 that the FIR camera can capture the river surface pattern clearly even during the night with low light condition, while a normal camera can recognize only the water surface just beneath street lamps at the bridge as indicated in lower left image in Figure 6. It can be considered that an FIR camera senses only the heat differences of objects, but the intensity of the received infrared ray changes depending on the slope of reflected surface even when the object temperature is the same. It should be noted that the original image size is only 320 by 240 pixels in FIR compared with the normal image size of 1920 by 1080 pixels. Although the surface pattern by FIR camera seems to be rather smoothed out in the night image as seen in Figure 6b, advection of the pattern in the downstream direction was observed in STI, which is a significant advantage over conventional cameras when developing a real-time measurement system using STIV.

(**a**) 10:00 (**b**) 20:00

(**c**) 10:00 (**d**) 20:00

Figure 6. Comparison of image quality by an FIR camera and a normal high definition video camera: (**a**) image by FIR camera in the daytime; (**b**) image by FIR camera at night; (**c**) image by normal HD camera in the daytime; (**d**) image by normal HD camera at night.

Figure 7 shows 25 search lines set on oblique and rectified images for the FIR camera A. The search lines were drawn parallel to the river bank at an equal spacing in a similar manner indicated in Figure 1. Its length is 17.8 m and the spacing is 5.7 m, covering the river width of about 140 m. It can be noted in Figure 7b that the image texture within a rectangle region becomes significantly blurred after the ortho-rectification. This is because the pixel image is greatly expanded for the region far from the camera location. As an example, the original spatial resolution along the farthest search line is 33.6 cm/pixel in this case, which is a very large value trying to reproduce the actual water surface roughness pattern. Because of this insufficient information of image texture in that region, it was difficult to apply LSPIV to the whole width since LSPIV uses a pattern matching technique to track the surface texture. As an example, Figure 8 provides the results by LSPIV applied to sequential ortho-rectified images such as shown in Figure 7b. The images ortho-rectified with a pixel size of 0.2 m and a time interval of 0.133 seconds were used for the analysis. The template size for the pattern matching in PIV was set at 33 by 33 pixels, which is 6.6 by 6.6 m in physical scale. In Figure 8, instantaneous vectors are superposed to show variations in data. It is obvious that variation in data increases significantly with the distance from the left bank where the FIR camera is located. Since the actual flow direction is almost unidirectional, the scatter in the data is unreliable. This is because the image is largely stretched in the transverse direction in the process to ortho-rectification. On the other hand, the data nearer to the left bank shows almost uniform variation, indicating that reliable data are obtained only until about half-width, i.e., about 70 m, from the left bank. This feature suggests that the flow rate cannot be analyzed from the LSPIV in the cases where the image resolution is low and the image stretching becomes large in the image transformation, such as in the present situation.

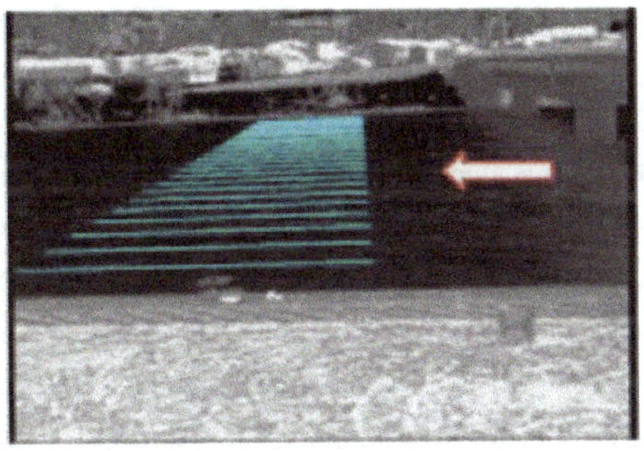

(**a**) original image

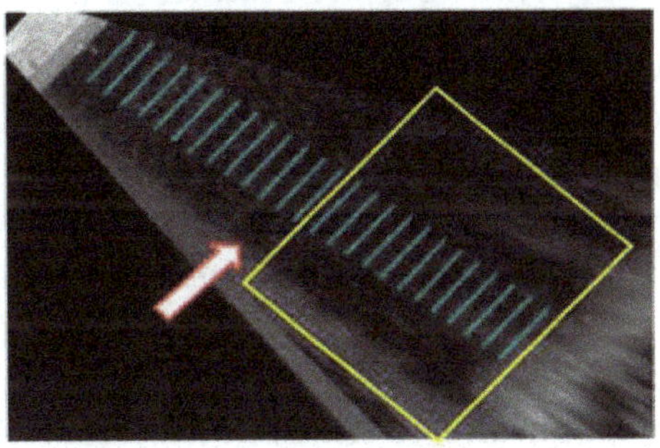

(**b**) ortho-rectified image

Figure 7. Images from FIR camera A at 10:00: (**a**) original image taken from the left bank with search lines, (**b**) ortho-rectified image of (**a**).

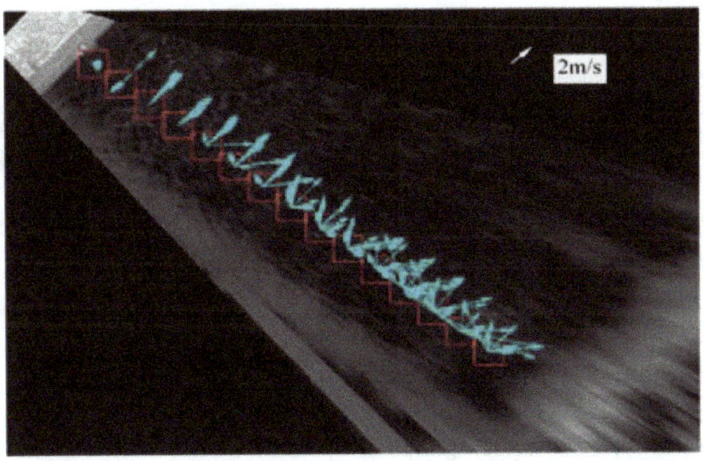

Figure 8. Superposed instantaneous velocity vectors measured by large-scale image velocimetry (LSPIV).

In order to show good sensitivity when using spatiotemporal images, Figure 9 shows the STI obtained by the FIR camera at almost the same position as the STI shown in Figure 2 obtained by a high resolution camera. The images are enhanced by the same technique as for Figure 2. Although the FIR camera used in the measurement have only one sixth of the resolution of the high definition camera, similar patterns appear in the resulting STIs even at the farther locations from the camera location. This feature of STI is the source of robustness in applying STIV to various types of flood flows [16].

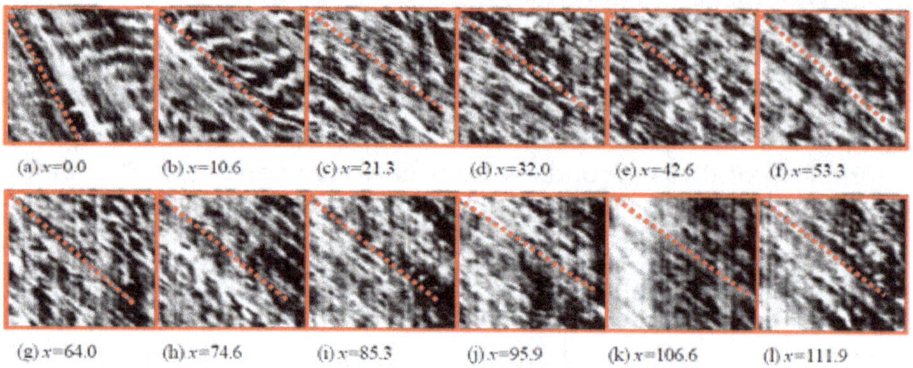

Figure 9. Space time images obtained by an FIR camera at several transverse locations (units in meters): horizontal scale is 17.6 m and downward vertical scale is ten seconds. Images are enhanced for a better recognition. The dotted line is the main orientation angle of the pattern.

3.4. Traceability of Water Surface Features

In using non-contact techniques for surface flow measurement such as RVM, LSPIV or STIV, water surface features generated by turbulence are assumed to be advected with the surface flow velocity as previously mentioned. In order to validate this assumption, an STI for the search line intentionally set along a trajectory of a float captured by camera B is shown in Figure 10 together with an STI by an FIR camera for the same location. The trajectory of the float is visible as a straight line as seen in Figure 10a, indicating that the float moved at a constant speed along the search line, from which its speed can be easily obtained from the slope of the trajectory. In Figure 10a, there appears a texture almost parallel to the float trajectory. An STI at the same location obtained by an FIR camera is shown in Figure 10b. The important point is that the gradient of the oblique patterns obtained by the two cameras is almost the same as that of the float trajectory. Referring to the four straight lines of the same gradient superposed in the figures, a good agreement of the speeds can be visually recognized, indicating the traceability of surface features to the surface flow.

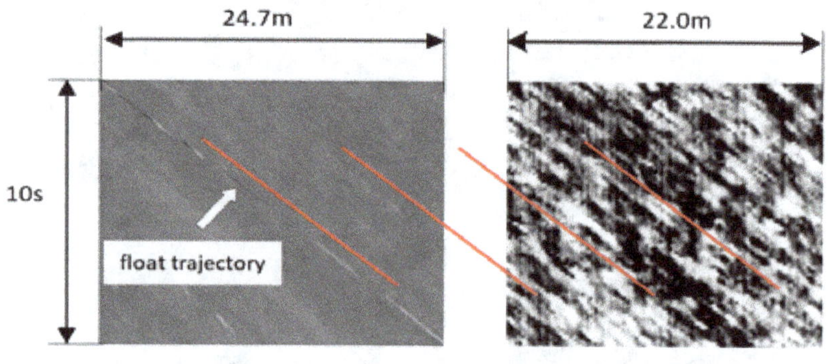

(**a**) normal camera (**b**) FIR camera

Figure 10. Comparison of space time image (STI) captured downstream of the bridge in the daytime: (**a**) STI captured by normal HD camera for a search line along a float trajectory; (**b**) STI captured by FIR camera for the same search line as (**a**).

Another example for examining traceability of surface features is shown in Figure 11, which compares a trajectory of a piece of driftwood and surface features. Since these STIs are obtained from the camera D located upstream of the bridge, surface features displays straight uniform patterns. This is because the water surface was not disturbed by the pier of the bridge as in the downstream region. The trajectory of driftwood coincides with the surface pattern gradient so that it seems to be embedded in the pattern. This suggests that measurements by STIV should be conducted on the upstream side of the bridge, without disturbance by the bridge pier

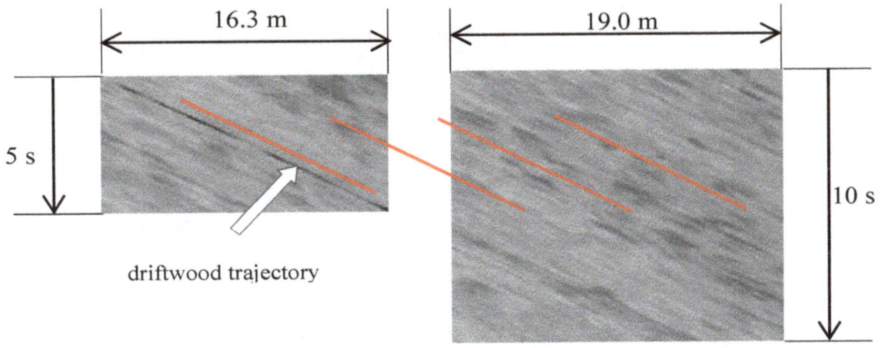

Figure 11. Comparison of STIs with and without a driftwood obtained at the upstream of the bridge from camera D shown in Figure 5.

Finally, Figure 12 compares the trajectory of a luminous float on the STI captured by a normal camera and STI by an FIR camera obtained at night. The luminous float is used conventionally in a dark condition in Japan. It has a fluorescent part at the top of the normal float so that an observer can easily detect the light from the river bank. Again, due to the low spatial resolution of the FIR camera used in the present study, oblique patterns displayed in Figure 12b do not have sharp outlines, but it is possible to determine the slope of the general pattern even in such a deteriorated image condition. Referring to the parallel lines drawn in the figure, the gradient of the light trajectory agrees quite well with the pattern slope by the FIR camera. Therefore, the conventional measurement during the night by using a luminous float can be replaced by STIV using an FIR camera. In this respect, STIV is much more robust than LSPIV in which a reliable matching of image pattern has to be established in a sub-pixel level. Alternatively, STI containing a float trajectory, such as indicated in Figure 12a, can be used for STIV measurement when an FIR camera is not available. The gradient of the pattern can be obtained either manually or automatically with the aid of an STIV algorithm [17].

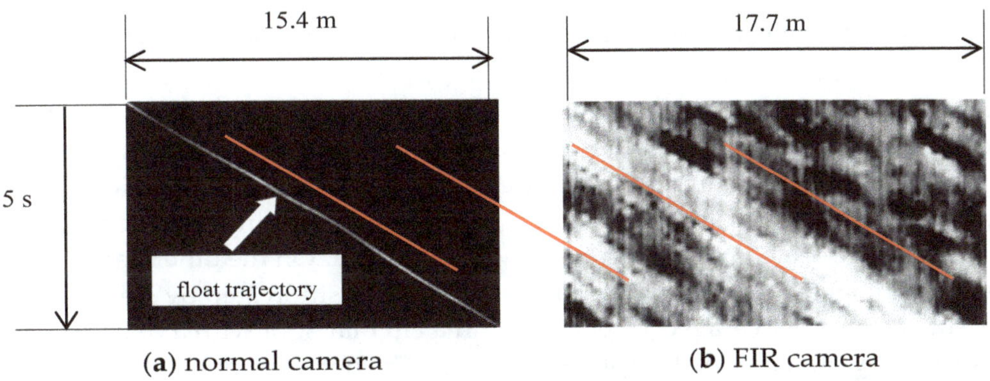

Figure 12. Comparison of STI obtained at night: (**a**) STI captured by normal HD camera for a search line along a luminous float trajectory; (**b**) STI captured by FIR camera for the same search line as (**a**).

4. Results and Discussion

4.1. Comparison with Other Measurement Methods

As mentioned previously, the measurements were conducted every hour from 10:00 a.m. to 8:00 p.m. During the measurements, the water level rose about 50 cm due to the melting of snow from the mountainous region. The water level repeated up and down changes due to the snowmelt in the daytime as has been indicated in Figure 5. Figure 13 shows surface velocity distributions downstream of the Negoya Bridge at 1:00 p.m. and 8:00 p.m. measured by various methods, i.e., STIV by the far infrared camera on the left bank, STIVs by normal camcorders on the left and the right banks, a radio-wave velocity meter, and an ADCP. The cross-sectional bottom shape measured by ADCP was also indicated in the figure.

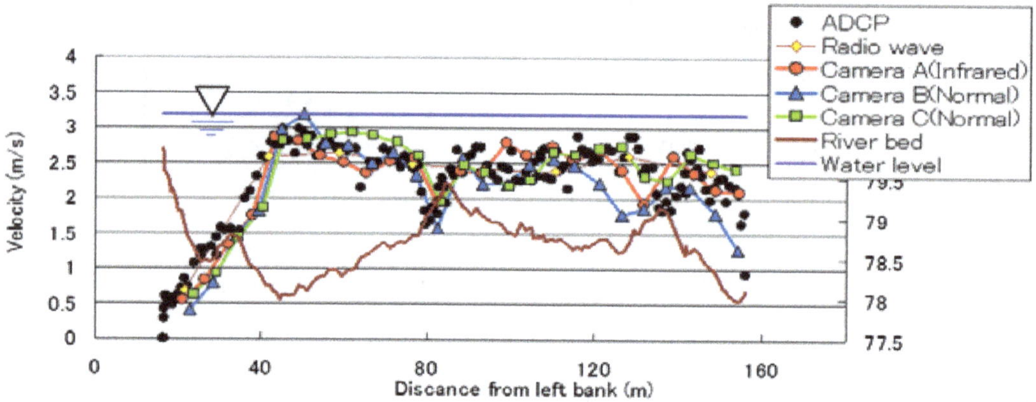

(**a**) 13:00

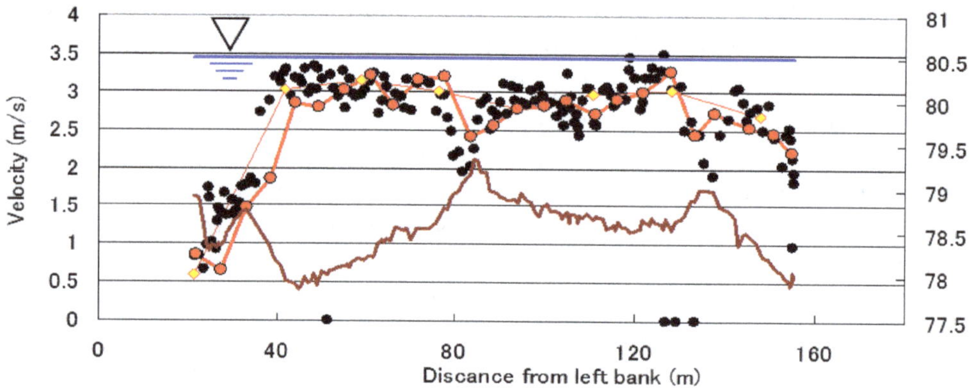

(**b**) 20:00

Figure 13. Comparison of surface velocity distributions by various techniques.

As for the ADCP data, the streamwise velocity component closest to the water surface, about 30 cm below it, were compared as representing the surface velocity. Since the ADCP measurement is conducted by tagging the boat-mounted sensor manually in a transverse direction, the data indicated are almost instantaneous values and have a great scatter especially in the region downstream of the bridge piers, where large-scale unsteady wake flow is generated. On the other hand, measurements by a radio-wave velocity meter were conducted by shifting its position step by step manually along the bridge. It measures a locally averaged velocity within a circular spot with a diameter of a few meters on the water surface by utilizing the Doppler shift effect. The obtained data falls into a scattered range of ADCP data.

Regarding the measurements by STIV, plots by the normal video cameras and the FIR camera show a large scatter within the wake flow region with the maximum relative error more than 0.5 m/s as shown in Figure 13a. Compared to the ADCP data, the FIR camera produced a favorable velocity distribution with a relative error of about 10%, which also agreed well with the RVM data. Since normal cameras are impossible to use at night, STIV by the FIR camera are compared with ADCP and RVM as shown in Figure 13b. Keeping in mind the scatter in ADCP data, data by the FIR camera agrees quite with the other data with a relative error of about 10%.

Comparing the STIV data obtained by normal video cameras and the FIR camera, normal cameras record the light reflection from the water surface roughness while the FIR camera mainly detects the temperature difference as well as the radiation intensity that varies with the roughness shape. Since the spatial resolution along a search line decreases with the distance from the camera, there should be some limitation in terms of the width of the river that STIV can be applicable. According to the author's experience, the present case with a width of about 150 m might be the maximum distance at which STIV can yield reliable data.

Regarding the transverse bed profile at the section downstream of the bridge shown in Figure 13, the river bed is deeper closer to the left bank and a ridge-like deposition occurs downstream of a bridge pier. Another lower deposition can be found closer to the right bank just after the other bridge pier. Generally speaking, water surface velocity distributions measured by the above methods agree well with each other in the entire cross-section with its distribution influenced by the bottom variation, i.e., faster in a deeper zone and slower in a shallower part. The maximum velocity occurred near the left bank increased from about 2.8 m/s at 1:00 p.m. to about 3.3 m/s at 8:00 p.m. due to the increase of discharge.

In order to examine the effect of the bridge piers that disturb the surface features, time variation of velocity distributions measured by STIV with Camera D just upstream of the Negoya Bridge is shown in Figure 14. The location of the Camera D is indicated in Figure 4. As can be seen from Figure 10 obtained at the upstream section, the oblique patterns that appeared on the STI show continuous straight features with little disturbance, indicating that surface features are advected steadily at a constant speed without the influence of hydraulic structures. As a result, the transverse velocity distributions are much smoother than that shown in Figure 13. This feature is important for improving the measurement accuracy by STIV. Unfortunately, we could not calculate the flow rate because the riverbed shape of this section was unknown, but once the topographic data is available, more reliable discharge data could be obtained.

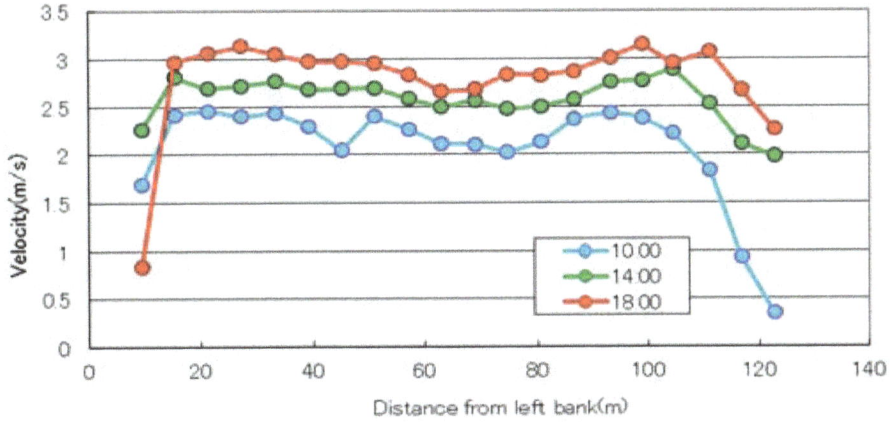

Figure 14. Surface velocity distribution measured by space time image velocimetry (STIV) at the upstream section of the bridge.

4.2. Discharge Measurement Accuracy

Figure 15 compares a discharge directly measured by ADCP and estimated discharges from surface velocity distributions by STIV and a radio-wave velocity meter. In integrating the velocity distribution measured by ADCP, the surface velocity U_s was assumed to be the same as the data 30 cm below the water surface. In using STIV, discharge was obtained by dividing the river width by the interval of the search line Δy_i, multiplying the flow velocity at the search line by a surface velocity ratio α, the local flow depth h_i, and finally summing the piecewise discharges. Therefore, the discharge can be calculated by the following formula:

$$Q_{ADCP/RVM} = \sum_{i=1}^{n} \alpha U_{si} h_i \Delta y_i. \tag{10}$$

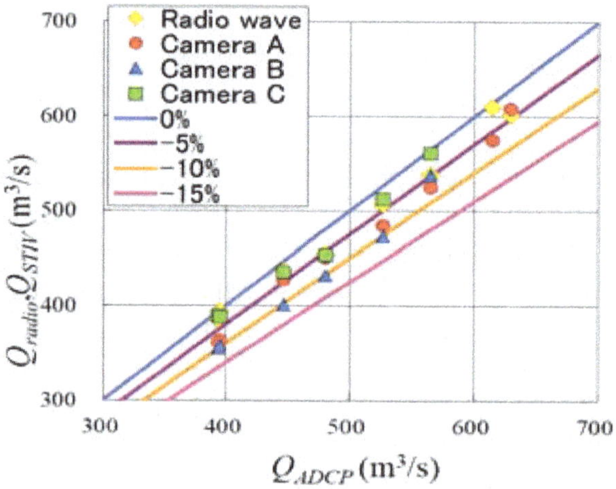

Figure 15. Comparison of measured discharges.

Similar flow rate calculation is also performed by RVM, but the number of measurement cross-sections by STIV was much larger than RVM; i.e., 25 sections in STIV and eight sections in RVM. In STIV, the number of sections can be increased easily, but it becomes physically difficult to conduct a simultaneous measurement in using RVM. In the present discharge calculation shown in Figure 14, the conventional value of 0.85 was used for α, and, as a result, underestimated values were obtained when compared to ADCP data. Therefore, using a value of 0.9 for α would increase the degree of coincidence at the section downstream of the bridge. As shown in Figure 14, even with a value of 0.85 for α, the relative error is less than ten percent, which is acceptable in the practical usage of STIV.

Figure 16 compares the discharge hydrographs obtained by the above methods. The discharge increased about 200 m^3/s due to the snowmelt. The discharge began to increase just after noon, reached its peak at 7:00 p.m. and decreased again after that, which indicates a typical feature of snowmelt flood in Japan.

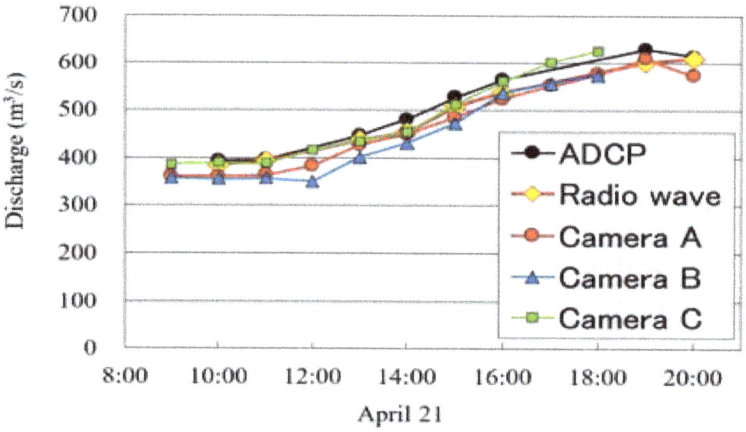

Figure 16. Discharge hydrographs obtained by various techniques.

5. Conclusions

Velocity fields of the snowmelt flood that occurred in the Uono River were measured by various cutting-edge methods such as imaging techniques, ADCP and radio-wave velocity meter. Among them, imaging techniques used to be considered difficult to apply during the night. However, this shortcoming was overcome by introducing a far-infrared ray camera in the present study for the first time for flood flow measurement with surface velocities more than 3 m/s. Puelo et al. [21] used a thermal infrared camera to measure small stream flows, but the velocity level was low with the maximum velocity less than 0.9 m/s, and it was not a flood measurement. Regarding the traceability of surface features, the assumption that surface features are advected with the surface velocity was verified by directly comparing it with the velocities of floating objects such as driftwood or floats in a space-time image. It was also made clear that the measurement accuracy of discharge by STIV was comparable to the direct measurement by ADCP with a relative error less than ten percent. In the actual situation of discharge measurements, measurement sections should have to be located upstream of a bridge because surface features at the downstream section can be destroyed by wake flows generated at the piers. Regarding the use of LSPIV, another widely-used image-based technique, it yielded erroneous values where the stretching of the image in the transverse direction is remarkable. On the other hand, STIV yielded stable results even where the image resolution and quality is low farther from the camera location. In that sense, STIV is a promising technique for measuring river discharges safely from a remote place even for an overtopping flood flow. However, the conventional surveillance camera such as closed-circuit television (CCTV) is difficult to use at night with little surrounding light. Therefore, an FIR camera is recommended for establishing a real-time measurement system that works all the time. Further investigation is required to examine the effects of strong wind or heavy rain on the accuracy of STIV measurements.

Acknowledgments: The study was supported by a research grant from the Foundation of River and Basin Integrated Communications (FRICS) Foundation in collaboration with ICHARM. The data other than imaging data were provided through ICHARM. I am grateful to their support of the present research.

Abbreviations

The following abbreviations are used in this manuscript:

ADCP	Acoustic Doppler Current Profiler
RVM	Radio-wave Velocity Meter
FIR	Far-Infrared Ray
ICHARM	International Centre for Water Hazard and Risk Management
LSPIV	Large Scale Particle Image Velocimetry
STIV	Space Time Image Velocimetry
STI	Space time image
CCTV	Closed-circuit television

References

1. Schiermeier, Q. Increased flood risk linked to global warming: Likelihood of extreme rainfall may have been doubled by rising greenhouse-gas levels. *Nature* **2011**, *470*, 316. [CrossRef] [PubMed]
2. Okazumi, T.; Nakasu, T. Lessons learned from two unprecedented disasters in 2011—Great East Japan Earthquake and Tsunami in Japan and Chao Phraya River flood in Thailand. *Int. J. Disaster Risk Reduct.* **2015**, *13*, 200–206. [CrossRef]
3. Xu, G.-Y. Climate change and hydrologic models: A review of existing gaps and recent research developments. *Water Resour. Manag.* **1999**, *13*, 369–382. [CrossRef]
4. Moradkhani, H.; Sorooshian, S. General review of rainfall-runoff modeling: Model calibration, data assimilation, and uncertainty analysis. *Hydrol. Model. Water Cycle* **2009**, *63*, 1–24.

5. Turnipseed, D.P.; Sauer, V.B. *Discharge Measurements at Gaging Stations. U.S. Geological Survey Techniques and Methods Book 3*; USGS: Reston, VA, USA, 2010; Chapter 8, p. 87.
6. Fukami, K.; Yamaguchi, T.; Imamura, H.; Tashiro, Y. Current status of river discharge observation using non-contact current meter for operational use in Japan. *World Environ. Water Resour. Congr.* **2008**, *2008*, 1–10.
7. Welber, M.; Le Coz, J.; Laronne, J.B.; Zolezzi, G.; Zamler, D.; Dramais, G.; Hauet, A.; Salvaro, M. Field assessment of non-contact stream gauging using portable surface velocity radars (SVR). *Water Resour. Res.* **2016**, *52*, 1108–1126. [CrossRef]
8. Fujita, I.; Komura, S. Application of video image analysis for measurements of river surface flows. *Proc. Hydraul. Eng. JSCE* **1994**, *38*, 733–738. (In Japanese). [CrossRef]
9. Fujita, I.; Muste, M.; Kruger, A. Large-scale particle image velocimetry for flow analysis in hydraulic engineering applications. *J. Hydraul. Res.* **1998**, *36*, 397–414. [CrossRef]
10. Muste, M.; Fujita, I.; Hauet, A. Large-scale particle image velocimetry for measurements in riverine environments. *Water Resour. Res.* **2008**, *44*, W00D19. [CrossRef]
11. Muste, M.; Hauet, A.; Fujita, I.; Legout, C.; Ho, H.-C. Capabilities of Large-scale Particle Image Velocimetry to characterize shallow free-surface flows. *Adv. Water Res.* **2014**, *70*, 160–171. [CrossRef]
12. Hauet, A.; Creutin, J.-D.; Belleudy, P. Sensitivity study of large-scale particle image velocimetry measurement of river discharge using numerical simulations. *J. Hydrol.* **2008**, *349*, 178–190. [CrossRef]
13. Jodeau, M.; Hauet, A.; Paquier, A.; Le Coz, J.; Dramais, G. Application and evaluation of LS-PIV technique for the monitoring of river surface velocities in high flow conditions. *Flow Meas. Instrum.* **2008**, *19*, 117–127. [CrossRef]
14. Dramais, G.; Le Coz, J.; Camenen, B.; Hauet, A. Advantages of a mobile LSPIV method for measuring flood discharges and improving stage–discharge curves. *J. Hydro-Environ. Res.* **2011**, *5*, 301–312. [CrossRef]
15. Le Coz, J.; Hauet, A.; Pierrefeu, G.; Dramais, G.; Camenen, B. Performance of image-based velocimetry (LSPIV) applied to flash-flood discharge measurements in Mediterranean rivers. *J. Hydrol.* **2010**, *394*, 42–52. [CrossRef]
16. Fujita, I.; Kumano, G.; Asami, K. Evaluation of 2D river flow simulation with the aid of image-based field velocity measurement techniques. In *River Flow 2014*; Schleiss, A.J., Cesare, G.D., Franca, M.J., Pfister, M., Eds.; Taylor & Francis Group: London, UK, 2014; pp. 1969–1977.
17. Fujita, I.; Watanabe, H.; Tsubaki, R. Development of a non-intrusive and efficient flow monitoring technique: The space time image velocimetry (STIV). *Int. J. River Basin Manag.* **2007**, *5*, 105–114. [CrossRef]
18. Tsubaki, R.; Fujita, I.; Tsutsumi, S. Measurement of the flood discharge of a small-sized river using an existing digital video recording system. *J. Hydro-Environ. Res.* **2011**, *5*, 313–321. [CrossRef]
19. Fujita, I.; Hara, H. Development of space time image velocimetry introduced fast Fourier transform for improving robustness in river surface flow measurement. *J. Hydrosci. Hydraul. Eng.* **2011**, *29*, 123–135.
20. Fujita, I.; Kosaka, Y.; Yorozuya, A.; Motonaga, Y. Surface flow measurement of snow melt flood using a far infrared camera. *J. Jpn. Soc. Civ. Eng. Ser. B1 (Hydraul. Eng.)* **2013**, *69*, I_703–I_708. (In Japanese) [CrossRef]
21. Puleo, J.A.; McKenna, T.E.; Holland, K.T.; Calanton, J. Quantifying riverine surface currents from time sequences of thermal infrared imagery. *Water Resour. Res.* **2012**, 48. [CrossRef]

6

Snow Precipitation Measured by Gauges: Systematic Error Estimation and Data Series Correction in the Central Italian Alps

Giovanna Grossi [1,*], Amerigo Lendvai [1], Giovanni Peretti [2] and Roberto Ranzi [1]

1 Department of Civil, Environmental, Architectural Engineering and Mathematics, Università degli Studi di Brescia-DICATAM, Via Branze, 42, 25123 Brescia BS, Italy; a.lendvai@studenti.unibs.it (A.L.); roberto.ranzi@unibs.it (R.R.)

2 ARPA Lombardia–Centro Nivometeorologico di Bormio (SO), 23032 Bormio SO, Italy; g.peretti@arpalombardia.it

* Correspondence: giovanna.grossi@unibs.it

Abstract: Precipitation measurements by rain gauges are usually affected by a systematic underestimation, which can be larger in case of snowfall. The wind, disturbing the trajectory of the falling water droplets or snowflakes above the rain gauge, is the major source of error, but when tipping-bucket recording gauges are used, the induced evaporation due to the heating device must also be taken into account. Manual measurements of fresh snow water equivalent (SWE) were taken in Alpine areas of Valtellina and Vallecamonica, in Northern Italy, and compared with daily precipitation and melted snow measured by manual precipitation gauges and by mechanical and electronic heated tipping-bucket recording gauges without any wind-shield: all of these gauges underestimated the SWE in a range between 15% and 66%. In some experimental monitoring sites, instead, electronic weighing storage gauges with Alter-type wind-shields are coupled with snow pillows data: daily SWE measurements from these instruments are in good agreement. In order to correct the historical data series of precipitation affected by systematic errors in snowfall measurements, a simple 'at-site' and instrument-dependent model was first developed that applies a correction factor as a function of daily air temperature, which is an index of the solid/liquid precipitation type. The threshold air temperatures were estimated through a statistical analysis of snow field observations. The correction model applied to daily observations led to 5–37% total annual precipitation increments, growing with altitude (1740 ÷ 2190 m above sea level, a.s.l.) and wind exposure. A second 'climatological' correction model based on daily air temperature and wind speed was proposed, leading to errors only slightly higher than those obtained for the at-site corrections.

Keywords: precipitation measurement; precipitation correction; snow water equivalent

1. Introduction

Snow precipitation is one of the key hydrometeorological variables, besides snow monitoring activities connected to hydrological and meteorological processes. Planning and managing water resources, as well as preventing floods and landslides, need precise and accurate precipitation data. However, precipitation measurements by rain gauges are normally affected by systematic errors that lead to an underestimation of the real value [1–3]. The wind, disturbing the trajectory of the falling water droplets or snowflakes above the rain gauge, is the major source of error [4–6]. When precipitation occurs in its liquid phase, the maximum underestimation, even in strong wind conditions, is generally below 15%. Errors are much higher when precipitation turns into snow: even 100% underestimation is possible in this case [7,8].

Evaporation induced by heating devices (necessary for tipping bucket rain gauges) adds to the wind effect [9]. Tipping-bucket rain gauges are the most used around the world [10], but they are very often installed in those areas where snowfall exceeds rainfall throughout the year. This happens despite their well-known low reliability in such conditions [11–13].

The errors clearly show up in the Alps, a densely-monitored area, where precipitation measurements are commonly underestimated, more and more as altitude increases. It is at higher elevations that precipitation is greater, the wind is stronger and more continuous, and snowfall is the predominant form of precipitation: these are the most critical issues in correctly measuring the water volume that falls onto the ground. The increasing automatization in meteorological monitoring systems poses new challenges in this field of work, but also pushes to develop new, and maybe crucial, technologies, including advanced methods for monitoring the snow water equivalent and snow properties with remote sensing [14].

In this study a dataset from snow-meteorological monitoring stations located in Northern Italy is analysed. The focus is on the Valtellina and Vallecamonica upper areas, belonging to the Lombardy Region, with the purpose of estimating systematic errors in solid precipitation measurements collected by rain gauges (1740 ÷ 2190 m a.s.l.). To do so, daily rain gauge observations are compared to fresh Snow Water Equivalent (SWE) data from monitoring sites close to the study area. The dataset was collected thanks to cooperation between Università degli Studi di Brescia and the Snow and Avalanches ARPA centre located in Bormio, which is the official snow, avalanche, and weather monitoring institution for mountain areas in Lombardy. Field surveys supported the choice of the most suitable sites, according to the study aim. This paper not only draws the reader's attention to the issues of solid precipitation-measuring methods, but it also suggests a practical way to reconstruct the time series of precipitation observations affected by underestimation errors. Therefore, a simple statistical one-parameter model is proposed here, useful to correct daily precipitation data and based on two meteorological variables, temperature, and precipitation.

2. Case Study and the Error Estimate

Lombardy Region owns a dense meteorological monitoring network, part of which is a legacy of the former National Hydrological Service SIMN (Servizio Idrografico e Mareografico Nazionale, decommissioned in the nineties) and it is still composed of the original manual and mechanical devices, such as the mechanical tipping-bucket rain gauges with paper ink recorders manufactured by SIAP (Villanova (BO), Italy) (Figure 1a,b). With the transfer to regional administrations of the responsibility of environmental monitoring services from the national level to the regional level (ARPA Lombardia, the regional Environment Protection Agency), electronic devices and digital recording systems were installed in addition to the old ones, sometimes adjacent to the old manual stations. In many cases rain gauges manufactured by CAE (San Lazzaro di Savena (BO), Italy) (PMB2 model, Figure 1c) were chosen, so these are now the newest instruments in Lombardy.

For each one of the sites shown in Figure 2 and listed in Table 1, selected from those that are still active in the area, freshly-fallen snow water equivalent (SWE) daily measurements were compared to daily precipitation data retrieved from rain gauges nearby (Figure 3). SWE data by depth and density measurements of freshly-fallen snow collected on snowboards are assumed to be reliable because they provide a direct estimate of the real precipitation fallen on the ground [15]. In fact SWE was previously used by Sevruk [11,16] to reconstruct solid precipitation data from rain gauges and also by [17] as a representative reference measure for quality control of snow pillow measurements by the Snow Telemetry (SNOTEL) Data Collection Network by the U.S Department of Agriculture (USDA).

(a) (b) (c)

Figure 1. (**a**) Precipitation gauge at Pantano d'Avio (Brescia province, see also Figure 7b) monitoring site. The tank is inside a heated housing made by SIAP (Italian weather instruments manufacturer). On the left, the heating device, activated by a thermostat, here set to 4 °C; (**b**) Cancano monitoring site, where a CAE PMB2 electronic tipping-bucket heated gauge (yellow circle) is installed besidea heated mechanical SIAP one (red circle); (**c**) Electronic tipping-bucket heated gauge model CAE PMB2 at the Aprica Magnolta (SO) monitoring site.

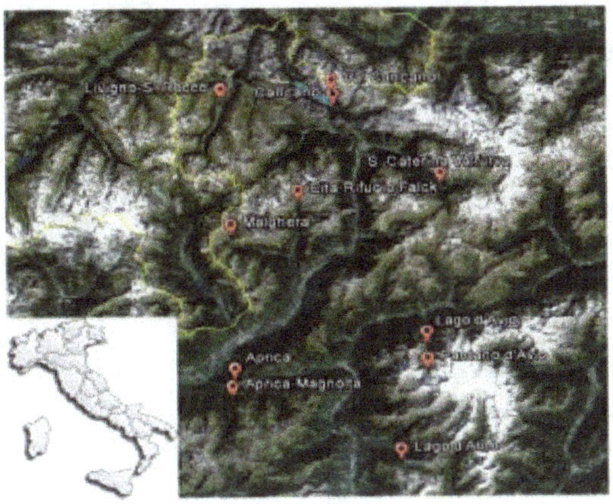

Figure 2. Map of the sites selected for the error estimate for solid precipitation measurements through precipitation gauges in the upper Valtellina and upper Valcamonica, in Northern Italy.

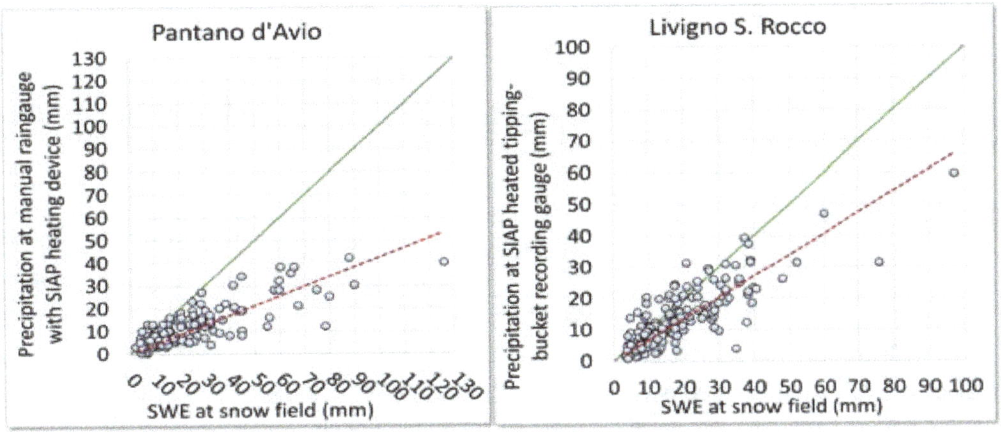

Figure 3. Comparison between melted snow measured daily by rain gauges and SWE of freshly-fallen snow manually measured at snow field sites. Plots refer to Pantano d'Avio (2390 m a.s.l., period 1987–2014) and Livigno S. Rocco (1865 m a.s.l., period 1989–2014).

Table 1. Comparison between melted snow measured by precipitation gauges and SWE of freshly-fallen snow manually measured at snow field sites nearby. Instruments abbreviation: SWE = manual measurement of fresh fallen snow with snow board and cylindrical coring device; P1 = manual rain gauge with 1000 cm^2 top hole and melting operation made at measuring time; P2 = manual rain gauge with SIAP heating device; P3 = mechanical SIAP heated tipping-bucket recording gauge; P4 = electronic heated tipping-bucket recording gauge CAE PMB2.

Sites	Position WGS84 (Lat; Long)	Elevation (m a.s.l.)	Period	Wind Exposure	Wind Data Availability	Site Context	Instruments Compared	Distance SWE/SP-Px (m)	No. of Compared Measurements	Accumulated SWE (mm)	Accumulated P (mm)	Bias (%)
S. Caterina Valfurva	46.410; 10.500 46.413; 10.493	1740	2001–2014 2003–2010	Low	2005–2010	Valley, open	SWE-P3 SWE-P4	590	75 28	866 359	649 177	−25 −51
Livigno S. Rocco	46.522; 10.125	1865	1989–2014	Low	none	Valley, open	SWE-P3	0	241	3641	2662	−27
Cancano	46.514; 10.317 46.513; 10.318	1950	1989–2013 2003–2010	High	2005–2010	Valley, open	SWE-P3 SWE-P4	120	301 73	5723 1579	2353 535	−59 −66
Aprica	46.154; 10.147	1180	1993–2014	Low	none	Valley, garden	SWE-P1	0	66	1452	912	−37
Aprica Magnolta	46.132; 10.140 46.129; 10.148	1865–1950	2009–2014	Low	2012–2014	Slope, forest	SWE-P4	740	38	744	631	−15
Lago d'Arno	46.048; 10.430	1830	1984–2014	Medium	none	Slope, open	SWE-P2	0	163	3429	2597	−24
Pantano d'Avio	46.167; 10.474	2390	1987–2014	High	none	Slope, open	SWE-P2	20	176	3956	1909	−52
Lago d'Avio	46.202; 10.474	1902	2003–2012	Low	none	Valley, open	SWE-P1	0	12	220	176	−20
Eita	46.388; 10.258	1950	2008–2014	Medium	none	Valley, open	SP-PT	0	195	2405	2554	6
Malghera	46.344; 10.142	1995	2008–2014	Medium	none	Valley, open	SP-PT	0	164	1835	1806	−2
Val Cancano	46.532; 10.313	2190	2008–2014	High	none	Slope, open	SP-PT	0	254	2113	2294	9

Starting with seasonal manual monitoring observations, only the precipitation events with the fresh snow density and depth correctly recorded by the observers were selected, in order to evaluate the snow water equivalent (SWE) with snow course data according to the following formula:

$$SWE = \frac{\rho \cdot H_N}{100} \left(mm = \frac{kg}{m^2} \right), \tag{1}$$

where the snow density ρ is expressed in kg/m^3 and the depth of fresh snow H_N is in cm and SWE in mm or kg m^{-2}.

To assess the underestimation of fresh fallen snow, no snow-mixed-to-rain events were considered (specific forms following the international standards are adopted in Italy [18–20]), nor were those with missing, incomplete, or inconsistent SWE measurements or rain gauge precipitation values. SWE manual measurements were taken between 8:00 and 10:00 am with a white wood snow board, a graduated scale and a snow sampler attached to a dynamometer (Figure 4 shows an example snow monitoring field). SWE measurements given by Equation (1) were assumed as the 'true' precipitation reference value. These data are affected by uncertainties in both depth of fresh snow and snow density measurements. In order to assess such uncertainty, results of a recent intercomparison of SWE measurement methods adopted in Italy were analysed [21]. They showed a 7% coefficient of variation of 92 SWE estimates conducted with combined depth and density measurements. This is a result of a 1 cm standard uncertainty of snow depth measurements and a typical 5% standard relative uncertainty of gravimetric snow density measurements with density cutters and portable dynamometers in the field [22]. Rain gauge types are: manual rain gauge with 1000 cm^2 hole and melting snow operation made at the measurement time (P1); manual rain gauge (1000 cm^2) in a heated housing SIAP model (P2); mechanical tipping-bucket recording gauge (1000 cm^2) SIAP or Salmoiraghi in a heated housing SIAP model (P3); CAE PMB2 electronic tipping-bucket recording heated gauge (1000 cm^2) (P4).

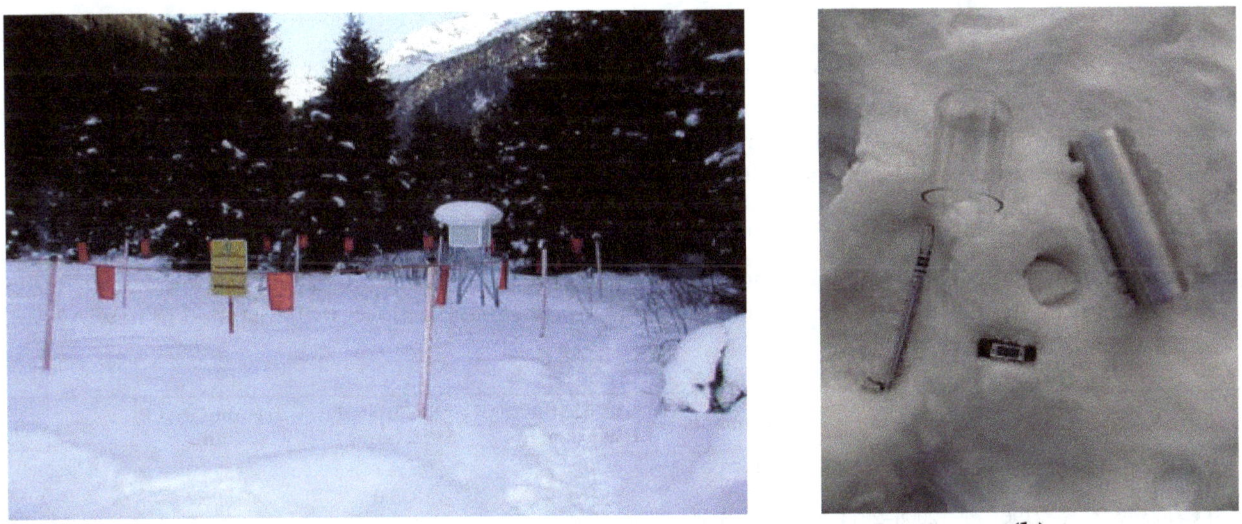

(a) (b)

Figure 4. (**a**) S. Caterina Valfurva manual monitoring site; and (**b**) instruments for the direct manual measurement of snow water equivalent: snow sampler and dynamometer.

In addition to the ARPA Lombardia monitoring network, A2A (an Italian energy generation and trading company) also manages several hydrometeorological stations, used in their management activity of hydroelectric power plants located in this area. Some of these stations are equipped with snow pillows and electronic weighing storage precipitation gauges with Alter-type wind-shields (Figure 5). Table 2 shows the comparison between daily precipitation measurements from weighing storage gauges (model Geonor T200B, with 200 cm^2 hole) and the respective snow pillows (model STS ATM/N; see also [23]). In both cases the total amount in 24 h and a 0.5 mm lower bound were considered to avoid high relative differences [24,25]. A 0 °C higher bound was also set for the daily average temperature, to leave out most of the snow-mixed-to-rain events. The obtained daily dataset was finally purged of all records with missing or incoherent data from one of the instruments. The literature [17] indicates that snow pillow SWE data are within 15% of the snow course reference value in 68% of the SNOTEL network sites and within 5% in 28% of the sites.

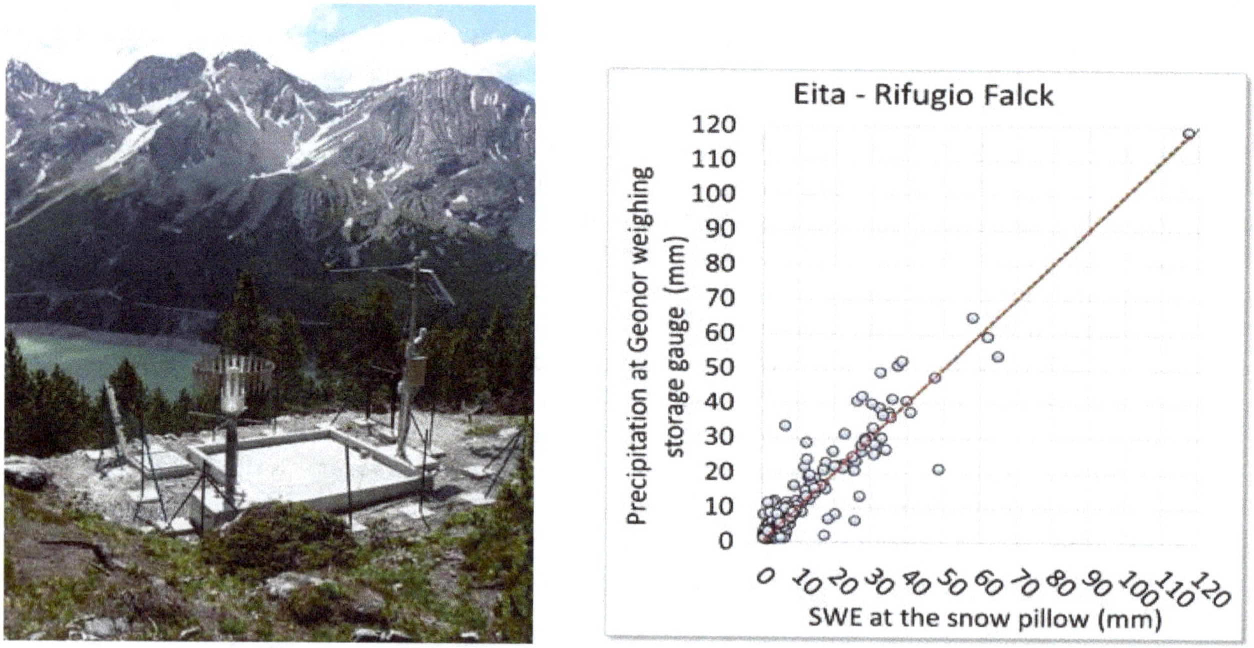

(a) (b)

Figure 5. (**a**) Val Cancano monitoring site managed by A2A equipped with Geonor T200B weighing storage gauge with an Alter-type wind-shield and snow pillow; and (**b**) a comparison of daily precipitation measurements: melted snow measured by the Geonor weighing storage gauge with an Alter-type wind-shield versus SWE of fresh fallen snow measured by snow pillows, for the period 2008–2014.

Table 2. Comparison between melted snow measured by Geonor weighing storage gauge with Alter windshield (PT) and SWE of fresh fallen snow measured by snow pillows (SP) installed in new A2A monitoring sites.

Sites	Elevation (m a.s.l.)	Period	Instruments	No. of Measurements	Accumulated SWE (mm)	Accumulated P (mm)	Bias (%)
Eita	1950	2008–2014	SP-PT	195	2405.4	2554.3	6
Malghera	1995	2008–2014	SP-PT	164	1835.0	1805.7	−2
Val Cancano	2190	2008–2014	SP-PT	254	2112.7	2294.3	9

Regarding the manual rain gauges and tipping-bucket heated recording gauges, the underestimation is systematic and varies from 15% to 66%; windy sites (Cancano and Pantano d'Avio) are those with the highest errors, above 50%, while in low wind conditions or wind-covered stations the average error is around 25%. The wind effect is responsible for about half of the total underestimation encountered for the Cancano mechanical tipping-bucket gauge SIAP model (Figure 1b) and Pantano d'Avio manual gauge in a SIAP heated housing (Figure 1a), since these instruments have no wind shielding device; in fact, the same instrument installed in low wind sites shows about halved errors. For the majority of the sites, the lack of wind data makes it difficult to comment more precisely on the influence of the wind. For this reason our basic correction procedures of rain gauge precipitation measurements are based on air temperature only, as presented in the next section.

However, at Cancano and S. Caterina, with two types of tipping bucket gauges, and Aprica Magnolta wind data are also available, although for a limited period and with some gaps. The results of the analysis of those data confirm the influence of the wind on the SIAP mechanical tipping-bucket heated gauge systematic errors, which in S. Caterina are about halved compared to Cancano, where the station is significantly more exposed to the wind. Using the SWE, precipitation, air temperature, and wind speed data available for these three sites a correction procedure of precipitation measurements based on both daily air temperature and wind speed data will be discussed in the fourth section.

Less significant, and apparently contradictory, are the results from the CAE PMB2 electronic tipping-bucket heated gauge. In fact, at S. Caterina, the errors are large despite the fact that the wind speed is even lower when compared to the Aprica Magnolta ones, where the same instrument shows only 15% underestimation, the lowest of all the examined ones, as shown in Table 1. This discrepancy may be justified by the fact that the Aprica Magnolta precipitation gauge is equipped with a heating device only from the end of 2012. Measurements considered for this comparison are mostly from the 2013–2014 season (only one winter season), in which exceptional temperature anomalies were recorded in Northern Italy (up to 4 °C above the average; see also [26]). At Aprica Magnolta the average temperature during the days taken into account for the 2013–2014 season was −1.7 °C; the remaining measurements are from the 2012–2013 season, in which days the average temperature was −3.7 °C. Considering only the data from the 2012–2013 season the gap in accumulated precipitation is −29.3%. At S. Caterina, where the same instrument shows an underestimate near 50%, the data refer to the 2005–2010 period, with an average temperature equal to −4.0 °C. This fact, added to an on-average lower intensity precipitation occurring at this site compared to Aprica, because of the interior Alps precipitation regime, could justify this different instrument performance, less accurate at lower temperatures and lower rainfall intensities. In fact, at S. Caterina, the SIAP mechanical tipping-bucket heated gauge shows fewer underestimations than the CAE PMB2 electronic one, probably due to the different heating devices of the two instruments, given that all other conditions are nearly identical.

Regarding the automated monitoring stations equipped with snow pillows and weighing storage gauges, there is a good correspondence between the increments of precipitation recorded by the gauge and daily increases of snow water equivalent on the snow pillow (Table 2). However there might still be a non-negligible underestimation of the measurements from the snow pillow because the considered sites are not fully wind protected and the Geonor T200B weighing precipitation gauge, equipped with an Alter-type wind shield, is known to underestimate snow precipitations by about 5–10% already for 1–2 m/s wind speed [27–30]. Therefore, it is reasonable to assume that there is an underestimation in the measurements from the snow pillow, because of the well-known problem caused by the internal cohesion of the snowpack. Unfortunately, none of these stations is equipped with wind sensors, so it was not possible to resolve these uncertainties. Results are consistent with those obtained in other similar studies referring to the Alpine region [11,31–34].

3. Correction of Precipitation Data

In order to reconstruct the precipitation data series, a simple model was developed and calibrated. It is based on the application of a correction factor to the daily measurement as a function of daily average temperature [31,35–41]. It appears thusly:

$$P_c = (1 + \alpha_s C_s)P, \tag{2}$$

$$\alpha_s = \begin{cases} 1 \text{ if } T_m < T_l \\ (T_h - T_m)/(T_h - T_l) \text{ if } T_l < T_m < T_h \\ 0 \text{ if } T_m > T_h \end{cases} \quad (3)$$

where the daily recorded precipitation P is corrected using a factor C_s as a function of the daily average temperature T_m through the coefficient α_s, which varies linearly between 0 and 1 as the temperature increases from the minimum threshold value for liquid precipitation T_h to the maximum threshold value for solid precipitation T_l (Figure 6a). In this way, theoretically, the snow-only precipitation measurements are increased, rain-only precipitation measurements remain the same and the snow-mixed-to-rain ones are partially corrected in a linear way according to the position of the daily mean temperature value between the two thresholds. This model is widely applicable because daily observations of air temperature are easy to find even in very old datasets. Wind speed data would be very useful for a more accurate correction, but they are seldom available and representative data samples are too hard to find, especially regarding the time extension. The nature of the area concerned does not even allow easy connections or interpolations with the data of other stations nearby, as the wind is influenced by the extremely complex orography. In order to determine the transition temperature between liquid and solid precipitation, a statistical analysis was carried out on the data collected by the observers at manual monitoring sites.

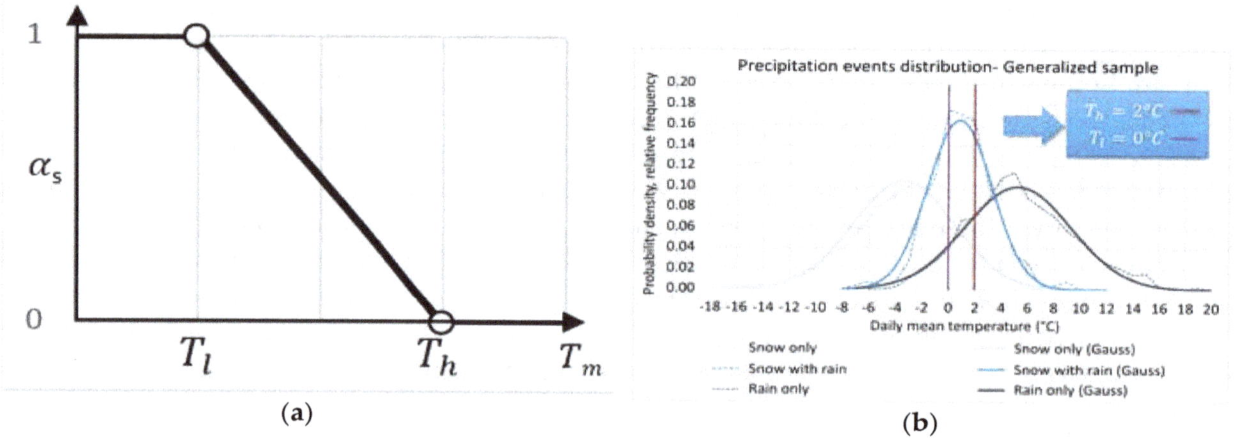

Figure 6. (**a**) Variability of the coefficient α_s as a function of mean daily temperature; (**b**) Probability density distribution of the precipitation events as a function of the daily mean temperature for the generalized sample, obtained joining many single station samples (S. Caterina Valfurva, Livigno S. Rocco, Cancano, Aprica, Lago d'Arno, Pantano d'Avio, for the time period shown in Table 1).

AINEVA (Associazione Interregionale di coordinamento e documentazione per i problemi inerenti la NEve e le VAlanghe), the Italian association for snow and avalanche studies, has set a procedure useful to clearly classify the precipitation events as: rain-only, snow-only, and sleet events. Given the availability of samples with a high abundance for each type of event and, very often, also for each individual station, it was possible to carry out a statistical analysis which can be assumed to be representative. Subdividing the samples of the three types of events mentioned above in temperature classes of 1 °C, it was possible to observe that they are well approximated by the normal distribution (Gauss). First, the samples of the three types of meteorological events selected for each station were taken into account; then, to obtain a generalized result, the samples of individual stations were also added.

The generalized sample analysis (Figure 6b) led to the setting of the following thresholds:

- 90% of snow-only events occur in days with average temperature below 1.5 °C
- 90% of rain-only events occur in days with average temperature above 0.1 °C
- 90% of sleet events occur in days with average temperature between −3.1 °C and 4.8 °C

The number of snow-mixed-to-rain events is significantly less than the number of rain-only or snow-only events (264 against 1284 and 2780, respectively). Therefore, it is logical to give more importance to the temperature thresholds identified by the rain-only and snow-only events. On the other hand the fact that the sleet events have a distribution centred at 1 °C was accounted for by choosing symmetrical temperature thresholds with respect to this value. In particular, the selected threshold temperatures are T_h = 2 °C and T_l = 0 °C. That is coherent with those specified by [31] and [35] for the application of the same model (based on a similar precipitation event analysis) to Malga Bissina and Pinzolo-Prà Rodont sites, in the Trentino-Alto Adige/Südtirol Region.

The correction factor C_s was calibrated for each monitoring site by minimizing the square bias of the individual measurements or the bias between the cumulative values (Table 3). The C_s value appeared to always be higher in the latter case. Subsequently, an intermediate value was chosen, fulfilling also the condition that the slope of the best-fit line of each measurement has to be lower than 1. In this way the resulting values of C_s were close to those obtained by minimizing the RMSE.

Table 3. Estimate of the correction factor C_s for each monitoring site and statistics of the resulting errors by considering the freshly-fallen snow water equivalent as the 'true' reference measurement.

Sites	Elevation (m a.s.l.)	Distance (m)	Instruments	No. of Measurements	Correction Factor C_s	New Bias (%)	Errors St. Dev (mm)
S. Caterina Valfurva	1740	590	SWE-P3	75	**0.30**	−3.6	4.9
			SWE-P4	28	**0.80**	−11.8	4.3
Livigno-S. Rocco	1865	0	SWE-P3	241	**0.35**	−2.6	6.6
Cancano	1950	120	SWE-P3	301	**1.15**	−17.0	13.1
			SWE-P4	73	**1.50**	−17.7	10.9
Aprica	1180	0	SWE-P1	66	**0.60**	−4.7	13.4
Aprica-Magnolta	1865/1950	740	SWE-P4	38	**0.11**	−6.1	5.8
Lago d'Arno	1830	0	SWE-P2	163	**0.25**	−6.9	10.7
Pantano d'Avio	2390	20	SWE-P2	176	**1.05**	−2.3	11.5
Lago d'Avio	1902	0	SWE-P1	12	**0.20**	−3.6	13.9

The model was applied to four precipitation data series: Pantano d'Avio, Lago d'Avio, S. Caterina and Cancano. In particular for the following years: Pantano d'Avio: 1990, 1994, 1996; Lago d'Avio: 1987–1994, 1996, 2002, 2003, 2005; S. Caterina Valfurva: 1968–1970, 1972–1976, 1978–1984, 1986, 1988–1992, 1994–1997, 1999–2007, 2009, 2012–2013; Cancano: 1979–1986; 1988–1989; 1997; 2002–1913.

An example of the correction effects on monthly precipitation averages and annual total value is reported in Figure 7a and Table 4. Corrected values for Cancano were compared to data from the nearby Val Cancano station, about 2 km apart and 250 metres higher than the first, but equipped with a Geonor T200B weighing precipitation gauge with a Alter-type wind-shield, considered a more reliable instrument in measuring solid precipitation. Regarding the 2009–2013 period, the average precipitation shows significant differences both in the annual values, and in the monthly ones, that cannot be explained only by the different exposure and the altitude gap. The corrected data show excellent consistency on the mean yearly total; at the monthly scale the variability is greater, probably due to an over-correction in the winter months but, overall, the precipitation regime (with the maximum in November) is properly reproduced.

Table 4. Mean yearly precipitation in some monitoring stations before and after the correction model application.

Sites	No. of Years	C_s	P_m (mm)	$P_{m,c}$ (mm)	Error ΔP (%)
S. Caterina Valfurva	40	0.3	860.3	923.6	**7%**
Cancano	23	1.15	788.3	1055.5	**34%**
Pantano d'Avio	3	1.05	1417.3	1897.0	**34%**
Lago d'Avio	12	0.2	1263.3	1324.8	**5%**

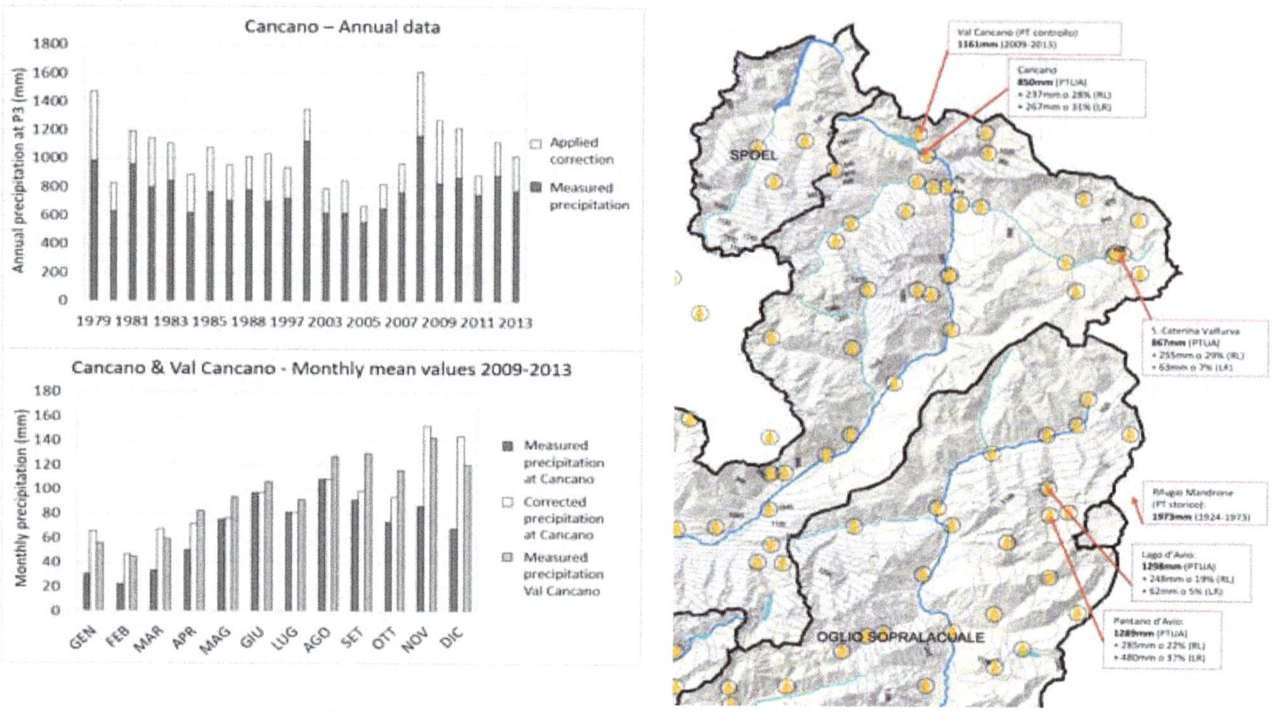

(a) (b)

Figure 7. (**a**) The effects of the application of the correction model to the precipitation data gathered by mechanical SIAP heated tipping-bucket recording gauges in Cancano. In the upper part, the annual values and, beneath, the monthly mean values for the 2009–2013 period also compared with the precipitation measured by the close weighing storage gauge with wind-shields at the Val Cancano monitoring site; (**b**) Probability part of the official Lombardy mean annual precipitation map, made in 2003 for the General Water Management and Protection Plan (PTUA). For the highlighted stations a more representative value using the correction model was estimated: the panels show a comparison between the official values with no correction (PTUA), the correction suggested by Regione Lombardia (RL), from a mean estimated SWE, and the correction suggested in this paper using the Lendvai-Ranzi model (LR). For a better comparison, the mean annual precipitation are also measured at two weighing storage gauges with wind-shields (PT) are shown.

4. Discussion

The results obtained for this and other stations can be compared in Figure 7b with those reported on the current official Lombardy Region precipitation distribution map (2003), here limited to the area of this study. To compensate the well-known underestimation of precipitation measurements at the high elevation sites, the Lombardy Region in its General Water Management and Protection Plan-PTUA (2006) uses, and proposes to use for other hydrological studies, a procedure in order to compensate the solid precipitation missing values. This consists in adding, to the yearly precipitation average value, a variable SWE value according to the different river basins and only to the basin area above the winter mean freezing level.

Figure 7b also shows the average precipitation measured by the manual storage gauge at Rifugio Mandrone (TN). Data published on Hydrological Yearbooks by SIMN show a mean yearly precipitation of 1973 mm calculated on the hydrological year and according to the available data for the period between 1924 and 1973. In the same period, starting from the daily measurements, the manual rain gauge at Lago d'Avio shows a yearly average of 1240 mm. Although the comparison is not as rigorous as the one made for Cancano site, since there is no coincidence in time period, nor the availability of such evolved instruments, the reconstruction performed on the data of Pantano d'Avio can still be considered consistent: it gives an yearly average of about 1900 mm for the three years 1990, 1994, and 1996.

A complete dataset of wind data was available for only three stations, and for an extensive and systematic solid precipitation data correction for the past, the simple temperature-based correction procedure was implemented for all of the stations. However, for the three stations equipped with five instruments for which wind data were available some variability with respect to wind speed in the systematic underestimation of precipitation was found.

To test the possibility to extend the information collected for these stations to other environments a temperature and wind-based data correction was tested. The value of the C_s (now $C_{s,w}$) factor in Equation (2) was assigned by minimising the RMSE between the corrected value P_c and the reference 'true' value of the fresh fallen SWE with the following criteria:

- if daily wind speed V is < 1.5 m/s, then $C_{s,w} = 0.4$ (S. Caterina, Aprica Magnolta)
- if $1.5 < V < 2.5$ m/s, then $C_{s,w} = 0.7$ (Cancano P3)
- if $V > 2.5$, then $C_{s,w} = 1.5$ (Cancano P4)

Since this second procedure is no more an at-site correction through a best fit value of $C_{s,w}$ for each station, but has a more general climatological value, a slight loss of accuracy (bias) and precision (RMSE) was observed, as expected, although it has to be kept in mind that the dataset used for the climatological correction is smaller than the one used for the at-site correction, because wind data are available only for a subset of measurements. The results are still acceptable, though, as it can be seen by comparing the error standard deviation in Tables 3 and 5 and, for this reason, this second method is suggested in case both temperature and wind data are available.

Table 5. Data used for the verification of the correction procedure based on both temperature and wind speed data. The coefficient $C_{s,w}$ is calibrated by minimising the RMSE of the corrected precipitation P_c vs. the 'true' value of fresh fallen snow SWE. Uncorrected rain gauge precipitation measurements are indicated by P.

Sites	Instrument	Number of Measurements	Accumulated SWE (mm)	Accumulated Measured Precipitation P (mm)	Corrected Precipitation P_c (mm)	Error St. Dev (mm)	Event Average Wind Speed (m/s)	Annual Average Wind Speed (m/s)
SCP/S.Caterina Valfurva	SWE-P3 SWE-P4	26 26	337.8 337.8	256.2 170.2	387.0 264.6	4.5 3.3	0.7	1.1 (2005–2009)
CAN/Cancano	SWE-P3 SWE-P4	82 71	1758.3 1566.8	650.4 530.0	1354.4 1106.1	13.4 9.7	2.3	2.6 (2005–2009)
AMA/Aprica Magnolta	SWE-P4	37	694.6	595.4	991.9	13.4	1.1	2.2 (2012–2014)

5. Conclusions

This study concerned Valtellina and Vallecamonica upper areas, located in the Lombardy Region, where ARPA Lombardia is officially in charge for the snow monitoring and avalanche risk warning service. Regarding the precipitation measurement in high-altitude areas, the current monitoring network is mostly equipped with mechanical and electronic tipping-bucket recording gauges, and only some of them have a heating device useful to melt the snow; moreover, there are still some manual precipitation gauges in use. During the snow season, there are several sites where human observers take fresh fallen snow water equivalent (SWE) measurements that are assumed as a reference 'true value' to correct systematic errors of precipitation measured by rain gauges nearby. Heated tipping-bucket gauges and manual gauges underestimate the real precipitation depth by between 15% and 66%; windy monitoring sites are those with the highest errors, above 50%, while in low wind or wind-covered stations the average error is around 25%.

In addition the A2A hydropower generation company had some hydrometeorological stations installed in 2008, equipped with snow pillows and electronic weighing storage precipitation gauges with Alter windshield. For each station, a comparison between measurements taken by these instruments showed to be in good agreement. This fact confirms the high reliability of electronic weighing gauges in solid precipitation measurement. However, it is likely that the snow pillows also underestimate the snowpack SWE when this is subject to significant internal cohesion, so that it creates a bridge effect that threatens the correct sensor reading.

The results obtained are, overall, consistent with those obtained from other similar studies conducted in the Alpine region and confirm the great variability of the underestimation of solid precipitation observations through rain gauges, depending on both site conditions and instrument type and model. In order to correct the precipitation datasets affected by systematic errors, first, a simple model was developed, which takes into account only daily average temperature and a site and instrument-dependent correction factor as parameters to set the correction. The model is then widely applicable since the air temperature is often the only meteorological information available, especially in old monitoring sites with similar characteristics. The model evaluates the constant correction factor C_s, ranging between 0.11 and 1.50, to daily precipitation measurements when they refer to a snow-only event: it does not modify the measurements of rain-only events. It also considers a linear variation within the temperature range in which snow-mixed-to-rain event occurred. The proposed model was applied to the precipitation datasets of four stations, with different results in terms of the impact on the monthly and yearly averages, which are increased between 5% and 37% as altitude above the sea level and the site's wind exposure increase. These results were compared to those reported on the Lombardy map of yearly precipitation distribution, which is the current official reference in this field of work, and with some interesting old data from the former National Hydrological Service (SIMN). A second correction procedure with more general climatological value, based on both daily temperature and wind speed data is then considered. The verification of this procedure provides only slightly worse performances than those obtained with the at-site and instrument-dependent correction and, therefore, can be used when complete meteorological data are available.

Acknowledgments: Special thanks for their cooperation in this study: ARPA Lombardia, especially the Centro Nivometeorologico di Bormio (SO) and the Servizio Idrografico di Milano; A2A S.p.A., especially the Uffici di Grosio (SO) for providing data from their stations.

Author Contributions: Amerigo Lendvai collected and analysed the data with the help of Giovanni Peretti and his collaborators, who did most of the fieldwork. Giovanna Grossi contributed in framing the study in the international research on snow monitoring. Roberto Ranzi indicated the methodologies and the statistical analyses to be performed.

References

1. Sevruk, B. Point precipitation measurement: Why are they not corrected? In *Water for the Future: Hydrology in Perspective*; IAHS Publication: Rome, Italy, 1987.
2. Immerzeel, W.; Wanders, N.; Lutz, A.; Shea, J.; Bierkens, M. Reconciling high altitude precipitation in the upper Indus basin with glacier mass balances and runoff. *Hydrol. Erath Syst. Sci.* **2015**, *19*, 4673–4687. [CrossRef]
3. Groot Zwaaftink, C.D.; Cagnati, A.; Crepaz, A.; Fierz, C.; Macelloni, G.; Valt, M.; Lehning, M. Event-driven deposition of snow on the Antarctic Plateau: Analyzing field measurements with SNOWPACK. *Cryosphere* **2013**, *7*, 333–347. [CrossRef]
4. Sevruk, B.; Hertig, J.; Spiess, R. Wind field deformation above precipitation gauge orifices. *IAHS* **1989**, *179*, 65–70.
5. Chvila, B.; Ondras, M.; Sevruk, B. The wind-induced loss of precipitation measurement of small time intervals as recorded in the field. In Proceedings of the WMO/CIMO Technical Conference, Bratislava, Slovaki, 25 September–3 October 2002.
6. Rasmussen, R.; Baker, B.; Kochendorfer, J.; Meyers, T.; Landolt, S.; Fischer, A.; Black, J.; Theriault, J.; Kucera, P.; Gochis, D.; et al. How well are we measuring snow? The NOAA/FAA/NCAR winter precipitation test bed. *BAMS* **2012**, *6*, 811–829. [CrossRef]

7. Duchon, C.; Cole, J.; Rasmussen, R. Measuring heavy snowfall using five different windshields and vibrating-wire precipitation gauges. In Proceedings of the 65th Eastern Snow Conference, Lake Morey, Fairlee, VT, USA, 28–30 May 2008.
8. Pan, X.; Yang, D.; Li, Y.; Barr, A.; Helgason, W.; Hayashi, M.; Marsh, P.; Janowicz, J.P.R. Bias Corrections of Precipitation Measurements across Experimental Sites in Different Ecoclimatic Regions of Western Canada. *Cryosphere Discuss.* **2016**, *10*, 2347. [CrossRef]
9. Sevruk, B. Distribution of precipitation in mountainous areas. In *Evaporation Losses from Storage Gauges*; World Meteorological Organization (WMO): Geneva, Switzerland, 1972; pp. 96–102.
10. Nitu, R.; Wong, K. *CIMO Survey on National Summaries of Methods and Instruments for Solid Precipitation Measurement at Automatic Weather Stations*; World Meteorological Organization (WMO): Geneva, Switzerland, 2010.
11. Sevruk, B. Correction of measured precipitation in the Alps using the water equivalent of new snow. *Nord. Hydrol.* **1983**, *1*, 49–58.
12. Savina, M.; Schaeppi, B.; Molnar, P.; Burlando, P.; Sevruk, B. Comparison of a tipping-bucket and electronic weighing precipitation gage for sbnowfall. *Atmos. Res.* **2012**, *103*, 45–51. [CrossRef]
13. Nitu, R.; Rasmussen, R.; Baker, B.; Lanzinger, E.; Joe, P.; Yang, D.; Smith, C.; Roulet, Y.; Goodison, B.; Liang, H.; et al. *WMO Intercomparison of Instruments and Methods for the Measurement of Solid Precipitation and Snow on the Ground: Organization of the Experiment*; World Meteorological Organization (WMO): Geneva, Switzerland, 2012.
14. Cagnati, A.; Crepaz, A.; Macelloni, G.; Pampaloni, P.; Ranzi, R.; Tedesco, M.; Tomirotti, M.; Valt, M. Study of the snow meltfreeze cycle using multi-sensor data and snow modelling. *J. Glaciol.* **2004**, *50*, 419–426. [CrossRef]
15. Bocchiola, D.; Rosso, R. The distribution of daily snow water equivalent in the central Italian Alps. *Adv. Water Resour.* **2007**, *30*, 135–147. [CrossRef]
16. Sevruk, B. Conversion of Snowfall Depths to Water Equivalents in the Swiss Alps. *Zürcher Geogr. Schr.* **1986**, *23*, 13–23.
17. Serreze, M.C.; Clark, M.P.; Armstrong, R.L.; McGinnis, D.A.; Pulwarty, R.S. Characteristics of the western United States snowpack from snowpack telemetry (SNOTEL) data. *Water Resour. Res.* **1999**, *35*, 2145–2160. [CrossRef]
18. Cagnati, A. *Strumenti di Misura e Metodi di Osservazione Nivometeorologici*; AINEVA: Trento, Italy, 2003. (In Italian)
19. Valt, M.; Cagnati, A.; Corso, T. Stima dell'equivalente in acqua della neve. *Neve e Valanghe* **2006**, *59*, 24–33. (In Italian)
20. Valt, M.; Moro, M. Average snowcover density values in eastern Alps mountain. In Proceedings of the EGU General Assembly 2009, Vienna, Austria, 19–24 April 2009.
21. D'Aosta, A.V. *Secondo Interconfronto SWE*; ARPA: Valle d'Aosta, Aosta, Italy, 2016.
22. Proksch, M.; Rutter, N.; Fierz, C.; Schneebeli, M. Intercomparison of snow density measurements: Bias, precision, and vertical resolution. *Cryosphere* **2016**, *10*, 371–384. [CrossRef]
23. Valgoi, P. Cuscinetto per la misura della densità della neve (snow pillow). *Neve e Valanghe* **2011**, *72*, 48–53. (In Italian)
24. Goodison, B.; Louie, P.; Yang, D. *WMO Solid Precipitation Measurement Intercomparison-Final Report*; World Meteorological Organization (WMO): Geneva, Switzerland, 1998.
25. Sevruk, B. Adjustment of tipping-bucket precipitation gauge measurements. *Atmos. Res.* **1996**, *42*, 237–246. [CrossRef]
26. ArCIS-Gruppo di lavoro Archivio climatologico dell'Italia centro-settentrionale. Il clima nell'inverno 2013–2014: Le eccezionali anomalie climatiche del Centro-Nord Italia. *Neve e Valanghe* **2014**, *81*, 4–9. (In Italian)
27. Colli, M.; Lanza, L.; Rasmussen, R.; Thériault, J. The collection efficiency of shielded and unshielded precipitation gauges. Part I: CFD airflow modeling. *J. Hydrometeorol.* **2016**, *17*, 231–243. [CrossRef]
28. Colli, M.; Lanza, L.; Rasmussen, R.; Thériault, J. The collection efficiency of shielded and unshielded precipitation gauges. Part II: Modeling particle trajectories. *J. Hydrometeorol.* **2016**, *17*, 245–255. [CrossRef]
29. Smith, C. Correcting the wind bias in snowfall measurements made with a Geonor T-200B precipitation gauge and alter wind shield. In Proceedings of the 15th Symposium on Meteorological Observation and Instrumentation, Atlanta, GA, USA, 17–21 January 2010.

30. Macdonald, J.; Pomeroy, J. Gauge undercatch of two common snowfall gauges in a prairie environment. In Proceedings of the 64th Eastern Snow Conference, St. John's, NL, Canada, 29 May–1 June 2007.
31. Ranzi, R.; Grossi, G.; Bacchi, B. Ten years of monitoring areal snowpack in the Southern Alps using NOAA-AVHRR imagery, ground measurements and hydrological data. *Hydrol. Process.* **1999**, *13*, 2079–2095. [CrossRef]
32. Zweifel, A.; Sevruk, B. Comparative accuracy of solid precipitation measurement using heated recording gauges in the Alps. In *Workshop on Determination of Solid Precipitation in Cold Climate Regions*; Fairbanks: Zürich, Switzerland, 2002.
33. Savina, M.; Schaeppi, B.; Molnar, P.; Burlando, P.; Sevruk, B. Comparison of a tipping-bucket and electronic weighing precipitation gage for snowfall. *Atmos. Res.* **2012**, *103*, 45–51. [CrossRef]
34. Cugerone, K.; Allamano, P.; Salandin, A.; Barbero, S. Stima della precipitazione in siti di alta quota. *Neve e Valanghe* **2012**, *77*, 36–43. (In Italian)
35. Eccel, E.; Cau, P.; Ranzi, R. Data reconstruction and homogenization for reducing uncertainties in high-resolution climate analysis in Alpine regions. *Theor. Appl. Climatol.* **2012**, *110*, 345–358. [CrossRef]
36. Ranzi, R.; Grossi, G.; Gitti, A.; Taschner, S. Energy and mass balance of the Mandrone Glacier. *Geogr. Fis. Din. Quat.* **2010**, *33*, 45–60.
37. Grossi, G.; Caronna, P.; Ranzi, R. Hydrologic vulnerability to climate change of the Mandrone glacier (Adamello-Presanella group, Italian Alps). *Adv. Water Resour.* **2013**, *55*, 190–203. [CrossRef]
38. World Meteorlolgical Organization. *Intercomparison of Models of Snowmelt Runoff*; WMO Operational Hydrology Report No. 23, WMO Publ. No. 646; World Meteorological Organization (WMO): Geneva, Switzerland, 1986.
39. Kienzle, S. A new temperature based method to separate rain and snow. *Hydrol. Process.* **2008**, *22*, 5067–5085. [CrossRef]
40. Forland, E.; Allerup, P.; Dahlström, B.; Elomaa, E.T.; Perälä, J.; Rissanen, P.; Vedin, H.; Vejen, F. Manual for operational correction of Nordic precipitation data. *DNMI Oslo* **1996**, *96*, 66.
41. Morin, S.; Lejeune, Y.; Lesaffre, B.; Panel, J.; Poncet, D.; David, P.; Sudul, M. An 18-yr long (1993–2011) snow and meteorological dataset from a mid-altitude mountain site (Col de Porte, France, 1325 m alt.) for driving and evaluating snowpack models. *Earth Syst. Sci. Data* **2012**, *4*, 13–21. [CrossRef]

7

Soil Moisture for Hydrological Applications: Open Questions and New Opportunities

Luca Brocca *, Luca Ciabatta, Christian Massari, Stefania Camici and Angelica Tarpanelli

Research Institute for Geo-Hydrological Protection, National Research Council, 06128 Perugia, Italy; luca.ciabatta@irpi.cnr.it (L.C.); christian.massari@irpi.cnr.it (C.M.); stefania.camici@irpi.cnr.it (S.C.); angelica.tarpanelli@irpi.cnr.it (A.T.)
* Correspondence: luca.brocca@irpi.cnr.it

Academic Editors: Roberto Ranzi and Tommaso Moramarco

Abstract: Soil moisture is widely recognized as a key parameter in the mass and energy balance between the land surface and the atmosphere and, hence, the potential societal benefits of an accurate estimation of soil moisture are immense. Recently, scientific community is making great effort for addressing the estimation of soil moisture over large areas through in situ sensors, remote sensing and modelling approaches. The different techniques used for addressing the monitoring of soil moisture for hydrological applications are briefly reviewed here. Moreover, some examples in which in situ and satellite soil moisture data are successfully employed for improving hydrological monitoring and predictions (e.g., floods, landslides, precipitation and irrigation) are presented. Finally, the emerging applications, the open issues and the future opportunities given by the increased availability of soil moisture measurements are outlined.

Keywords: soil moisture; hydrology; in situ measurements; remote sensing; floods; landslides; precipitation; irrigation

---

1. Introduction

The importance of soil moisture in the hydrological cycle has been stressed in a number of papers [1–4] and scientific projects (European Space Agency Climate Change Initiative Soil Moisture, ESACCISM [5], Soil Moisture Active and Passive mission, SMAP [6], International Soil Moisture Network, ISMN [7], and Cosmic-ray Soil Moisture Observing System, COSMOS [8]). Soil moisture governs the partitioning of the mass and energy fluxes between the land and the atmosphere, thus playing a key role in the assessment of the different components of the water and energy balance. Soil moisture is an important variable for flood and landslide modelling and prediction [9–12], for drought assessment and forecasting [13–15], and for numerical weather prediction [16–19], to cite a few. Soil moisture is also the water source for plants and, hence, its knowledge is required for irrigation management and agricultural studies [20]. The scientific community has well recognized the very important role of soil moisture in Earth Science applications and in the last 40 years new approaches and techniques for monitoring, modelling and using soil moisture data have been developed.

In this review, an overview of the potential of soil moisture observations for improving hydrological applications is presented. Firstly, we briefly describe the main techniques employed for soil moisture estimation at different spatial scales. Secondly, the understanding that we gained in the assessment of soil moisture spatial-temporal variability is illustrated. Thirdly, we provide a list of the hydrological applications that have benefited (and will benefit) from the use of soil moisture observations. Finally, the open questions and the novel opportunities that would be important to address in future investigations are described.

We underline that this paper reflects the opinion and the experience of the authors on this research topic and we mainly focus on the applications and the research questions that we investigated in the past. This paper should not be considered as a review paper. Indeed, several recent reviews already investigated different hydrological aspects related to soil moisture, e.g., the estimation of soil moisture through remote sensing [3,21,22], the different techniques for in situ soil moisture monitoring at different scales [23,24], the use of satellite soil moisture data in hydrological and climatic applications [25,26], and the assessment and modelling of soil moisture spatial-temporal variability [2,27,28]. Therefore, we believe there is no need of an additional review on soil moisture. Differently, the manuscript is conceived to be a comprehensive summary of our research activity on soil moisture and our view for future opportunities to be analysed. We wanted to identify the main knowledge gaps that need to be filled in future investigations for improving the monitoring, and the use, of soil moisture observations in hydrological (and others) applications.

2. How Do We Estimate Soil Moisture?

Notwithstanding soil moisture is largely recognized as a fundamental physical variable, e.g., soil moisture is included at the 2nd place among the Essential Climate Variable by the Global Climate Observing System (GCOS), its monitoring over large areas is still an open research activity [23,29]. Indeed, differently from precipitation, in situ monitoring networks for soil moisture are much less developed and only in few countries (e.g., United States) a good coverage of in situ stations is present (see Figure 1). This should be attributed to the very recent availability, with respect to precipitation, of ground sensors measuring soil moisture, and partly to the lower knowledge of the large benefits that we can obtain from soil moisture observations. Currently, three different approaches are used for providing soil moisture estimates: (1) in situ observations; (2) remote sensing; and (3) modelled data. In the following, we describe the main features of each approach in order to underline their pros and cons. Overall, it is widely accepted that the integration of in situ, satellite and modelled data is the best approach to exploit the potential of soil moisture in hydrological applications [25,30].

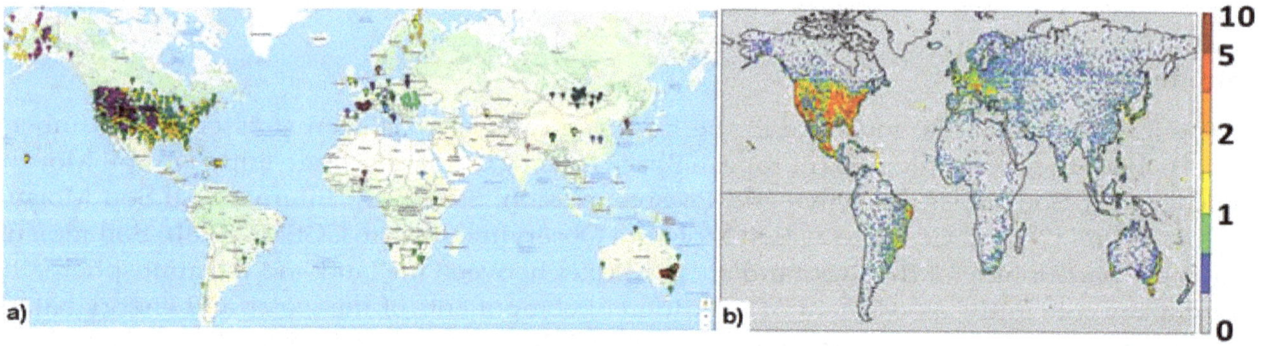

Figure 1. (**a**) Global distribution of soil moisture stations available at the International Soil Moisture Network from January 2000 ([7]); different colours refer to different networks; (**b**) Density of rain gauges underlying the Climate Prediction Center Unified Gauge-Based Analysis of Global Daily Precipitation (number of gauges/0.5°). Even though the two maps are not directly comparable, it is evident that the number of soil moisture stations is much lower than precipitation gauges, and, except in USA and Europe, soil moisture data are nearly absent.

2.1. *In Situ Measurements*

The first approaches for monitoring soil moisture through in situ observations are based on gravimetric, tensiometric and nuclear techniques [24]. Although invasive and time consuming, the gravimetric technique is still the reference method to which the other techniques and new methods are calibrated and tested. In the 1980s, Topp et al. [31] proposed the Time Domain Reflectometry

(TDR) technique. TDR was found to provide accurate measurements for a wide range of soils and settings thus becoming the new standard approach for measuring soil moisture and replacing the gravimetric approach (TDR is less time consuming and invasive). More recently, the emergence of low-cost capacitance sensors, based on Frequency Domain Reflectometry (FDR) technique, as well as impedance, time domain transmission sensors, etc., strongly promoted the usage of soil moisture sensors in environmental research [32]. Currently, FDR is likely the most used approach for in situ monitoring of soil moisture thanks to its lower cost with respect to TDR, even though at the expense of lower accuracy. Moreover, recent advances for the calibration and testing of electromagnetic sensors were proposed by Bogena et al. [33], thus further improving the accuracy and reliability of soil moisture sensors.

However, all the approaches listed above provide only point measurements that are representative of a small volume of soil [34]. The emergence of wireless sensor networks made it possible to cover larger areas with low-budget soil moisture sensors [35], but still this technology is only used for scientific studies. In the last decade, novel technologies were developed for providing soil moisture measurements over larger areas [29]. Currently, cosmic ray neutron sensors, Global Positioning System (GPS), and geophysical measurements (e.g., electrical resistivity and electromagnetic induction) are the most promising techniques [33,36–38]. Indeed, they are able to provide measurements with a support scale much larger than few cubic centimetres as point scale measurements. Cosmic ray and GPS techniques are well developed and some networks are already established in various countries (COSMOS [8], COSMOS-UK [39], and Plate Boundary Observatory to study the water cycle: PBO H2O [40]). However, these techniques are very recent, e.g., COSMOS network in UK was set up starting from 2016, and their use as established technique still requires further research and technical investigations.

Point in situ measurements of soil moisture (e.g., gravimetric, TDR, and FDR) are surely the most accurate techniques; however, they suffer from low spatial representativeness. The new techniques (e.g., cosmic ray and GPS) partly address this issue. Specifically, several papers demonstrated that the accuracy of the cosmic ray is similar to that point sensors, even under unfavourable conditions (e.g., [41–43]). Moreover, the maintenance of in situ soil moisture networks requires significant economic and human resources (even though lower than climatic stations) and, frequently, it is a non-trivial task to make a network operational for several (e.g., >5) years [34]. On this basis, it is evident that the development of established techniques providing measurements at 0.1–1 km spatial scale is vital [29].

2.2. Remote Sensing

Remote sensing is definitely the most appropriate technique to obtain large scale soil moisture measurements. Different methods were developed in the last 40 years for the retrieval of soil moisture from microwave, optical and thermal satellite sensors. Several review papers on this topic were published recently [21,22,44] that well describe the algorithms developed for the retrieval of soil moisture from the different bands of the electromagnetic spectrum. In this paper, we focus only on active and passive microwave-based products as they are the most used methods, also providing operational soil moisture products [45]; the reader is referred to Rahimzadeh-Bajgiran and Berg [44] for a review on recent advances in thermal and optical remote sensing for soil moisture estimation. Specifically, it is important to distinguish among active microwave sensors, i.e., between Synthetic Aperture Radars (SARs) and scatterometers. The former (SAR) can provide high spatial resolution (<1 km) but coarse temporal resolution (i.e., >10 days). The latter (scatterometer) provides coarse spatial resolution (~20 km) and high temporal coverage (~daily). Passive microwave radiometers are characterized by nearly the same spatial-temporal resolution of scatterometers. We note that the temporal coverage of satellite data can be significantly improved by using multiple sensors or even a constellation of sensors [46,47]. Another important aspect is the specific band in the microwave range of the electromagnetic spectrum. Usually, X-, C- and L-band are used for estimating soil moisture with the latter theoretically more suitable [48]. However, other aspects such as the radiometric accuracy and the radio frequency interferences can significantly affect the final quality of soil moisture retrievals, even more than the microwave band [45].

At the time of writing (January 2017), four quasi-operational, i.e., available either in near real time (NRT) or few days after sensing, coarse resolution satellite surface soil moisture products are available: (1) the Soil Moisture Active and Passive (SMAP) mission (L-band radiometer) starting from April 2015 with ~36 km/2-day spatial/temporal resolution [49]; (2) the Advanced Microwave Scanning Radiometer 2 (AMSR2) onboard the Global Change Observation Mission for Water, GCOM-W, satellite (C- and X-band radiometers) starting from July 2012 with ~25 km/1-day spatial/temporal resolution [50]; (3) the Soil Moisture and Ocean Salinity (SMOS) mission product (L-band radiometer) starting from January 2010 with ~50 km/2-day spatial/temporal resolution [51]; and (4) the Advanced SCATterometer (ASCAT) onboard Metop-A and Metop-B satellites (C-band scatterometer) starting from January 2007 with ~25 km/1-day spatial/temporal resolution [45]. Recently, higher resolution (~1 km) soil moisture products are becoming available based on the disaggregation of coarse resolution products [52–54] and, in the near future, from Sentinel-1 satellites (e.g., [55]). However, thus far a high-resolution soil moisture product from SAR sensors has never been made available.

The most striking benefit of remote sensing observations is related to their spatial-temporal coverage, and the relatively lower costs for large scale applications. The accuracy of satellite soil moisture products is surely lower than in situ observations, even though the recent products listed above have achieved a high level of reliability and maturity, as demonstrated in several validation studies comparing satellite data with in situ observations and land surface/hydrological modelling ([56–59] to cite a few). Moreover, the number of applications already using these products is an indirect assessment of their good quality level, as it was demonstrated their utility for improving and supporting hydrological and climatic predictions (see below and [26,60,61]). The three major limitations of satellite soil moisture products are: (1) the very shallow soil layer (2–7 cm) that is sensed from satellites; (2) the coarse spatial resolution (~20 km) of the currently available products; and (3) the very low quality under certain surface conditions (dense vegetation, frozen soils, snow, mountainous terrain). While these limitations may prevent their use in several hydrological applications [25], some approaches were already developed to overcome these issues, i.e., spatial downscaling techniques (e.g., [53]) and simplified methods for root-zone soil moisture estimation from surface measurements (e.g., [62,63]).

2.3. Hydrological and Land Surface Modelling

Hydrological and land surface models are largely used for estimating soil moisture at different spatial and temporal scales (e.g., [64]). Basically, both hydrological and land surface models use the same set of equations for simulating the water and energy balance, and hence soil moisture [65]. Indeed, different models use different structures for simulating each of the components (e.g., Richard's equation or variable infiltration curve for infiltration), but the main governing balance equations are the same. The models differ with respect to the spatial (horizontal and vertical) and temporal discretization, to the simulated physical processes and to the corresponding parameterization. More frequently, hydrological models are discretized at basin level and land surface model considers regular grids, but it is not always the case. Potentially, through modelling we are able to obtain soil moisture estimates at the desired temporal and spatial resolution (e.g., sub-hourly and 100 m, [66]).

The accuracy of modelled soil moisture data is strongly dependent on the employed model. However, besides to the model accuracy, the quality of meteorological observations used as input data also plays a very important role. Even a perfect model will fail if the quality, or the density, of meteorological inputs is low. Similarly, also the spatial and temporal discretization depends on the resolution of meteorological forcings, as well as of static information such as land use and soil texture maps [67]. Sometimes, high resolution modelled data (e.g., 500 m) are obtained by using as inputs coarse scale meteorological forcings (e.g., precipitation data at 10–100 km resolution). In these cases, the effective resolution of modelled data should be the one of the meteorological inputs (i.e., 10–100 km and not 500 m). A common issue faced by modeller is related to the parameterization of complex modelling structure and of the model parameter values. For instance, the parameterization of soil hydraulic parameters is a very difficult task, even in well gauged or experimental sites [68], and mainly over large areas [69]. Additionally, many key hydrologic processes are extremely difficult to parameterize (e.g., irrigation, dam operation, snow melting, and interception), especially in challenging regions (deserts, pluvial forests, and high altitudes). Therefore, modelled soil moisture data surely represents an important dataset that, however, needs to be used with caution [70,71].

3. How Does the Soil Moisture Vary in Space and Time?

In the last 40 years, a large number of studies were dedicated to the understanding of the spatial and temporal variability of soil moisture from the local to the regional and global scale (e.g., [2,72–75]). Indeed, after the development of the first measurement techniques, several researchers carried out detailed field campaigns in different geomorphological and climatic settings to assess soil moisture variability [24]. Different methods were applied including statistical [2,75] and geostatistical approaches [76], temporal stability analysis [73,77], and regression and wavelet techniques [78]. In this section, we provide a short overview of the major results obtained in the different studies trying to summarize the status of the knowledge we gained about the soil moisture spatial-temporal variability.

A large body of scientific literature analysed the statistical properties of soil moisture datasets obtained by field campaigns. The values of the first two statistical moments of absolute soil moisture data (in volumetric units, m^3/m^3), i.e., mean and variance, as well as their mutual relationship, were largely investigated (e.g., [79]). In most cases, a convex upward relationship is obtained between mean and variance thus concluding that soil moisture variability peaks for intermediates wetness conditions (e.g., [2,75,80,81]). Recently, Mittelbach and Seneviratne [82] proposed a new framework to decompose the soil moisture variability of the temporal mean and temporal anomalies. Interestingly, they obtained that the variability of soil moisture anomalies is larger for wet and dry conditions, with the minimum occurring at intermediate wetness conditions (see also [83,84]). Therefore, the behaviour of absolute and anomaly soil moisture values is not consistent and additional investigations are required.

A fundamental concept was introduced by [73], i.e., the soil moisture temporal stability. Vachaud et al. [73] found that the spatial patterns of soil moisture are stable over time and they named it as "temporal stability" (see [77] for a recent review). In other words, while the mean soil moisture for a given area might be changing over time, the spatial location of wetter and drier areas does not change. Figure 2 shows an example of soil moisture spatial patterns obtained by portable soil moisture measurements in the Colorso experimental plot in central Italy [2]. As can be seen, the different maps show the same pattern with wetter areas in the centre and southern part of the plot, and drier values in the northern part. The patterns are also quite consistent with the topography [2]. The simple concept of temporal stability has tremendous implications in hydrological and climate applications. It means that the monitoring of soil moisture with in situ observations over large areas (>100–200 km^2) can be carried out with a limited number of stations [75,77,83,85], with evident economic and time savings. Similarly, a limited number of stations can be used for the validation of coarse scale satellite soil moisture products, as indirectly demonstrated by the high correlations that are frequently obtained between coarse scale (~20 km) and point scale (~5 cm) measurements (e.g., [57–59]). Vice versa, coarse scale measurements, as obtained by remote sensing, can be used for regional or local studies (e.g., [11,86]). In summary, the temporal stability concept allows to mitigate the spatial discrepancies between point measurements and the scale needed in the applications.

A series of publications investigated the factors influencing soil moisture variability, both in time [87] and in space [88]. In time, soil moisture is mainly driven by precipitation and evapotranspiration, and its temporal variability is also a function of soil characteristics, vegetation, topography and groundwater [87,89]. Several modelling studies obtained very good performances in simulating point- (grid-) scale soil moisture temporal evolution [65,67,90], thus showing that a good knowledge has been gained about the factors influencing point scale soil moisture temporal variability. Figure 3 shows a comparison between observed and modelled data at two sites in central Italy using the Soil Water Balance Model [67]. The agreement with observations highlights the good model performances, and similar results were obtained at several sites in Europe (e.g., [57,91]). In space, the same meteorological factors, i.e., precipitation and evapotranspiration, have a clear impact on soil moisture patterns at large scales (>500–1000 km^2). At smaller scales, static factors such as land cover, topography and soil texture/structure affects soil moisture spatial variability (see Figure 1 in [75]). Several authors compared soil moisture spatial patterns with these static factors (e.g., [2,92]) obtaining moderate to low predictive capability, largely varying across sites and climates. Therefore, our knowledge about the factors influencing spatial variability of soil moisture is still limited. This important point is evident if we compare the capability of hydrological models in reproducing soil moisture spatial variability with respect to temporal variability. Frequently, the magnitude of variability is severely underestimated [90,93,94], and at regional scale soil moisture spatial patterns from modelling and satellite observations are found to be not consistent [95]. In this context, some recent modelling studies are demonstrating the value of soil moisture data for a more in-depth validation of distributed hydrological models [96,97].

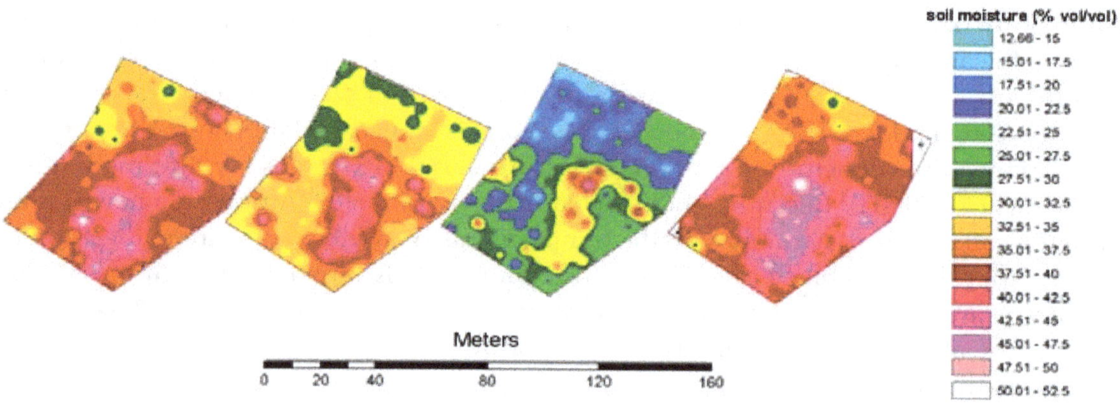

Figure 2. Spatial soil moisture maps (% vol/vol) obtained for the Colorso experimental plot in Italy at fuor different dates (from left to right): 21 April 2005, 28 April 2005, 5 May 2005, and 2 December 2005 (data from [2]). Even though the mean soil moisture conditions are changing over time, the spatial pattern of wetter and drier areas remains the same.

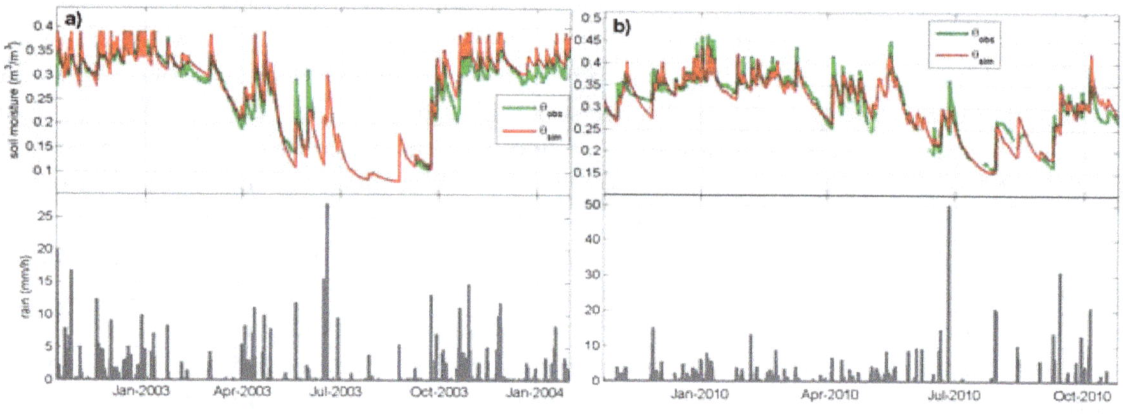

Figure 3. Modelled, θ_{sim}, through the Soil Water Balance Model [67], versus observed, θ_{obs}, soil moisture for: Colorso (**a**); and Ingegneria (**b**) sites in central Italy. The agreement between observed and modelled data reveals the good capability we achieved in simulating point scale soil moisture temporal evolution.

4. Which Hydrological Applications Are Benefiting (and Will Benefit) from Soil Moisture Data?

As evidenced in Section 2, several in situ, satellite and modelled soil moisture datasets are currently available, with large spatial and temporal coverage, and with appropriate spatial and temporal resolutions for hydrological, climatic and agricultural applications. Drought monitoring, runoff modelling and flood forecasting, numerical weather prediction, land surface and climate models assessment, wildfire risk assessment, agricultural monitoring and crop yield forecast [25,26,60,61] are among the most important applications benefiting from in situ and satellite soil moisture observations.

In this section, we mainly focus on a single hydrological application that was the object of our previous and on-going studies, i.e., runoff modelling and flood forecasting. Moreover, we list a number of "emerging" applications in which soil moisture data are being used in the last couple of years: (1) landslides and erosion prediction; (2) epidemic risk monitoring; (3) rainfall estimation; and (4) irrigation detection and assessment.

The purpose of this section is to briefly describe the main outcomes of previous studies on the selected applications in order to highlight the open issues to be addressed.

4.1. *Runoff Modelling*

Soil moisture is the key variable for the partitioning of rainfall into infiltration and runoff, thus playing a fundamental role in runoff modelling and flood forecasting [9,10]. On the one hand, several studies employed in situ soil moisture observations for enhancing: (1) the estimation of initial conditions in flood modelling [98,99]; (2) the calibration of hydrological models [100,101]; and (3) flood simulation through data assimilation approaches [86,102,103]. Overall, these studies demonstrated the value of in situ soil moisture observations for improving flood simulation even though a comprehensive assessment was not possible due to the very limited spatial and temporal coverage of in situ observations. On the other hand, due the recent availability of satellite soil moisture products, the number of studies considering satellite soil moisture observations in this topic is significantly increasing in the last five years. For instance, under the Climate Change Initiative (CCI)—Soil Moisture of the European Space Agency (ESA), a global-scale and long-term soil moisture product is currently available from 1979 to 2015 (i.e., 36 years [46]), and next year it is foreseen that it will be available in near real-time [104]. The global-scale ASCAT soil moisture product is already available in near real-time (130 min after satellite pass) through the Satellite Application Facility on support to Operational Hydrology and Water Management (H SAF) project of EUMETSAT (European Organization for the Exploitation of Meteorological Satellites) since 2007. These new products have large potential to be employed for long-term and real-time flood forecasting. Similarly to in situ observations, satellite data have been used for model initialization [105–109], for hydrological model calibration [110,111], and in a data assimilation framework [9,86,112–115]. By way of example, Figure 4 shows the results of the assimilation of the ESA CCI soil moisture product [46] in MISDc (Modello Idrologico Semi-Distribuito in continuo) rainfall-runoff model [116] for the Niccone basin in central Italy in the period 1995–2010. For the latest period 2007–2010, in which the temporal density of satellite observations is higher, the assimilation provides a significant increase in the model performance with a Nash-Sutcliffe efficiency value from 0.79 to 0.88 when the ESA CCI soil moisture product is assimilated. The improvement in discharge simulation is more evident for larger flood events.

However, the analysis of the scientific literature on this topic reveals that the actual added-value in using satellite soil moisture for runoff modelling is still unclear [86,113]. Some authors obtained moderate to significant improvement through the assimilation of satellite data soil moisture in hydrological modelling (e.g., [9,112,117]) while other studies obtained a deterioration of the performances (e.g., [118]). These contrasting results have to be attributed to the inherent uncertainties and issues involved in the use of satellite (and in situ) soil moisture data in hydrological modelling. By considering the example of data assimilation, Massari et al. [113] clarified well that several choices should be made (i.e., the "cooking techniques") in a data assimilation study and they might have a significant impact on final results, with even the same relevance of the considered hydrological model, observation to be assimilated, and assimilation approach (i.e., the "ingredients"). These choices include the assessment of the magnitude and the structure of the errors in the hydrological model and in the observations, and the selection of the rescaling (e.g., linear and non-linear bias correction, triple collocation, [119]) and/or filtering (e.g., the Soil Water Index method [62]) techniques. Moreover, several inherent issues are related to the spatial mismatch between observations, i.e., point scale for in situ data and coarse scale (plus shallow soil depth) for satellite data, and modelled data, to the structure of hydrological models [103,112], and to the error characterization of (satellite) observations. For that, a strong effort is required from the hydrologic and remote sensing communities to overcome these issues. On the one hand, hydrologists should modify hydrological models that usually consider a single soil layer and are spatially lumped (basin scale). Both in situ and satellite data requires a model tailored to their use. It would mean including a shallow soil layer close to the surface and a distributed spatial discretization for fully exploiting the potential of satellite soil moisture observations [112,120]. On the other hand, remote sensing scientists should improve the characterization of the errors associated to the soil moisture products, increase their spatial and temporal resolution, their temporal coverage and temporal consistency. Indeed, long-term soil moisture datasets (e.g., >10–15 years) are needed not only in the climate community but also from hydrologists to assess the impact of using such datasets through robust analyses.

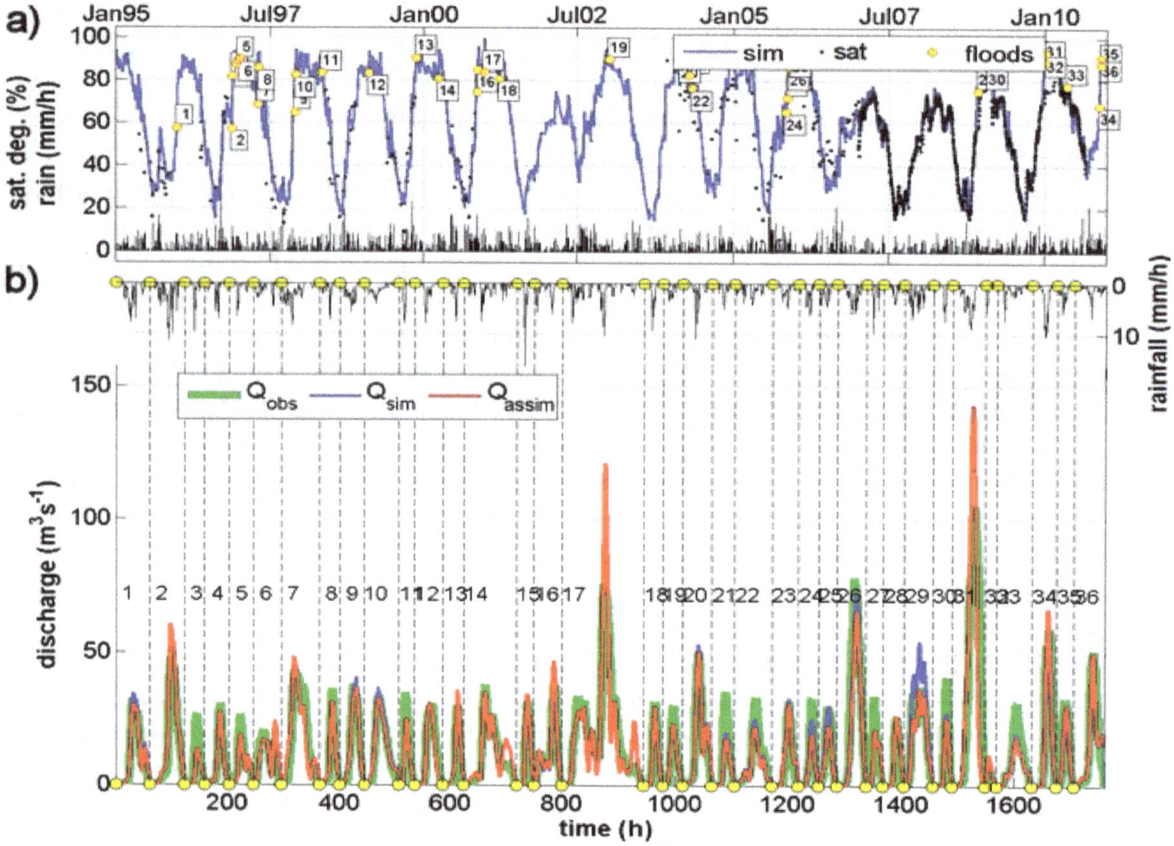

Figure 4. MISDc model results without and with the assimilation of ESA CCI soil moisture product in the Niccone river basin (central Italy). (**a**) Root zone soil moisture simulated by the model, sim, and obtained by ESA CCI soil moisture product, sat, for the period 1995–2010. The filled circles, floods, represent the initial conditions of the selected flood events simulated by MISDc model and identified by the numbers in the lower panel; (**b**) Observed, Q_{obs}, versus simulated, Q_{sim}, and simulated with assimilation, Q_{assim}, discharge for the sequence of the flood events occurred in the study period. The corresponding hourly rainfall is also shown in both panels (note that time axis is different in the upper and lower panel).

4.2. Emerging Applications

Soil moisture observations play an important role for the mitigation of many natural hazards other than floods, i.e., for landslide and erosion prediction [121,122]. Several studies used in situ observations for studying the linkage between soil moisture conditions and landslide occurrence [12,123]. For instance, Hawke and McConchie [124] analysed this relationship in a small area of New Zealand affected by landslide and obtained that the slope failure occurred when maximum soil moisture conditions were observed. Based on these studies, the use of soil moisture stations for monitoring slope stability conditions is becoming a standard (e.g., [125], United States Geological Survey website [126]). Differently from in situ observations, satellite soil moisture data were rarely employed in this context and only three studies were published [11,127,128]. Brocca et al. [11] demonstrated that the ASCAT soil moisture product is able to improve the performances of a statistical regression model aimed at predicting the temporal evolution of landslide movement for a small landslide in central Italy. Similarly, in the context of erosion prediction, Todisco et al. [122] recently suggested that ASCAT soil moisture data might be used for predicting event soil losses at plot scale in central Italy. The very small number of studies employing satellite soil moisture data for these applications is related firstly to the spatial mismatch between the targeted areas, typically a hillslopes or a small catchment, and the spatial resolution of satellite products, and secondly, to the difficulties in the exploitation of these technologies from the hydrological community [129,130]. Based on the studies mentioned above, we think that satellite soil moisture products have a high potential for improving the prediction of landslide and erosion processes, therefore these aspects deserve further investigations [26].

Soil moisture information might be helpful for the assessment and the monitoring of the epidemic risk. In the scientific literature, meteorological observations have been widely employed for determining spatial and temporal occurrence of epidemic risk (e.g., [131]), while only recently soil moisture data from modelling and observations have been considered in this respect [45,132,133]. For instance, Montosi et al. [132] developed an ecohydrological model for identifying the factors influencing malaria dynamics and highlighted the important role played by soil moisture. By using published data of malaria incidence rates in Mpumalanga and Botswana regions (Africa), we performed a simple correlation analysis between European Remote Sensing Scatterometer (ESCAT) and ASCAT satellite soil moisture products and malaria incidence rates (Figure 5). Although mainly driven by the seasonal cycle, a moderate agreement is observed between the monthly time series in the two regions, thus highlighting the potential of using satellite soil moisture data in this application that is extremely important in developing countries.

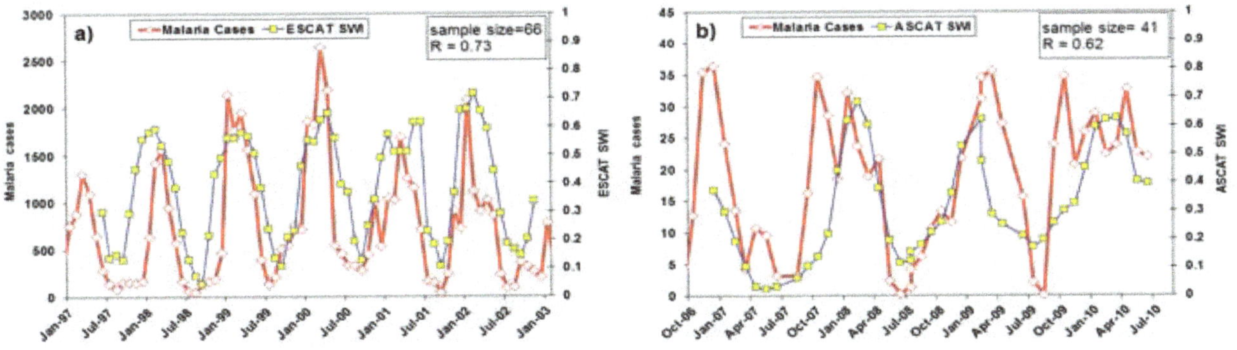

Figure 5. Comparison between malaria cases and soil moisture time series obtained by ESCAT Soil Water Index: SWI, in Mpumalanga (**a**); and ASCAT SWI in Botswana (**b**). The data on malaria cases are obtained by: Montosi et al. [132] (**a**); and by Chirebvu et al. [131] (**b**). R is the Pearson's correlation coefficient.

Pellarin et al. [134] and Crow et al. [135] made the first studies using satellite soil moisture observations for correcting satellite precipitation estimates. Thanks to the strong relationship between the temporal evolution of rainfall and soil moisture, the latter can be employed for improving the quality of satellite precipitation products that are affected by several issues (e.g., [136]). More recently, the number of publications on this topic is remarkably increasing likely due to the work by Brocca et al. [137,138] who developed a "bottom-up" approach, called SM2RAIN, for directly estimating precipitation rates from soil moisture observations only. The method has been applied on a local scale with in situ observations [137,139] and on a regional/global scale with satellite data [138,140,141]. Moreover, the "bottom-up" approach was integrated with state-of-the-art rainfall products (i.e., "top-down" approach) for obtaining a superior rainfall product by Brocca et al. [141] and Ciabatta et al. [142] in Australia and Italy. Finally, the precipitation product corrected through soil moisture data were used in several recent studies for improving flood prediction [98,143,144]. This new application is receiving more and more attention in the recent time, as also confirmed by a number of research projects funded by ESA, EUMETSAT, and NASA on this topic. Therefore, we believe that its further development and widespread application is highly beneficial for contributing to efforts in global precipitation estimation [140].

Soil moisture observations can be highly important for the detection and the quantification of irrigation, and specifically satellite datasets thanks to their large scale coverage [145,146]. For instance, Singh et al. [147] used Advanced Microwave Scanning Radiometer-Earth Observing System (AMSR-E) soil moisture data for discerning the shifting over time in the irrigation practices in north western India while Qiu et al. [146] highlighted that the ESA CCI SM product can be used to detect irrigated areas in eastern China by comparing trends in satellite precipitation and soil moisture. For instance, by using the SM2RAIN method described above, the quantification of irrigation from soil moisture data can be also performed. Figure 6 shows an example in Nebraska in which soil moisture data from ASCAT satellite and European Re Analysis -Interim (ERA-Interim) reanalysis were considered. The higher values of satellite soil moisture data compared to ERA-Interim reanalysis during summer are clearly visible in the Figure 6a (highlighted by grey areas). As ERA-Interim reanalysis does not incorporate irrigation data, the higher values obtained through satellite observations identify the occurrence and the amount of irrigation (Figure 6b,c). As the amount of water used for irrigation is largely unknown on a global scale, the capability of satellite soil moisture data to estimate irrigation might have a tremendous impact in future applications. We note, however, that dedicated studies still need to be carried out to demonstrate the feasibility of this application.

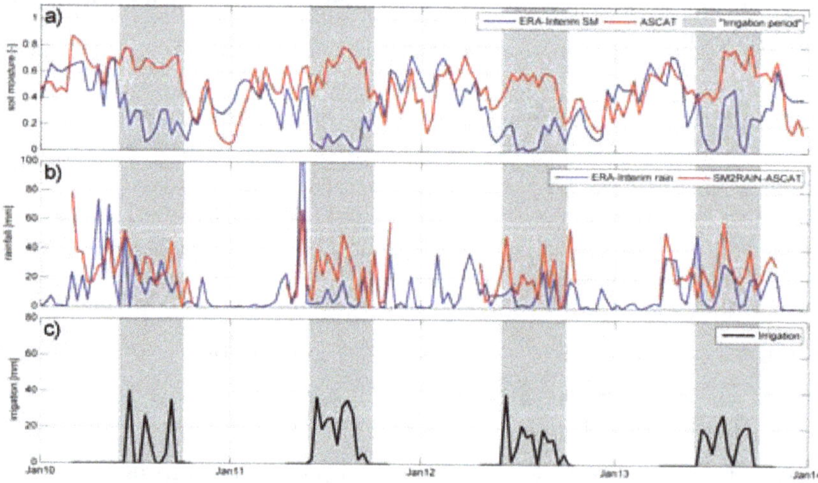

Figure 6. Soil moisture and rainfall time series for a pixel in Nebraska (Longitude, Latitude = −98 E, 40 N). (**a**) Temporal evolution of ASCAT and ERA-Interim soil moisture, SM, data (10-day moving average). Grey areas represent the months during which irrigation takes place; (**b**) Temporal evolution of ERA-Interim and SM2RAIN-ASCAT rainfall (10-day accumulations); (**c**) Temporal evolution of irrigation obtained as the difference between SM2RAIN-ASCAT and ERA-Interim rainfall (10-day accumulations).

5. Soil Moisture for Hydrological Applications: A Scientific Roadmap

In the previous three sections, we provided a short overview on the techniques used for soil moisture monitoring, on the spatial-temporal variability of soil moisture and on (some of) the hydrological applications in which soil moisture data are beneficial. Specifically, we highlighted the main limitations and research issues to be addressed together with the new hydrological applications that, in our opinion, should be investigated in the near future. We want to underline here that some of the open issues we raised might be considered as not doable, at least from the theoretical viewpoint. Theoretically, coarse resolution satellite soil moisture observations, or point scale in situ observations, cannot be used for catchment-scale applications; but several studies demonstrated the opposite in contrast with expectations (e.g., [9,11,86,113,122]). Similarly, it was not expected that soil moisture data could be used for rainfall estimation (as shown in [138]) or that C-band scatterometer data (i.e., ASCAT) could provide an accurate and usable satellite soil moisture product as it was shown in [26] and [45]. Research activities in the recent years demonstrated that our a priori, or theoretical, knowledge might fail when we started using actual observations. We believe that, as scientists, we have to experimentally demonstrate our hypothesis, and it was the rationale behind our past and on-going research activities. Bearing that in mind, we have defined three open issues to be addressed in the near future and the corresponding research opportunities that come from the attempt to answer to these scientific questions.

5.1. High Spatial-Temporal Resolution Soil Moisture Measurements

Currently, we are able to estimate accurately soil moisture at the point scale through in situ sensors. Moreover, less accurate measurements can be obtained at coarse scale (~20 km) using satellite sensors. Daily and sub-daily temporal resolutions can be obtained through these techniques. However, soil moisture measurements at high spatial (i.e., <5 km) and temporal (sub-daily) resolution are not available at the time of writing (January 2017). To get this target, we must improve, test and integrate the new ground-based monitoring techniques such as GPS, cosmic ray and geophysical methods with more accurate point measurements techniques (e.g., TDR). At the same time, satellite sensors are rapidly increasing their capabilities in terms of spatial and temporal resolution, and accuracy. By way of example, in the near future (2020), the launch of next generation scatterometer sensors by EUMETSAT [148] will provide 5 km/daily soil moisture data by using a well-established technology and a soil moisture retrieval algorithm largely tested with ASCAT data. Moreover, Sentinel-1 is expected to deliver soon high spatial resolution (1 km) soil moisture measurements every 6 days [55] and new technologies such as cubesat and nano-satellites, if well designed, have the potential to provide high spatial-temporal resolution at low costs [130]. Finally, spatial downscaling/upscaling approaches can be used to integrate the different techniques, as well as observations with modelling. Data assimilation and merging methods can be also considered to optimally integrate in situ, satellite and modelled data.

On this basis, an exciting future is foreseen in a couple of years from now in which high resolution soil moisture datasets are expected to be developed. The availability of these datasets to the hydrological (and climate) communities and users, also in an operational context (e.g., data availability in real-time) will open a number of new and unexplored research opportunities.

5.2. Soil Moisture Modelling in Space

As highlighted in Section 3, modelling soil moisture in space is an important issue that is largely unexplored. Currently, this issue is magnified by: (1) the unavailability (only for small regions and limited time periods) of spatially detailed soil moisture datasets; (2) the difficulties in obtaining static (e.g., soil type) and dynamic (e.g., precipitation) inputs at high spatial resolution for modelling; and (3) the larger interest of hydrological and climate communities to simulate soil moisture in time with respect to in space. In our opinion, we should improve our capability in modelling soil moisture both in time and in space.

Several research questions, and corresponding opportunities, should be addressed to achieve this target. Firstly, the spatial comparison between satellite and modelled data would reveal the similarities and the differences between the derived patterns. Mainly looking at the differences, we would gain knowledge on where (and when) satellite and modelled datasets need improvements. Secondly, a stronger link between studies that analysed soil moisture spatial variability and studies performing satellite datasets validation and/or using these datasets in hydrological and climatic applications should be carried out. For instance, lower (higher) performances in the comparison between in situ and satellite datasets can be expected when the spatial variance is higher (lower). Similarly, the error characterization of satellite-based soil moisture measurements (with coarse resolution) should be related to soil moisture spatial variability. We believe that detailed investigations on these two points would provide new important insights and, consequently, new research chance.

5.3. *Comprehensive Assessment of the Value of Soil Moisture Data*

The robust assessment of the effective value of in situ and satellite soil moisture data in hydrological applications is still missing. Theoretically, we are fully aware of the fundamental role of soil moisture but, with actual observations, contrasting results have been obtained. Open questions are: (1) What is the required accuracy of soil moisture data to be of benefit in hydrological applications? (2) What is the needed spatial and temporal resolution? Do we need higher spatial or temporal resolution? (3) What is the added-value of soil moisture data with respect to other physical quantities (e.g., river discharge, precipitation)? The answers to these questions are not trivial as, they are application-, model- and scale-dependent, and require dedicated studies encompassing a wide range of climates, models, etc. In this context, we strongly support the investigation of soil moisture data assimilation into flood modelling for different basins, climates, hydrological models and datasets (e.g., from different satellite sensors). For instance, we still need to assess if it is preferable to assimilate in situ or satellite soil moisture data (or alternatively, soil moisture or river discharge data). Moreover, another interesting research topic is related to the estimation of hydraulic soil properties using data assimilation of soil moisture (e.g., [149]). As mentioned above, the benefit in assimilating soil moisture data should also be related to soil moisture spatial-temporal variability studies.

Besides a comprehensive assessment in flood applications, we foster the use of soil moisture data in emerging applications. Indeed, we obtained that in situ and satellite soil moisture data can be extremely useful for landslides and erosion prediction, for malaria risk monitoring and for rainfall and irrigation assessment. Each of these new applications deserves further investigations in order to test the effective usability of soil moisture data on these topics. If successful, these new research activities can potentially open a number of new scientific and practical opportunities. For instance, the quantitative irrigation assessment from soil moisture data would be highly important for food security and management on a global scale, which is becoming more and more challenging under climate changing scenarios.

Acknowledgments: The study was partly supported by the EUMETSAT Satellite Application Facility in Support of Operational Hydrology and Water Management (H-SAF) project, the European Space Agency SMOS+rainfall and WACMOS-MED projects, the Italian Civil Protection Department. This work was partly funded by The Project of Interest NextData (MIUR-CNR). The authors wish to thank Umbria Region, Department of Environment, Planning and Infrastructure, for providing Tiber River basin data.

Author Contributions: All authors contributed extensively to the work presented in this paper. Luca Brocca conceived and designed the paper. Luca Ciabatta, Christian Massari, Stefania Camici and Angelica Tarpanelli contributed to the acquisition and processing of ground and satellite-based datasets, to the analysis and the elaboration of the data, and to the preparation and discussion of results. All co-authors contributed to the editing of the manuscript and to the discussion and interpretation of the results.

References

1. Seneviratne, S.I.; Corti, T.; Davin, E.L.; Hirschi, M.; Jaeger, E.B.; Lehner, I.; Orlowsky, B.; Teuling, A.J. Investigating soil moisture-climate interactions in a changing climate: A review. *Earth-Sci. Rev.* **2010**, *99*, 125–161. [CrossRef]

2. Brocca, L.; Morbidelli, R.; Melone, F.; Moramarco, T. Soil moisture spatial variability in experimental areas of central Italy. *J. Hydrol.* **2007**, *333*, 356–373. [CrossRef]
3. Wagner, W.; Blöschl, G.; Pampaloni, P.; Calvet, J.C.; Bizzarri, B.; Wigneron, J.P.; Kerr, Y. Operational readiness of microwave remote sensing of soil moisture for hydrologic applications. *Hydrol. Res.* **2007**, *38*, 1–20. [CrossRef]
4. Corradini, C. Soil moisture in the development of hydrological processes and its determination at different spatial scales. *J. Hydrol.* **2014**, *516*, 1–5. [CrossRef]
5. ESA Climate Change Initiative. Available online: http://www.esa-soilmoisture-cci.org/ (accessed on 13 February 2017).
6. Soil Moisture Active and Passive Mission. Available online: http://smap.jpl.nasa.gov/ (accessed on 13 February 2017).
7. International Soil Moisture Network. Available online: http://ismn.geo.tuwien.ac.at/ (accessed on 13 February 2017).
8. Cosmic-ray Soil Moisture Observing System. Available online: http://cosmos.hwr.arizona.edu/ (accessed on 13 February 2017).
9. Brocca, L.; Melone, F.; Moramarco, T.; Wagner, W.; Naeimi, V.; Bartalis, Z.; Hasenauer, S. Improving runoff prediction through the assimilation of the ASCAT soil moisture product. *Hydrol. Earth Syst. Sci.* **2010**, *14*, 1881–1893. [CrossRef]
10. Koster, R.D.; Mahanama, S.P.P.; Livneh, B.; Lettenmaier, D.; Reichle, R.H. Skill in Streamflow Forecasts Derived from Large-Scale Estimates of Soil Moisture and Snow. *Nat. Geosci.* **2010**, *3*, 613–616. [CrossRef]
11. Brocca, L.; Ponziani, F.; Moramarco, T.; Melone, F.; Berni, N.; Wagner, W. Improving landslide forecasting using ASCAT-derived soil moisture data: A case study of the Torgiovannetto landslide in central Italy. *Remote Sens.* **2012**, *4*, 1232–1244. [CrossRef]
12. Bittelli, M.; Valentino, R.; Salvatorelli, F.; Rossi Pisa, P. Monitoring soil-water and displacement conditions leading to landslide occurrence in partially saturated clays. *Geomorphology* **2012**, *173–174*, 161–173. [CrossRef]
13. Rahmani, A.; Golian, S.; Brocca, L. Multiyear monitoring of soil moisture over Iran through satellite and reanalysis soil moisture products. *Int. J. Appl. Earth Obs. Geoinf.* **2016**, *48*, 85–95. [CrossRef]
14. Enenkel, M.; Steiner, C.; Mistelbauer, T.; Dorigo, W.; Wagner, W.; See, L.; Atzberger, C.; Schneider, S.; Rogenhofer, E. A combined satellite-derived drought indicator to support humanitarian aid organizations. *Remote Sens.* **2016**, *8*, 340. [CrossRef]
15. Sánchez, N.; González-Zamora, Á.; Piles, M.; Martínez-Fernández, J. A New Soil Moisture Agricultural Drought Index (SMADI) Integrating MODIS and SMOS products: A case of study over the Iberian Peninsula. *Remote Sens.* **2016**, *8*, 287. [CrossRef]
16. De Rosnay, P.; Drusch, M.; Vasiljevic, D.; Balsamo, G.; Albergel, C.; Isaksen, L. A simplified Extended Kalman Filter for the global operational soil moisture analysis at ECMWF. *Q. J. R. Meteorol. Soc.* **2013**, *139*, 1199–1213. [CrossRef]
17. De Rosnay, P.; Balsamo, G.; Albergel, C.; Muñoz-Sabater, J.; Isaksen, L. Initialisation of land surface variables for Numerical Weather Prediction. *Surv. Geophys.* **2014**, *35*, 607–621. [CrossRef]
18. Dharssi, I.; Bovis, K.; Macpherson, B.; Jones, C. Operational assimilation of ASCAT surface soil wetness at the Met Office. *Hydrol. Earth Syst. Sci.* **2011**, *15*, 2729–2746. [CrossRef]
19. Capecchi, V.; Brocca, L. A simple assimilation method to ingest satellite soil moisture into a limited-area NWP model. *Meteorol. Z.* **2014**, *23*, 105–121.
20. Rodríguez-Iturbe, I.; Porporato, A. *Ecohydrology of Water-Controlled Ecosystems: Soil Moisture and Plant Dynamics*; Cambridge University Press: Cambridge, UK, 2007; p. 460.
21. Fang, B.; Lakshmi, V. Soil moisture at watershed scale: Remote sensing techniques. *J. Hydrol.* **2014**, *516*, 258–272. [CrossRef]
22. Muñoz-Sabater, J.; Al Bitar, A.; Brocca, L. Soil moisture retrievals based on active and passive microwave data: State-of-the-art and operational applications. In *Satellite Soil Moisture Retrievals: Techniques and Applications*; Petropoulos, G.P., Srivastava, P., Kerr, Y., Eds.; Elsevier: Amsterdam, The Netherlands, 2016; Volume 18, pp. 351–378.
23. Ochsner, T.E.; Cosh, M.H.; Cuenca, R.H.; Dorigo, W.A.; Draper, C.S.; Hagimoto, Y.; Kerr, Y.H.; Njoku, E.G.; Small, E.E.; Zreda, M. State of the art in large-scale soil moisture monitoring. *Soil Sci. Soc. Am. J.* **2013**, *77*, 1888–1919. [CrossRef]

24. Romano, N. Soil moisture at local scale: Measurements and simulations. *J. Hydrol.* **2014**, *516*, 6–20. [CrossRef]
25. Tebbs, E.; Gerard, F.; Petrie, A.; De Witte, E. Emerging and Potential Future Applications of Satellite-Based Soil Moisture Products. In *Satellite Soil Moisture Retrievals: Techniques and Applications*; Petropoulos, G.P., Srivastava, P., Kerr, Y., Eds.; Elsevier: Amsterdam, The Netherlands, 2016; Volume 19, pp. 379–400.
26. Brocca, L.; Crow, W.T.; Ciabatta, L.; Massari, C.; de Rosnay, P.; Enenkel, M.; Hahn, S.; Amarnath, G.; Camici, S.; Tarpanelli, A.; et al. A review of the applications of ASCAT soil moisture products. *IEEE J. Sel. Top. Appl. Earth Obs. Remote Sens.* **2017**. [CrossRef]
27. Vereecken, H.; Huisman, J.A.; Pachepsky, Y.; Montzka, C.; van der Kruk, J.; Bogena, H.; Weihermüller, L.; Herbst, M.; Martinez, G.; Vanderborght, J. On the spatiotemporal dynamics of soil moisture at the field scale. *J. Hydrol.* **2014**, *516*, 76–96. [CrossRef]
28. Vereecken, H.; Schnepf, A.; Hopmans, J.; Javaux, M.; Or, D.; Roose, T.; Vanderborght, J.; Young, M.; Amelung, W.; Aitkenhead, M.; et al. Modelling Soil Processes: Key challenges and new perspectives. *Vadose Zone J.* **2016**, *15*. [CrossRef]
29. Bogena, H.R.; Huisman, J.A.; Güntner, A.; Hübner, C.; Kusche, J.; Jonard, F.; Vey, S.; Vereecken, H. Emerging methods for noninvasive sensing of soil moisture dynamics from field to catchment scale: A review. *WIREs Water* **2015**, *2*, 635–647. [CrossRef]
30. Crow, W.T.; Yilmaz, M.T. The Auto-Tuned Land Assimilation System (ATLAS). *Water Resour. Res.* **2014**, *50*, 371–384. [CrossRef]
31. Topp, G.C.; David, J.L.; Annan, A.P. Electromagnetic Determination of Soil Water Content: Measurement in Coaxial Transmission Lines. *Water Resour. Res.* **1980**, *16*, 574–582. [CrossRef]
32. Robinson, D.A.; Campbell, C.S.; Hopmans, J.W.; Hornbuckle, B.K.; Jones, S.B.; Knight, R.; Ogden, F.; Selker, J.; Wendroth, O. Soil moisture measurement for ecological and hydrological watershed-scale observatories: A review. *Vadose Zone J.* **2008**, *7*, 358–389. [CrossRef]
33. Bogena, H.; Huisman, J.A.; Schilling, B.; Weuthen, A.; Vereecken, H. Effective calibration of low-cost soil water content sensors. *Sensors* **2017**, *17*, 208. [CrossRef] [PubMed]
34. Dorigo, W.A.; Xaver, A.; Vreugdenhil, M.; Gruber, A.; Hegyiová, A.; Sanchis-Dufau, A.D.; Zamojski, D.; Cordes, C.; Wagner, W.; Drusch, M. Global automated quality control of in situ soil moisture data from the International Soil Moisture Network. *Vadose Zone J.* **2013**, *12*. [CrossRef]
35. Bogena, H.R.; Herbst, M.; Huisman, J.A.; Rosenbaum, U.; Weuthen, A.; Vereecken, H. Potential of wireless sensor networks for measuring soil water content variability. *Vadose Zone J.* **2010**, *9*, 1002–1013. [CrossRef]
36. Calamita, G.; Perrone, A.; Brocca, L.; Onorati, B.; Manfreda, S. Field test of a multi-frequency electromagnetic induction sensor for soil moisture monitoring in southern Italy test sites. *J. Hydrol.* **2015**, *529*, 316–329. [CrossRef]
37. Franz, T.E.; Wang, T.; Avery, W.; Finkenbiner, C.; Brocca, L. Combined analysis of soil moisture measurements from roving and fixed cosmic-ray neutron probes for multi-scale real-time monitoring. *Geophys. Res. Lett.* **2015**, *42*, 3389–3396. [CrossRef]
38. Larson, K.M.; Small, E.E.; Gutmann, E.D.; Bilich, A.L.; Braun, J.J.; Zavorotny, V.U. Use of GPS receivers as a soil moisture network for water cycle studies. *Geophys. Res. Lett.* **2008**, *35*, L24405. [CrossRef]
39. Cosmic-ray Soil Moisture Monitoring Network-UK. Available online: http://cosmos.ceh.ac.uk/ (accessed on 13 February 2017).
40. PBO H_2O Data Portal. Available online: http://xenon.colorado.edu/portal/ (accessed on 13 February 2017).
41. Bogena, H.R.; Huisman, J.A.; Baatz, R.; Hendricks Franssen, H.-J.; Vereecken, H. Accuracy of the cosmic-ray soil water content probe in humid forest ecosystems: The worst case scenario. *Water Resour. Res.* **2013**, *49*, 5778–5791. [CrossRef]
42. Baatz, R.; Bogena, H.; Hendricks Franssen, H.-J.; Huisman, J.A.; Wei, Q.; Montzka, C.; Vereecken, H. Calibration of a catchment scale cosmic-ray soil moisture network: A comparison of three different methods. *J. Hydrol.* **2014**, *516*, 231–244. [CrossRef]
43. Heidbüchel, I.; Güntner, A.; Blume, T. Use of Cosmic-Ray Neutron Sensors for Soil Moisture Monitoring in Forests. *Hydrol. Earth Syst. Sci.* **2016**, *20*, 1269–1288. [CrossRef]
44. Rahimzadeh-Bajgiran, P.; Berg, A. Soil Moisture Retrievals Using Optical/TIR Methods. In *Satellite Soil Moisture Retrievals: Techniques & Applications*; Petropoulos, G.P., Srivastava, P., Kerr, Y., Eds.; Elsevier: Amsterdam, The Netherlands, 2016; Volume 3, pp. 47–72.

45. Wagner, W.; Hahn, S.; Kidd, R.; Melzer, T.; Bartalis, Z.; Hasenauer, S.; Figa, J.; de Rosnay, P.; Jann, A.; Schneider, S.; et al. The ASCAT soil moisture product: A review of its specifications, validation results, and emerging applications. *Meteorol. Z.* **2013**, *22*, 5–33.
46. Liu, Y.Y.; Parinussa, R.M.; Dorigo, W.A.; De Jeu, R.A.M.; Wagner, W.; van Dijk, A.I.J.M.; McCabe, M.F.; Evans, J.P. Developing an improved soil moisture dataset by blending passive and active microwave satellite-based retrievals. *Hydrol. Earth Syst. Sci.* **2011**, *15*, 425–436. [CrossRef]
47. Brocca, L.; Massari, C.; Ciabatta, L.; Wagner, W.; Stoffelen, A. Remote sensing of terrestrial rainfall from Ku-band scatterometers. *IEEE J. Sel. Top. Appl. Earth Obs. Remote Sens.* **2016**, *9*, 533–539. [CrossRef]
48. Kerr, Y.H.; Waldteufel, P.; Wigneron, J.-P.; Martinuzzi, J.; Font, J.; Berger, M. Soil moisture retrieval from space: The Soil Moisture and Ocean Salinity (SMOS) mission. *IEEE Trans. Geosci. Remote Sens.* **2001**, *39*, 1729–1735. [CrossRef]
49. Entekhabi, D.; Njoku, E.G.; Neill, P.E.; Kellogg, K.H.; Crow, W.T.; Edelstein, W.N.; Entin, J.K.; Goodman, S.D.; Jackson, T.J.; Johnson, J.; et al. The soil moisture active passive (SMAP) mission. *Proc. IEEE* **2010**, *98*, 704–716. [CrossRef]
50. Kim, S.; Liu, Y.Y.; Johnson, F.M.; Parinussa, R.M.; Sharma, A. A global comparison of alternate AMSR2 soil moisture products: Why do they differ? *Remote Sens. Environ.* **2015**, *161*, 43–62. [CrossRef]
51. Kerr, Y.H.; Waldteufel, P.; Richaume, P.; Wigneron, J.P.; Ferrazzoli, P.; Mahmoodi, A.; Juglea, S.E.; Leroux, D.; Mialon, A.; Delwart, S. The SMOS soil moisture retrieval algorithm. *IEEE Trans. Geosci. Remote Sens.* **2012**, *50*, 1384–1403. [CrossRef]
52. Malbéteau, Y.; Merlin, O.; Molero, B.; Rüdiger, C.; Bacon, S. DisPATCh as a tool to evaluate coarse-scale remotely sensed soil moisture using localized in situ measurements: Application to SMOS and AMSR-E data in Southeastern Australia. *Int. J. Appl. Earth Obs. Geoinf.* **2016**, 221–234. [CrossRef]
53. Piles, M.; Sánchez, N.; Vall-llossera, M.; Camps, A.; Martinez-Fernandez, J.; Martinez, J.; Gonzalez-Gambau, V. A downscaling approach for SMOS land observations: Evaluation of high-resolution soil moisture maps over the Iberian Peninsula. *IEEE J. Sel. Top. Appl. Earth Obs. Remote Sens.* **2014**, *7*, 3845–3857. [CrossRef]
54. Santi, E.; Paloscia, S.; Pettinato, S.; Brocca, L.; Ciabatta, L. Robust assessment of an operational algorithm for the retrieval of soil moisture from AMSR-E data in central Italy. *IEEE J. Sel. Top. Appl. Earth Obs. Remote Sens.* **2016**, *9*, 2478–2492. [CrossRef]
55. Paloscia, S.; Pettinato, S.; Santi, E.; Notarnicola, C.; Pasolli, L.; Reppucci, A. Soil moisture mapping using Sentinel-1 images: Algorithm and preliminary validation. *Remote Sens. Environ.* **2013**, *134*, 234–248. [CrossRef]
56. Al-Yaari, A.; Wigneron, J.-P.; Ducharne, A.; Kerr, Y.H.; Wagner, W.; De Lannoy, G.; Reichle, R.H.; Al-Bitar, A.; Dorigo, W.; Richaume, P.; et al. Global-scale comparison of passive (SMOS) and active (ASCAT) satellite based microwave soil moisture retrievals with soil moisture simulations (MERRA-Land). *Remote Sens. Environ.* **2014**, *152*, 614–626. [CrossRef]
57. Brocca, L.; Hasenauer, S.; Lacava, T.; Melone, F.; Moramarco, T.; Wagner, W.; Dorigo, W.; Matgen, P.; Martínez-Fernández, J.; Llorens, P.; et al. Soil moisture estimation through ASCAT and AMSR-E sensors: An intercomparison and validation study across Europe. *Remote Sens. Environ.* **2011**, *115*, 3390–3408. [CrossRef]
58. Dorigo, W.A.; Gruber, A.; De Jeu, R.A.M.; Wagner, W.; Stacke, T.; Loew, A.; Albergel, C.; Brocca, L.; Chung, D.; Parinussa, R.M.; et al. Evaluation of the ESA CCI soil moisture product using ground-based observations. *Remote Sens. Environ.* **2015**, 380–395. [CrossRef]
59. Paulik, C.; Dorigo, W.; Wagner, W.; Kidd, R. Validation of the ASCAT Soil Water Index using in situ data from the International Soil Moisture Network. *Int. J. Appl. Earth Obs. Geoinf.* **2014**, *30*, 1–8. [CrossRef]
60. Escobar, V.M.; Srinivasan, M.; Arias, S.D. Improving NASA's Earth Observation Systems and Data Programs through the Engagement of Mission Early Adopters. In *Earth Science Satellite Applications*; Hossain, F., Ed.; Springer International Publishing: Argovia, Switzerland, 2016; Volume 9, pp. 223–267.
61. Moran, M.S.; Doorn, B.; Escobar, V.; Brown, M.E. Connecting NASA science and engineering with earth science applications. *J. Hydrometeorol.* **2015**, *16*, 473–483. [CrossRef]
62. Wagner, W.; Lemoine, G.; Rott, H. A method for estimating soil moisture from ERS scatterometer and soil data. *Remote Sens. Environ.* **1999**, *70*, 191–207. [CrossRef]
63. Manfreda, S.; Brocca, L.; Moramarco, T.; Melone, F.; Sheffield, J. A physically based approach for the estimation of root-zone soil moisture from surface measurements. *Hydrol. Earth Syst. Sci.* **2014**, *18*, 1199–1212. [CrossRef]

64. Balsamo, G.; Albergel, C.; Beljaars, A.; Boussetta, S.; Brun, E.; Cloke, H.; Dee, D.; Dutra, E.; Muñoz-Sabater, J.; Pappenberger, F.; et al. ERA-Interim/Land: A global land surface reanalysis data set. *Hydrol. Earth Syst. Sci.* **2015**, *19*, 389–407. [CrossRef]
65. Famiglietti, J.S.; Wood, E.F. Multiscale modeling of spatially variable water and energy balance processes. *Water Resour. Res.* **1994**, *11*, 3061–3078. [CrossRef]
66. Bierkens, M.F.; Bell, V.A.; Burek, P.; Chaney, N.; Condon, L.E.; David, C.H.; de Roo, A.; Döll, P.; Drost, N.; Famiglietti, J.S.; et al. Hyper-resolution global hydrological modelling: What is next? Everywhere and locally relevant. *Hydrol. Process.* **2015**, *29*, 310–320. [CrossRef]
67. Brocca, L.; Camici, S.; Melone, F.; Moramarco, T.; Martinez-Fernandez, J.; Didon-Lescot, J.-F.; Morbidelli, R. Improving the representation of soil moisture by using a semi-analytical infiltration model. *Hydrol. Process.* **2014**, *28*, 2103–2115. [CrossRef]
68. Morbidelli, R.; Corradini, C.; Govindaraju, R.S. A field-scale infiltration model accounting for spatial heterogeneity of rainfall and soil saturated hydraulic conductivity. *Hydrol. Process.* **2006**, *20*, 1465–1481. [CrossRef]
69. Hopmans, J.W.; Nielsen, D.R.; Bristow, K.L. *How Useful are Small-Scale Soil Hydraulic Property Measurements for Large-Scale Vadose Zone Modeling? In Environmental Mechanics: Water, Mass and Energy Transfer in the Biosphere: The Philip Volume*; Raats, P.A.C., Smiles, D., Warrick, A.W., Eds.; American Geophysical Union: Washington, DC, USA, 2002; pp. 247–258.
70. Juglea, S.; Kerr, Y.; Mialon, A.; Lopez-Baeza, E.; Braithwaite, D.; Hsu, K. Soil moisture modelling of a SMOS pixel: Interest of using the persiann database over the Valencia anchor station. *Hydrol. Earth Syst. Sci.* **2010**, *14*, 1509–1525. [CrossRef]
71. Montzka, C.; Bogena, H.R.; Zreda, M.; Monerris, A.; Morrison, R.; Muddu, S.; Vereecken, H. Validation of Spaceborne and Modelled Surface Soil Moisture Products with Cosmic-Ray Neutron Probes. *Remote Sens.* **2017**, *9*, 103. [CrossRef]
72. Bell, K.R.; Blanchard, B.J.; Schmugge, T.J.; Witczak, M.W. Analysis of surface moisture variations within large field sites. *Water Resour. Res.* **1980**, *16*, 796–810. [CrossRef]
73. Vachaud, G.; Passerat de Silans, A.; Balabanis, P.; Vauclin, M. Temporal stability of spatially measured soil water probability density function. *Soil Sci. Soc. Am. J.* **1985**, *49*, 822–828. [CrossRef]
74. Vinnikov, K.Y.; Robock, A.; Speranskaya, N.A.; Schlosser, C.A. Scales of temporal and spatial variability of midlatitude soil moisture. *J. Geophys. Res.* **1996**, *101*, 7163–7174. [CrossRef]
75. Crow, W.T.; Berg, A.A.; Cosh, M.H.; Loew, A.; Mohanty, B.P.; Panciera, R.; Rosnay, P.D.; Ryu, D.; Walker, J.P. Upscaling sparse ground-based soil moisture observations for the validation of coarse-resolution satellite soil moisture products. *Rev. Geophys.* **2012**, *50*, RG2002. [CrossRef]
76. Western, A.W.; Zhou, S.; Grayson, R.B.; McMahon, T.A.; Blöschl, G.; Wilson, D.J. Spatial correlation of soil moisture in small catchments and its relationship to dominant spatial hydrological processes. *J. Hydrol.* **2004**, *286*, 113–134. [CrossRef]
77. Vanderlinden, K.; Vereecken, H.; Hardelauf, H.; Herbst, M.; Martínez, G.; Cosh, M.H.; Pachepsky, Y.A. Temporal stability of soil water contents: A review of data and analyses. *Vadose Zone J.* **2012**, *11*. [CrossRef]
78. Biswas, A. Scaling analysis of soil water storage with missing measurements using the second-generation continuous wavelet transform. *Eur. J. Soil Sci.* **2014**, *65*, 594–604. [CrossRef]
79. Pan, F.; Peters-Lidard, C.D. On the Relationship between Mean and Variance of Soil Moisture Fields. *J. Am. Water Resour. Assoc.* **2008**, *44*, 235–242. [CrossRef]
80. Teuling, A.J.; Troch, P.A. Improved understanding of soil moisture variability dynamics. *Geophys. Res. Lett.* **2005**, *32*, L05404. [CrossRef]
81. Lawrence, J.E.; Hornberger, G.M. Soil moisture variability across climate zones. *Geophys. Res. Lett.* **2007**, *34*, L20402. [CrossRef]
82. Mittelbach, H.; Seneviratne, S.I. A new perspective on the spatio-temporal variability of soil moisture: Temporal dynamics versus time invariant contributions. *Hydrol. Earth Syst. Sci.* **2012**, *16*, 2169–2179. [CrossRef]
83. Brocca, L.; Zucco, G.; Mittelbach, H.; Moramarco, T.; Seneviratne, S.I. Absolute versus temporal anomaly and percent of saturation soil moisture spatial variability for six networks worldwide. *Water Resour. Res.* **2014**, *50*, 5560–5576. [CrossRef]

84. Hu, W.; Si, B.C. Estimating spatially distributed soil water content at small watershed scales based on decomposition of temporal anomaly and time stability analysis. *Hydrol. Earth Syst. Sci.* **2016**, *20*, 571–587. [CrossRef]
85. Cosh, M.H.; Jackson, T.J.; Starks, P.; Heathman, G. Temporal stability of surface soil moisture in the Little Washita River watershed and its applications in satellite soil moisture product validation. *J. Hydrol.* **2006**, *323*, 168–177. [CrossRef]
86. Matgen, P.; Fenicia, F.; Heitz, S.; Plaza, D.; de Keyser, R.; Pauwels, V.R.N.; Wagner, W.; Savenije, H. Can ASCAT-derived soil wetness indices reduce predictive uncertainty in well-gauged areas? A comparison with in situ observed soil moisture in an assimilation application. *Adv. Water Resour.* **2012**, *44*, 49–65. [CrossRef]
87. Porporato, A.; Daly, E.; Rodriguez-Iturbe, I. Soil water balance and ecosystem response to climate change. *Amer. Nat.* **2004**, *164*, 625–632. [CrossRef] [PubMed]
88. Riley, W.J.; Shen, C. Characterizing coarse-resolution watershed soil moisture heterogeneity using fine-scale simulations. *Hydrol. Earth Syst. Sci.* **2014**, *18*, 2463–2483. [CrossRef]
89. Rosenbaum, U.; Bogena, H.R.; Herbst, M.; Huisman, J.A.; Peterson, T.J.; Weuthen, A.; Western, A.; Vereecken, H. Seasonal and event dynamics of spatial soil moisture patterns at the small catchment scale. *Water Resour. Res.* **2002**, *48*, W10544. [CrossRef]
90. Chaney, N.W.; Roundy, J.K.; Herrera-Estrada, J.E.; Wood, E.F. High-resolution modeling of the spatial heterogeneity of soil moisture: Applications in network design. *Water Resour. Res.* **2015**, *51*, 619–638. [CrossRef]
91. Lacava, T.; Matgen, P.; Brocca, L.; Bittelli, M.; Moramarco, T. A first assessment of the SMOS soil moisture product with in-situ and modelled data in Italy and Luxembourg. *IEEE Trans. Geosci. Remote Sens.* **2012**, *50*, 1612–1622. [CrossRef]
92. Western, A.W.; Grayson, R.B.; Blöschl, G.; Willgoose, G.R.; McMahon, T.A. Observed spatial organization of soil moisture and its relation to terrain indices. *Water Resour. Res.* **1999**, *35*, 797–810. [CrossRef]
93. Li, B.; Rodell, M. Spatial variability and its scale dependency of observed and modeled soil moisture over different climate regions. *Hydrol. Earth Syst. Sci.* **2013**, *17*, 1177–1188. [CrossRef]
94. Cornelissen, T.; Diekkrüger, B.; Bogena, H.R. Significance of scale and lower boundary condition in the 3D simulation of hydrological processes and soil moisture variability in a forested headwater catchment. *J. Hydrol.* **2014**, *516*, 140–153. [CrossRef]
95. Polcher, J.; Piles, M.; Gelati, E.; Barella-Ortiz, A.; Tello, M. Comparing surface-soil moisture from the SMOS mission and the ORCHIDEE land-surface model over the Iberian Peninsula. *Remote Sens. Environ.* **2016**, *174*, 69–81. [CrossRef]
96. Koch, J.; Stisen, S.; Fang, Z.; Bogena, H.R.; Cornelissen, T.; Diekkrüger, B.; Kollet, S. Inter-comparison of three distributed hydrological models with respect to the seasonal variability of soil moisture patterns at a small forested catchment. *J. Hydrol.* **2016**, *533*, 234–249. [CrossRef]
97. Fang, Z.; Bogena, H.R.; Kollet, S.; Koch, J.; Vereecken, H. Spatio-temporal validation of long-term 3D hydrological simulations of a forested catchment using empirical orthogonal functions and wavelet coherence analysis. *J. Hydrol.* **2015**, *529*, 1754–1767. [CrossRef]
98. Massari, C.; Brocca, L.; Moramarco, T.; Tramblay, Y.; Didon Lescot, J.-F. Potential of soil moisture observations in flood modelling: Estimating initial conditions and correcting rainfall. *Adv. Water Resour.* **2014**, *74*, 44–53. [CrossRef]
99. Tramblay, Y.; Bouvier, C.; Martin, C.; Didon-Lescot, J.F.; Todorovik, D.; Domergue, J.M. Assessment of initial soil moisture conditions for event-based rainfall-runoff modelling. *J. Hydrol.* **2010**, *387*, 176–187. [CrossRef]
100. Wooldridge, S.A.; Kalma, J.D.; Walker, J.P. Importance of soil moisture measurements for inferring parameters in hydrologic models of low-yielding ephemeral catchments. *Environ. Model. Softw.* **2003**, *18*, 35–48. [CrossRef]
101. Koren, V.; Moreda, F.; Smith, M. Use of soil moisture observations to improve parameter consistency in watershed calibration. *Phys. Chem. Earth* **2008**, *33*, 1068–1080. [CrossRef]
102. Aubert, D.; Loumagne, C.; Oudin, L. Sequential assimilation of soil moisture and streamflow data in a conceptual rainfall-runoff model. *J. Hydrol.* **2003**, *280*, 145–161. [CrossRef]
103. Chen, F.; Crow, W.T.; Starks, P.J.; Moriasi, D.N. Improving hydrologic predictions of a catchment model via assimilation of surface soil moisture. *Adv. Water Resour.* **2011**, *34*, 526–536. [CrossRef]

104. Enenkel, M.; Reimer, C.; Dorigo, W.; Wagner, W.; Pfeil, I.; Parinussa, R.; De Jeu, R. Combining satellite observations to develop a global soil moisture product for near-real-time applications. *Hydrol. Earth Syst. Sci.* **2016**, *20*, 4191–4208. [CrossRef]
105. Brocca, L.; Melone, F.; Moramarco, T.; Morbidelli, R. Antecedent wetness conditions based on ERS scatterometer data. *J. Hydrol.* **2009**, *364*, 73–87. [CrossRef]
106. Massari, C.; Brocca, L.; Ciabatta, L.; Moramarco, T.; Gabellani, S.; Albergel, C.; de Rosnay, P.; Puca, S.; Wagner, W. The use of H-SAF soil moisture products for operational hydrology: Flood modelling over Italy. *Hydrology* **2015**, *2*, 2–22. [CrossRef]
107. Beck, H.E.; Jeu, R.A.M.D.; Schellekens, J.; Dijk, A.I.J.M.V.; Bruijnzeel, L.A. Improving curve number based storm runoff estimates using soil moisture proxies. *IEEE J. Sel. Top. Appl. Earth Obs. Remote Sens.* **2009**, *2*, 250–259. [CrossRef]
108. Tramblay, Y.; Bouaicha, R.; Brocca, L.; Dorigo, W.; Bouvier, C.; Camici, S.; Servat, E. Estimation of antecedent wetness conditions for flood modelling in northern Morocco. *Hydrol. Earth Syst. Sci.* **2012**, *16*, 4375–4386. [CrossRef]
109. Zhuo, L.; Dai, Q.; Han, D. Evaluation of SMOS soil moisture retrievals over the central United States for hydro-meteorological application. *Phys. Chem. Earth* **2015**, *83*, 146–155. [CrossRef]
110. Parajka, J.; Naemi, V.; Bloschl, G.; Wagner, W.; Merz, R.; Scipal, K. Assimilating scatterometer soil moisture data into conceptual hydrologic models at coarse scales. *Hydrol. Earth Syst. Sci.* **2006**, *10*, 353–368. [CrossRef]
111. Silvestro, F.; Gabellani, S.; Rudari, R.; Delogu, F.; Laiolo, P.; Boni, G. Uncertainty reduction and parameter estimation of a distributed hydrological model with ground and remote-sensing data. *Hydrol. Earth Syst. Sci.* **2015**, *19*, 1727–1751. [CrossRef]
112. Brocca, L.; Moramarco, T.; Melone, F.; Wagner, W.; Hasenauer, S.; Hahn, S. Assimilation of surface and root-zone ASCAT soil moisture products into rainfall-runoff modelling. *IEEE Trans. Geosci. Remote Sens.* **2012**, *50*, 2542–2555. [CrossRef]
113. Massari, C.; Brocca, L.; Tarpanelli, A.; Moramarco, T. Data assimilation of satellite soil moisture into rainfall-runoff modelling: A complex recipe? *Remote Sens.* **2015**, *7*, 11403–11433. [CrossRef]
114. Lievens, H.; Tomer, S.K.; Al Bitar, A.; De Lannoy, G.J.M.; Drusch, M.; Dumedah, G.; Hendricks Franssen, H.-J.; Kerr, Y.H.; Martens, B.; Pan, M.; et al. SMOS soil moisture assimilation for improved hydrologic simulation in the Murray Darling Basin, Australia. *Remote Sens. Environ.* **2015**, *168*, 146–162. [CrossRef]
115. Alvarez-Garreton, C.; Ryu, D.; Western, A.W.; Crow, W.T.; Robertson, D.E. The impacts of assimilating satellite soil moisture into a rainfall–runoff model in a semi-arid catchment. *J. Hydrol.* **2014**, *519*, 2763–2774. [CrossRef]
116. Masseroni, D.; Cislaghi, A.; Camici, S.; Massari, C.; Brocca, L. A reliable rainfall-runoff model for flood forecasting: Review and application to a semiurbanized watershed at high flood risk in Italy. *Hydrol. Res.* **2017**. [CrossRef]
117. Cenci, L.; Laiolo, P.; Gabellani, S.; Campo, L.; Silvestro, F.; Delogu, F.; Boni, G.; Rudari, R. Assimilation of H-SAF soil moisture products for flash flood early warning systems. Case study: Mediterranean catchments. *IEEE J. Sel. Top. Appl. Earth Obs. Remote Sens.* **2016**, *9*, 5634–5646. [CrossRef]
118. Alvarez-Garreton, C.; Ryu, D.; Western, A.W.; Su, C.-H.; Crow, W.T.; Robertson, D.E.; Leahy, C. Improving operational flood ensemble prediction by the assimilation of satellite soil moisture: Comparison between lumped and semi-distributed schemes. *Hydrol. Earth Syst. Sci.* **2015**, *19*, 1659–1676. [CrossRef]
119. Gruber, A.; Su, C.-H.; Zwieback, S.; Crow, W.; Dorigo, W.; Wagner, W. Recent advances in (soil moisture) triple collocation analysis. *Int. J. Appl. Earth Obs. Geoinf.* **2016**, *45*, 200–211. [CrossRef]
120. Zhuo, L.; Han, D. Could operational hydrological models be made compatible with satellite soil moisture observations? *Hydrol. Process.* **2016**, *30*, 1637–1648. [CrossRef]
121. Godt, J.W.; Baum, R.L.; Chleborad, A.F. Rainfall characteristics for shallow landsliding in Seattle, Washington, USA. *Earth Surf. Process. Landf.* **2006**, *31*, 97–110. [CrossRef]
122. Todisco, F.; Brocca, L.; Termite, L.; Wagner, W. Use of satellite and modelled soil moisture data for predicting event soil loss at plot scale. *Hydrol. Earth Syst. Sci.* **2015**, *19*, 3845–3856. [CrossRef]
123. Ponziani, F.; Pandolfo, C.; Stelluti, M.; Berni, N.; Brocca, L.; Moramarco, T. Assessment of rainfall thresholds and soil moisture modeling for operational hydrogeological risk prevention in the Umbria region (central Italy). *Landslides* **2012**, *9*, 229–237. [CrossRef]

124. Hawke, R.; McConchie, J. In situ measurement of soil moisture and pore-water pressures in an 'incipient' landslide: Lake Tutira, New Zealand. *J. Environ. Manag.* **2011**, *92*, 266–274. [CrossRef] [PubMed]
125. Baum, R.L.; Godt, J.W. Early warning of rainfall-induced shallow landslides and debris flows in the USA. *Landslides* **2010**, *7*, 259–272. [CrossRef]
126. USGS Real-Time Monitoring of Landslides. Available online: http://landslides.usgs.gov/research/rtmonitoring/ (accessed on 13 February 2017).
127. Ray, R.L.; Jacobs, J.M.; Cosh, M.H. Landslide susceptibility mapping using downscaled AMSR-E soil moisture: A case study from Cleveland Corral, California, US. *Remote Sens. Environ.* **2010**, *114*, 2624–2632. [CrossRef]
128. Ray, R.L.; Jacobs, J.M.; Ballestero, T.P. Regional landslide susceptibility: Spatiotemporal variations under dynamic soil moisture conditions. *Nat. Hazards* **2011**, *59*, 1317–1337. [CrossRef]
129. AghaKouchak, A.; Farahmand, A.; Melton, F.S.; Teixeira, J.; Anderson, M.C.; Wardlow, B.D.; Hain, C.R. Remote sensing of drought: Progress, challenges and opportunities. *Rev. Geophys.* **2015**, *53*, 452–480. [CrossRef]
130. McCabe, M.F.; Rodell, M.; Alsdorf, D.E.; Miralles, D.G.; Uijlenhoet, R.; Wagner, W.; Lucieer, A.; Houborg, R.; Verhoest, N.E.C.; Franz, T.E.; et al. The future of earth observation in hydrology. *Hydrol. Earth Syst. Sci.* **2017**. [CrossRef]
131. Chirebvu, E.; Chimbari, M.J.; Ngwenya, B.N.; Sartorius, B. Clinical malaria transmission trends and its association with climatic variables in tubu village, Botswana: A retrospective analysis. *PLoS ONE* **2016**, *11*, e0139843. [CrossRef] [PubMed]
132. Montosi, E.; Manzoni, S.; Porporato, A.; Montanari, A. An ecohydrological model of malaria outbreaks. *Hydrol. Earth Syst. Sci.* **2012**, *16*, 2759–2769. [CrossRef]
133. Peters, J.; Conte, A.; Verhoest, N.E.; De Clercq, E.; Goffredo, M.; De Baets, B.; Hendrickx, G.; Ducheyne, E. On the relation between soil moisture dynamics and the geographical distribution of Culicoides imicola. *Ecohydrology* **2014**, *7*, 622–632. [CrossRef]
134. Pellarin, T.; Ali, A.; Chopin, F.; Jobard, I.; Bergs, J.-C. Using spaceborne surface soil moisture to constrain satellite precipitation estimates over West Africa. *Geophys. Res. Lett.* **2008**, *35*, L02813. [CrossRef]
135. Crow, W.T.; Huffman, G.F.; Bindlish, R.; Jackson, T.J. Improving satellite rainfall accumulation estimates using spaceborne soil moisture retrievals. *J. Hydrometeorol.* **2009**, *10*, 199–212. [CrossRef]
136. Trenberth, K.E.; Asrar, G.R. Challenges and opportunities in water cycle research: WCRP contributions. *Surv. Geophys.* **2014**, *35*, 515–532. [CrossRef]
137. Brocca, L.; Melone, F.; Moramarco, T.; Wagner, W. A new method for rainfall estimation through soil moisture observations. *Geophys. Res. Lett.* **2013**, *40*, 853–858. [CrossRef]
138. Brocca, L.; Ciabatta, L.; Massari, C.; Moramarco, T.; Hahn, S.; Hasenauer, S.; Kidd, R.; Dorigo, W.; Wagner, W.; Levizzani, V. Soil as a natural rain gauge: Estimating global rainfall from satellite soil moisture data. *J. Geophys. Res.* **2014**, *119*, 5128–5141. [CrossRef]
139. Brocca, L.; Massari, C.; Ciabatta, L.; Moramarco, T.; Penna, D.; Zuecco, G.; Pianezzola, L.; Borga, M.; Matgen, P.; Martínez-Fernández, J. Rainfall estimation from in situ soil moisture observations at several sites in Europe: An evaluation of SM2RAIN algorithm. *J. Hydrol. Hydromech.* **2015**, *63*, 201–209. [CrossRef]
140. Koster, R.D.; Brocca, L.; Crow, W.T.; Burgin, M.S.; De Lannoy, G.J.M. Precipitation Estimation Using L-Band and C-Band Soil Moisture Retrievals. *Water Resour. Res.* **2016**, *52*, 7213–7225. [CrossRef]
141. Brocca, L.; Pellarin, T.; Crow, W.T.; Ciabatta, L.; Massari, C.; Ryu, D.; Su, C.-H.; Rudiger, C.; Kerr, Y. Rainfall estimation by inverting SMOS soil moisture estimates: A comparison of different methods over Australia. *J. Geophys. Res.* **2016**, *121*, 12062–12079. [CrossRef]
142. Ciabatta, L.; Brocca, L.; Massari, C.; Moramarco, T.; Puca, S.; Rinollo, A.; Gabellani, S.; Wagner, W. Integration of satellite soil moisture and rainfall observations over the Italian territory. *J. Hydrometeorol.* **2015**, *16*, 1341–1355. [CrossRef]
143. Ciabatta, L.; Brocca, L.; Massari, C.; Moramarco, T.; Gabellani, S.; Puca, S.; Wagner, W. Rainfall-runoff modelling by using SM2RAIN-derived and state-of-the-art satellite rainfall products over Italy. *Int. J. Appl. Earth Obs. Geoinf.* **2016**, *48*, 163–173. [CrossRef]
144. Alvarez-Garreton, C.; Ryu, D.; Western, A.W.; Crow, W.T.; Su, C.-H.; Robertson, D.R. Dual assimilation of satellite soil moisture to improve streamflow prediction in data-scarce catchments. *Water Resour. Res.* **2016**, *52*, 5357–5375. [CrossRef]

145. Kumar, S.V.; Peters-Lidard, C.D.; Santanello, J.A.; Reichle, R.H.; Draper, C.S.; Koster, R.D.; Nearing, G.; Jasinski, M.F. Evaluating the utility of satellite soil moisture retrievals over irrigated areas and the ability of land data assimilation methods to correct for unmodeled processes. *Hydrol. Earth Syst. Sci.* **2015**, *19*, 4463–4478. [CrossRef]
146. Qiu, J.; Gao, Q.; Wang, S.; Su, Z. Comparison of temporal trends from multiple soil moisture data sets and precipitation: The implication of irrigation on regional soil moisture trend. *Int. J. Appl. Earth Obs. Geoinf.* **2016**, *48*, 17–27. [CrossRef]
147. Singh, D.; Gupta, P.K.; Pradhan, R.; Dubey, A.K.; Singh, R.P. Discerning shifting irrigation practices from passive microwave radiometry over Punjab and Haryana. *J. Water Clim. Chang.* **2016**. [CrossRef]
148. Fois, F.; Lin, C.C.; Loiselet, M.; Scipal, K.; Stoffelen, A.; Wilson, J.J.W. The Metop second generation scatterometer. In Proceeding of the International Geoscience and Remote Sensing Symposium, Quebec City, QC, Canada, 13–18 July 2014; pp. 1–4.
149. Montzka, C.; Moradkhani, H.; Weihermuller, L.; Franssen, H.-J.H.; Canty, M.; Vereecken, H. Hydraulic parameter estimation by remotely-sensed top soil moisture observations with the particle filter. *J. Hydrol.* **2011**, *399*, 410–421. [CrossRef]

8

Daily Precipitation Changes over Large River Basins in China, 1960–2013

Bo Qu [1,2], Aifeng Lv [1,*], Shaofeng Jia [1] and Wenbin Zhu [1]

1 Key Laboratory of Water Cycle and Related Land Surface Processes, Institute of Geographic Sciences and Natural Resources Research, CAS, Beijing 100101, China; geo_qb@163.com (B.Q.); jiasf@igsnrr.ac.cn (S.J.); bfdh198612@163.com (W.Z.)

2 University of Chinese Academy of Sciences, Beijing 100049, China

* Correspondence: lvaf@163.com

Academic Editors: Tommaso Moramarco and Roberto Ranzi

Abstract: Based on a high-quality dataset of 713 daily precipitation series, changes in daily precipitation events during 1960–2013 were observed in China's ten largest river basins. Specifically, the amount of precipitation in four categories defined by fixed thresholds and their proportion on total precipitation were analyzed on annual and seasonal time scales. Results showed annual precipitation increased by 1.10 mm/10yr in China, but with obvious spatial differences. Regionally, annual precipitation increased significantly in northwestern rivers, upstream areas of the Yangtze River, the Yellow River, southwestern rivers (due to increase in light and moderate precipitation); and in southeastern rivers, downstream areas of the Yangtze River, and the Pearl River (due to increase in heavy and extreme precipitation). Annual precipitation decreased significantly in the mid-Yangtze River and upstream Pearl River (due to decrease in light, moderate, and heavy precipitation). Seasonally, precipitation decreased only in autumn; this was attributable to a decrease in light and moderate precipitation. Results show that the distribution of precipitation intensity over China has shifted to intense categories since the 1960s, there has been an increase in moderate precipitation in Northwestern and Northern China, and an increase in extreme precipitation in Southeastern China. This shift was detected in all seasons, especially in summer. Precipitation extremes were investigated in the categories of extreme precipitation and results show that the risk of flood has been exacerbated over the past half-century in the Huaihe River, the mid- and lower Yangtze River, the Pearl River, and southeastern rivers.

Keywords: daily precipitation; precipitation distribution; precipitation extremes; river basin

1. Introduction

As global climate experiences significant change with increasing concentrations of greenhouse gasses [1], assessing changes in climatic variables has become a key issue for researchers around the world [2]. Although it is widely recognized that global temperature is increasing significantly, there are large uncertainties concerning precipitation change because of variabilities in time and space [3]. Revealing changes in precipitation receives great interest because of its importance to agriculture, water supply, and natural hazards, such as droughts and floods.

Precipitation can be observed and analyzed in terms of long-term mean state and the intensity and frequency of precipitation events. Global land precipitation has increased by about 2% since 1900 [4], but this increase has been neither spatially nor temporally uniform [3].

A study of the US also showed a precipitation increase during the 20th century; the increase was most pronounced during the warm season [5]. To detect changes in the frequency and intensity of precipitation, daily precipitation is usually categorized using two methods: percentile thresholds and fixed thresholds. They are both widely used in many studies [1,5–7]. Brunetti *et al.* observed changes in daily precipitation within percentile categories across Italy and found that the significant decrease in frequency of precipitation was due to a decrease in low-intensity precipitation events [6]. Using fixed thresholds, Karl *et al.* observed that the proportion of total precipitation contributed by extreme precipitation had increased significantly within the US [8]. In the hydrological cycle, the frequency and intensity of precipitation directly affect rainfall-runoff processes and further determine regional water resources. For example, in forest stands, the proportion of interception loss could be up to 100% in light precipitation, while the loss could be 25% in heavy precipitation [9]. Percentile categories are generally defined by specific quantiles of the precipitation distribution at a single station, such as the top 5% for extreme precipitation [6] and bottom 50% for light precipitation [7]. Li *et al.* showed extreme events by the top 5% of precipitation distribution was about 10~20mm/d in Western China [10]. Additionally, the mean intensity of precipitation in the same percentile at different stations, therefore, could be different. Although percentile categories benefit the detection of precipitation changes across regions and seasons [1,6], fixed categories could be more useful in hydrological models, as rainfall-runoff processes generally derive from the amount of precipitation events across a river basin.

In recent years, changes in precipitation characteristics have also been widely discussed in China. Liu *et al.* found that the frequency of precipitation decreased by a significant 10% over the same period for most of China, except the northwest [1]. Precipitation extremes have also been studied extensively in China. A more recent study by Wu *et al.* found that heavy precipitation events had higher rates of change than did mean precipitation [7]. The river basin is an important unit in hydrological processes. Water resource management is principally practiced at the basin scale [11]. It is, therefore, valuable to identify precipitation changes at that scale. Precipitation extremes are one focus in studies over basins in China. Chen *et al.* studied precipitation extremes in large basins across China [12] and, similarly, studies were also performed in the Yangtze River [13], Pearl River [14,15], Lancang River [16], and Yellow River [17]. Zhang and Cong analyzed the mean intensity and frequency of precipitation events in large river basins across China from 1956 to 2005. Their results indicated that precipitation intensity had significantly increased, while precipitation frequency had decreased in all basins except for rivers in the northwest [11]. Changes in the distribution of precipitation intensity, however, are rarely explored in river basins across China. In hydrology, if the distribution of precipitation intensity shifts, even when there is no change in total precipitation, runoff quantities would also significantly change due to changes in rainfall-runoff processes. A relative study is, therefore, urgently needed in China.

Climatic variability research, however, is not an easy task. One of the greatest problems in examining climate change is a lack of high-quality long-term data [6] because of non-climatic noise in climate signals [18]. Although researchers have understood that developing reliable datasets is essential in climate-change studies, checking prospective datasets and adjusting for biases has received little attention around the world and in China.

This study focused on changes in daily precipitation of different intensities over large river basins in China and attempted to reveal changes in the distribution of precipitation intensity in these basins. The research first considered issues of quality control, homogeneity, and completeness and a high-quality dataset of daily precipitation was produced. Second, changes in daily precipitation were analyzed on annual and seasonal time scales during the past half-century. Specifically, fixed thresholds were used to categorize precipitation events. The amount and proportion of precipitation by category and annual or seasonal precipitation were calculated for every station each year, and then the trends were analyzed across China and its ten largest river basins. This paper is organized as follows. Section 2 describes the data, data-processing methods, and analysis methods used in this study. Results and analysis are presented in Section 3, and the discussion and conclusion are given in Section 4.

2. Data and Methodology

2.1. Data and Pre-Processing

The observed daily precipitation dataset obtained from the Chinese National Meteorological Center was used in our study. It contained 819 stations for the period from 1 January 1951 to 31 December 2013. This dataset is well distributed across China, including the Tibetan plateau [1], and is widely used in all kinds of studies [19–24]. In the 1950s, however, there were only about 160 to 400 stations with available data [25]. We, therefore, restricted our study to the period 1960–2013.

2.1.1. Quality Control

Quality control is an essential part of climatic variability research [26]. The dataset used in this study was compiled with quality control procedures including the extreme value test, consistency check, and spatial outliers test [11]. A detailed description of these procedures can be found in the work of Feng *et al.* and Qian and Lin. The study by Feng *et al.* applied quality control tests to daily meteorological data for 1951–2000 from the same source we used, and only 0.02% stations showed low quality [22]. We, therefore, considered the precipitation data to be of good quality.

2.1.2. Data Homogeneity

The homogeneity of meteorological data has been highlighted in many studies; climatic variability research is not possible without clear knowledge about it [18,27–33]. A number of non-climatic factors, including station relocation and changes in instruments, observing regulations, and algorithms for the calculation of means, can introduce bias into climatological time series [6]. These biases have disturbed the record of actual climatic variations over time [30].

A variety of direct and indirect methods has been developed to identify and adjust for inhomogeneities [18]. A comprehensive review of such methods was given by Peterson *et al.* [30]. In China, only 60 stations have metadata available [22], which the direct correction techniques for inhomogeneities require. Relatively indirect methods were, therefore, chosen for this study. Most of the indirect methods are based on the comparison of a candidate series with a reference series formed from nearby or correlative stations; these include Potter's ratio test [34], the standard normal homogeneity test (SNHT) [32], Bayesian procedures [20], and multiple linear regression [31].

No complete protocol, however, is available for processing daily precipitation datasets [27] and most methods were developed for monthly or annual data [30]. For daily precipitation, precipitation frequency is an important factor that affects the homogeneity of a time series [6,29]. Brunetti *et al.* attempted to check the homogeneity of daily precipitation records. They tested two kinds of heterogeneities in daily precipitation records: (1) the precipitation amount and (2) the number of rainy days [6]. This study followed their methods in that both precipitation amount and rainy days per month were used to evaluate the homogeneity in dataset. One of the most-used methods to identify inhomogeneities, the standard normal homogeneity test (SNHT) developed by Alexandersson was mostly used [27,35]. In China, Feng *et al.* also applied the SNHT method to identify inhomogeneity in daily meteorological data. Therefore, SNHT was used in this study.

After identifying inhomogeneities of our dataset, 11 stations showed homogeneity problems in precipitation amount and more stations, 57 stations, displayed homogeneity problems in rainy days. Among them, four stations were overlapped. At the end, a set of 755 stations constitute the newly-homogenized dataset in this study.

2.1.3. Missing Values

Missing values could also introduce bias in a climatological time series, so we attempted to fill in any missing values before proceeding. At first, we filtered the dataset to exclude stations with more than five missing years (nearly 10% of study period) between 1960 and 2013. Specifically, following Lucie *et al.*, the monthly and, consequently, annual, values were considered to be missing when data for more than three consecutive days or more than five random days within a month were missing [36]. In the end, 713 of 755 stations with more than 48 years of complete data remained in our study (Figure 1).

Each candidate station subjected to the SNHT was compared to a reference series created to identify its inhomogeneities. The reference series is a weighted average of a number of the nearest available stations [30]. The SNHT reference series was used to fill in the missing values from the 713 stations.

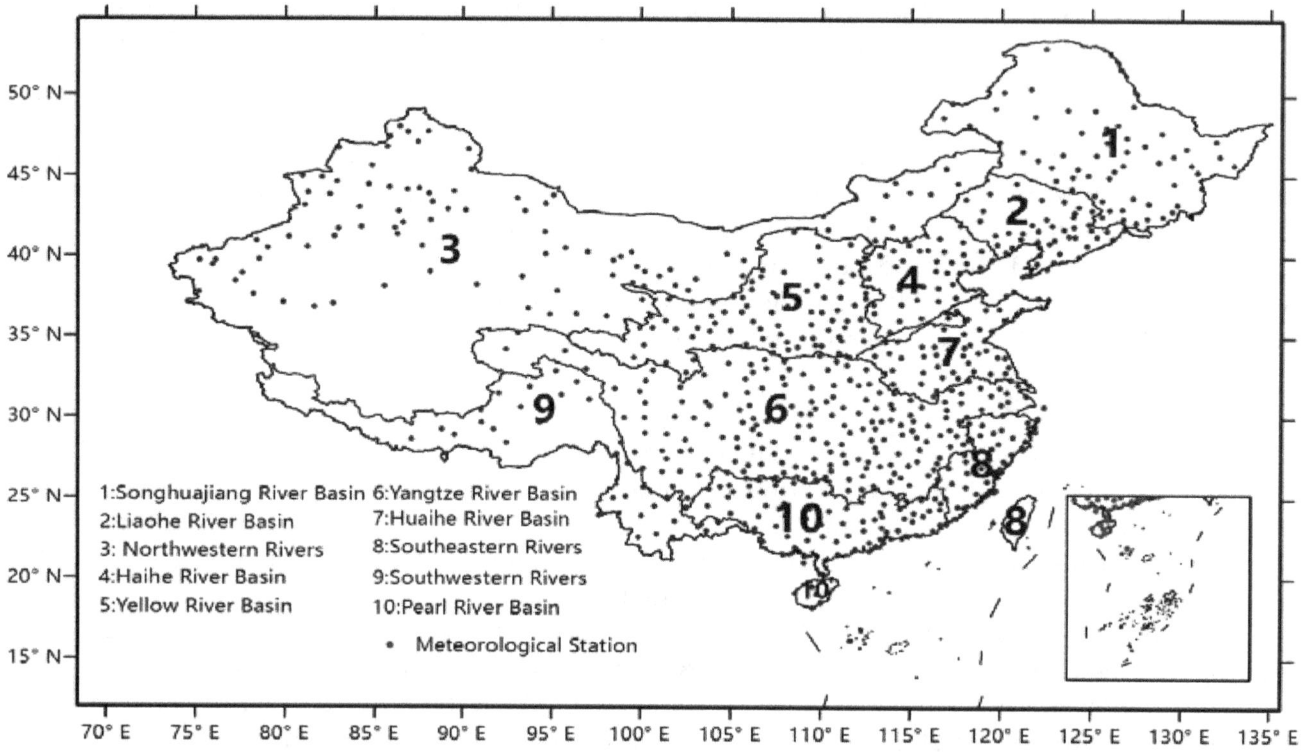

Figure 1. Large river basins and the locations of the 713 meteorological stations evaluated in China.

2.2. Methodology

The China Meteorological Administration (CMA) categorizes daily precipitation (P) into five intensity groups: $P < 10$ mm/d, $10 \leqslant P < 25$ mm/d, $25 \leqslant P < 50$ mm/d, $50 \leqslant P < 100$ mm/d, and $P \geqslant 100$ mm/d. Since the $P \geqslant 100$ mm/d group occurs infrequently in Northern China [13], following Gong *et al.*, we combined the last two groups and defined four categories: light precipitation (<10 mm/d), moderate precipitation (10–25 mm/d), heavy precipitation (25–50 mm/d), and extreme precipitation (≥50 mm/d). The amount and proportion of daily precipitation falling into these four categories were calculated for each station. We then analyzed the trends over China and in its ten large river basins (Figure 1).

2.2.1. Trends Detection

In this study, the trends were analyzed using the Mann-Kendall test, which is well suited to non-normally distributed data [37,38] and has been widely used in many previous studies [6,7]. To remove lag-1 autocorrelation in a time series, which could overestimate significance of the Mann-Kendall test, the trend-free pre-whitening (TFPW) method developed by Yue *et al.* was adopted before applying the Mann-Kendall test [38].

The Mann-Kendall test represents a ranked-based approach. The test statistic S is given as follows:

$$S = \sum_{i=1}^{n-1} \sum_{j=i+1}^{n} sgn\left(x_j - x_i\right) \tag{1}$$

where:

$$\operatorname{sgn}(x) = \begin{cases} 1, for\ x > 0 \\ 0, for\ x = 0 \\ -1, for\ x < 0 \end{cases} \tag{2}$$

x_i and x_j are the sequential data values, and n is the length of the dataset.

A positive value of S indicates an upward trend and a negative value of S indicates a downward trend. The test is conducted using the normal distribution, with the expectation (E) and variance (V) as follows:

$$E(S) = 0 \tag{3}$$

$$V(S) = \frac{1}{18}\left(n(n-1)(2n+5) - \sum_{k=1}^{h} t_k(t_k-1)(2t_k+5)\right) \tag{4}$$

where t_k is the number of data points in the kth tied group and h is the number of tied groups in the dataset.

The Mann-Kendall Z is then calculated by:

$$Z = \begin{cases} \frac{S-1}{\sqrt{V(s)}}\ for\ S > 0 \\ 0\ for\ S = 0 \\ \frac{S+1}{\sqrt{V(s)}}\ for\ S < 0 \end{cases} \tag{5}$$

The Z value can be related to a p-value of a specific trend. In a two-sided test for trend, the null hypothesis of no trend H_0 is accepted if $-Z_{1-\alpha/2} \leqslant Z \leqslant Z_{1-\alpha/2}$, where α is the significance level [39]. The null hypothesis of no trend is rejected if |Z| > 1:65, |Z| > 1:96 and |Z| > 2:57 at the 10%, 5%, and 1% significance levels, respectively.

The magnitudes of tendency were estimated using Theil-Sen approach (TSA), which is also a non-parametric statistical technique [39,40]. In this study, the TSA slope in precipitation was multiplied by 10 to express trends per decade. The TSA slope β is given by:

$$\beta = \text{median}\left[\frac{x_j - x_i}{j-i}\right]\ for\ all\ i < j \tag{6}$$

2.2.2. Spatial Interpolation

Meteorological stations can provide reliable and accurate precipitation data, but their spatial distribution is generally uneven in China [7]. In order to explore the temporal variation of ten large river basins, Kriging interpolation was chosen to generate grid data due to its good performance with geographic data [41]. The spherical semi-variogram model was used as the weighting function in interpolation [11]. Trends across the whole country and each basin were then derived from regional averaged data from 1960 to 2013.

3. Results and Analysis

3.1. Annual Character of Precipitation Tendencies

The Mann-Kendall trends for annual precipitation during 1960–2013 are shown in Figure 2. Stations with a significance level of 0.05 are labeled by pluses (positive) or triangles (negative). The averaged trends over China and its ten large river basins are showed in Table 1.

Over the past half-century, annual precipitation increased in China at a magnitude of 1.10 mm/10yr, but this trend was not statistically significant (Table 1). Regionally, significantly increasing trends occurred in the eastern coastal region, northwestern China, and the Tibetan Plateau, while significantly decreasing trends occurred in southwestern and northern China (Figure 2). In the ten large river basins, significant uptrend was observed in southwestern rivers. Large downtrends were observed in the Huaihe, Liaohe, and Haihe Rivers, while large uptrends were observed in southeastern rivers and northwestern rivers, although these findings were not statistically significant (Table 1).

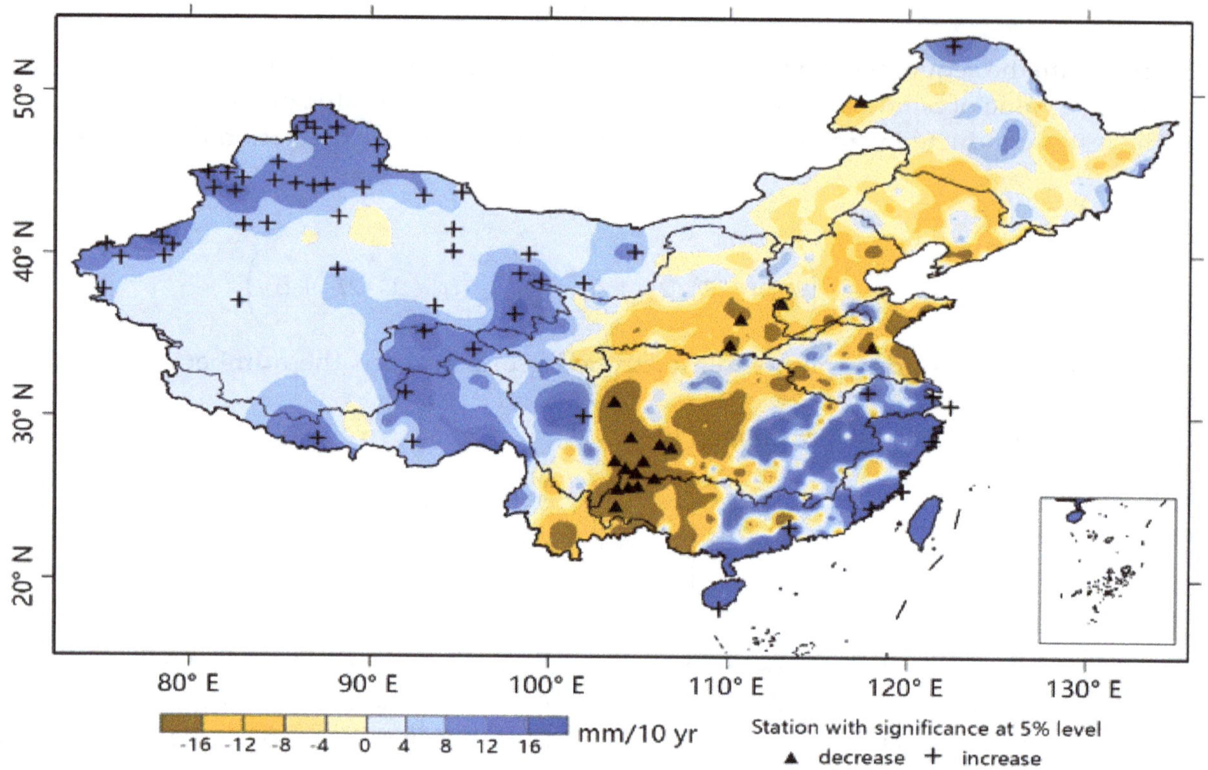

Figure 2. Trends in annual precipitation for 1960–2013. A plus/triangle shows a positive/negative trend for each station with statistical significance at the 5% level.

Table 1. Trends in amounts and proportion of precipitation in each category in the large river basins in China during 1960–2013.

Basin	Amount (mm/10yr)					Proportion (%/10yr)			
	Light Rain	**Moderate Rain**	**Heavy Rain**	**Extreme Rain**	**Total**	**Light Rain**	**Moderate Rain**	**Heavy Rain**	**Extreme Rain**
Songhuajiang River	−0.01	−1.65	−0.57	0.20	−2.89	0.25	−0.10	−0.06	0.00
Liaohe River	−1.68	−1.48	−1.80	−2.48	−8.60	0.13	0.12	0.05	−0.29
Northwestern Rivers	1.44[a]	0.99[a]	0.41	0.32	3.69[a]	−0.68[a]	0.37[b]	0.13	0.02
Haihe River	−1.65	0.86	−0.45	−4.39[b]	−5.59	−0.05	0.53[b]	0.08	−0.56[b]
Yellow River	−1.73	−1.44	0.61	−0.31	−2.69	−0.45	0.12	0.33[b]	−0.05
Yangtze River	−2.28[b]	−3.30[b]	0.61	3.88[b]	−1.13	−0.22[b]	−0.28[a]	0.08	0.45[a]
Huaihe River	−3.08[a]	−1.24	−1.80	−1.10	−7.20	−0.31[a]	−0.01	0.04	0.47
Southeastern Rivers	−3.94	−3.71	9.35[b]	15.63[a]	22.21	−0.48[a]	−0.52	0.34[a]	0.71[a]
Southwestern Rivers	1.44	1.04	2.14	1.84[b]	5.29	−0.43[a]	0.19	0.12	0.13[a]
Pearl River	−5.44[a]	−4.48	0.39	5.00	−5.71	−0.23[a]	−0.24[a]	0.04	0.34[b]
All China	−0.41	−0.65	0.47	1.76[b]	1.10	−0.39[a]	0.06	0.15[a]	0.15[a]

[a] Significance at 95% level; [b] Significance at 90% level; Unit "mm/10yr" and "%/10yr" means mm and % per decade.

Light precipitation showed a slight decrease (−1.20 mm/10yr) during 1960–2013 (Table 1), but a contrasting pattern of trends emerged across China (Figure 3a). Northwestern rivers, the headwaters of the Yangtze River and Yellow River, and southwestern rivers showed a significant increase, whereas other regions showed a significant decrease (Figure 3a). In the ten river basins, northwestern rivers and southwestern rivers showed an increase in light precipitation over the past half-century, and the trend in northwestern rivers was statistically significant (Table 1). Light precipitation in the Yangtze, Huaihe, and Pearl Rivers showed significant decreases. Data shown in Table 1 further suggest decreases in light precipitation largely contributes to the decrease in annual precipitation in these regions.

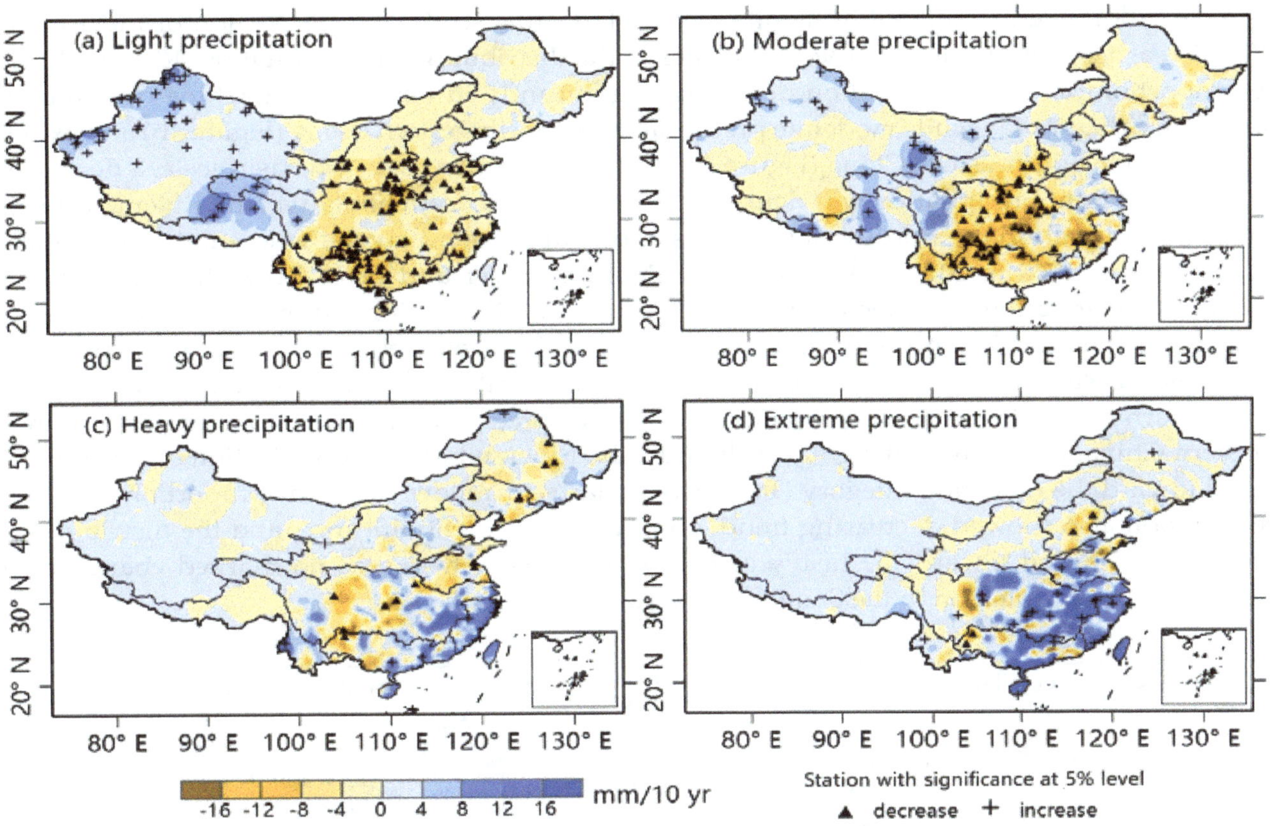

Figure 3. Trends in precipitation of each category for 1960–2013. (**a**) Light precipitation; (**b**) moderate precipitation; (**c**) heavy precipitation; and (**d**) extreme precipitation.

A decreasing trend for moderate precipitation was also detected across China (−0.65 mm/10yr, Table 1). Significant decreases occurred from Central to Southern China, including the mid-Yellow River, mid- and downstream Yangtze River, Pearl River, and southeastern rivers (Figure 3b). In the Yangtze River, moderate precipitation showed a significant decrease at a 0.1 significance level (−3.30 mm/10yr), which had a strong influence on the downtrend of annual precipitation (Table 1). Southeastern rivers and Pearl River also displayed large decreases in moderate precipitation (−3.71 mm/10yr and −4.48 mm/10yr, respectively). Significant increases mainly occurred in Northern and Western China (Figure 3b). Northwestern rivers were dominated by a significant increasing trend (0.99 mm/10yr). In the Haihe River, only moderate precipitation showed a positive trend (0.86 mm/10yr), which would partly alleviate the drier conditions over the past half-century.

Heavy precipitation was mainly detected in monsoon regions of China, and stations with significant increases were mostly located in the lower reaches of the Yangtze and Pearl Rivers and the southeastern rivers (Figure 3c). Table 1 shows that heavy precipitation significantly increased in northwestern rivers (0.82 mm/10yr) and southeastern rivers (9.35 mm/10yr), and fluctuated largely in Liaohe River, Huaihe River, and southwestern rivers, although these trends were not statistically significant. Extreme precipitation was mainly detected in the eastern and southeastern coastal regions, and stations with significant increases stretched from southeastern coastal areas to the middle reaches of the Yangtze and Huaihe Rivers (Figure 3d). In the whole country, on average, extreme precipitation showed a significant increase from 1960 to 2013 (1.76 mm/10yr). Regionally, in northwestern rivers, Yangtze River, southeastern rivers, and southwestern rivers, extreme precipitation significantly increased, while in Haihe River a significant decrease was detected (Table 1). The southeastern rivers displayed a large increase in total precipitation (22.21 mm/10yr), which was primarily the result of an increase in extreme precipitation (15.63mm/10yr).

Karl *et al.* noted that even when there is no change in the total precipitation, the distribution of precipitation events might shift significantly [8]. In this study, the proportion of precipitation falling in each category was sensitive to changes in the distribution of precipitation. Results showed the proportion of light precipitation decreased significantly in nearly whole China over the past half-century (Figure 4a), while moderate precipitation in Northwestern China (Figure 4b) and extreme precipitation in Southeastern China (Figure 4d) significantly increased. On basin scale, a decrease in proportion of light precipitation was detected in most basins except Songhuajiang River and Liaohe River, even though in the northwestern rivers and southwestern rivers, where light precipitation increased from 1960 to 2013 (Table 1). In the Yangtze River and Pearl River, the proportion of moderate precipitation also decreased significantly, mainly resulting from larger percentage of extreme precipitation. The proportion of heavy precipitation increased in nearly all basins except the Songhuajiang River, and the trend was significant in the northwestern rivers, Yellow River, and the southeastern rivers (Table 1). The proportion of extreme precipitation increased in basins locating in Southern China, but decreased significantly in Haihe River, where the distribution of precipitation shifted toward the moderate category (Table 1). Similarly, in Yellow River, the proportion of extreme precipitation also showed decreasing trend, although without significance, and the distribution of precipitation shifted toward the heavy category (Table 1). There was no marked change in the distribution of precipitation in the Songhuajiang River and Liaohe River.

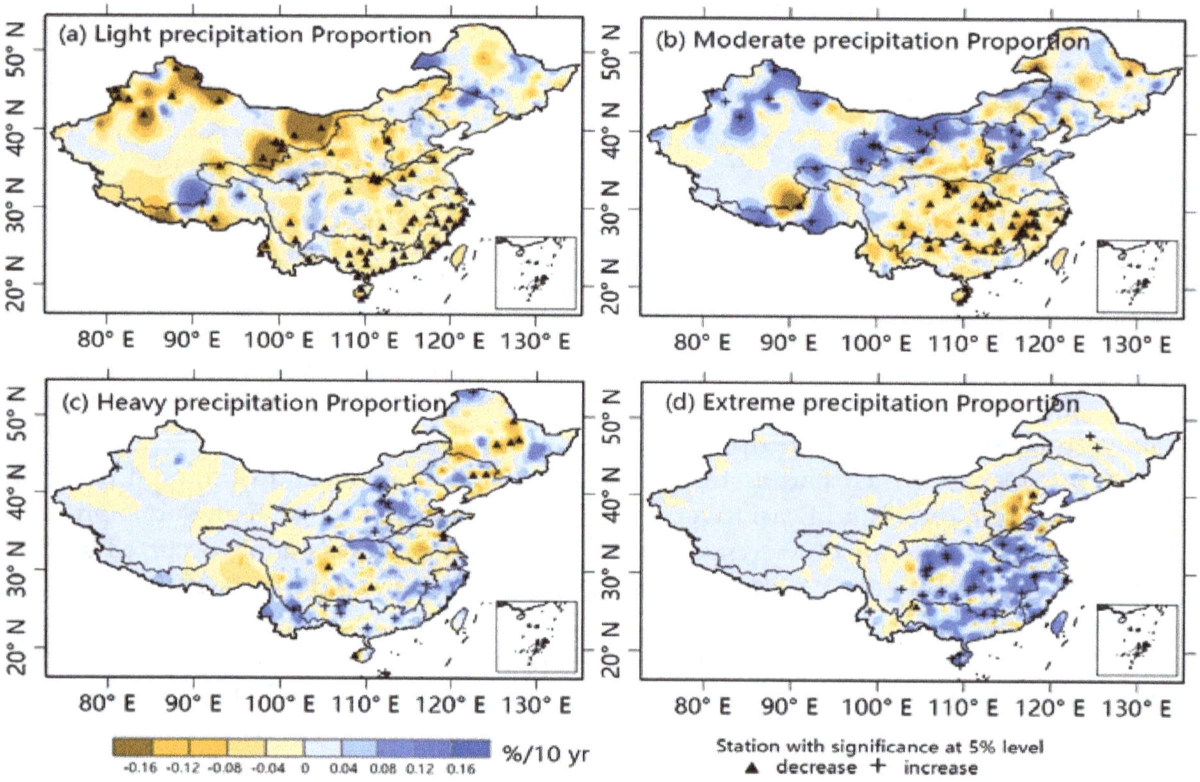

Figure 4. Trends in the proportion of each precipitation category relative to annual precipitation for 1960–2013. (**a**) Light precipitation; (**b**) moderate precipitation; (**c**) heavy precipitation; and (**d**) extreme precipitation.

3.2. Seasonal Characteristics of Precipitation Tendencies

Changes in seasonal precipitation over China during 1960–2013 are shown in Figure 5 and the averaged trends in each season over China and its ten large river basins are given in Table 2. In spring (Figure 5a), precipitation increased in Northern and Western China, especially over the southwestern rivers and the headwaters of the Yellow River and Yangtze River. Significant decreases were observed over the middle and lower reaches of the Yangtze River and the middle reaches of the Yellow River (Figure 5a).

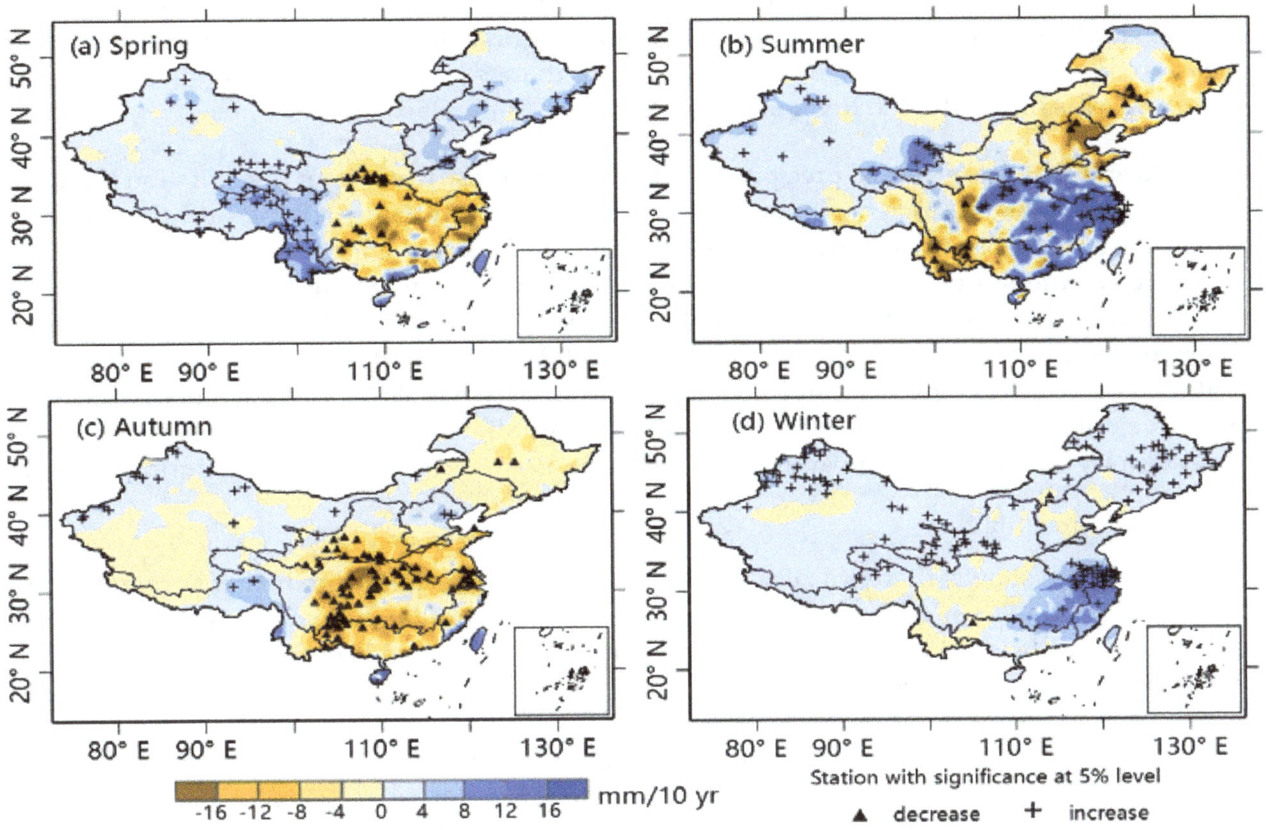

Figure 5. Trends in seasonal precipitation during 1960–2013. (**a**) Spring; (**b**) summer; (**c**) autumn; and (**d**) winter.

Table 2. Seasonal precipitation trends over China's large river basins during 1960–2013 (mm/10yr).

Basin	Spring	Summer	Autumn	Winter	Annual
Songhuajiang River	2.03	−4.90	−2.16[a]	1.33[a]	−2.89
Liaohe River	2.21	−7.75	−3.42	0.17	−8.60
Northwestern rivers	0.97[b]	1.36	0.33	0.75[a]	3.69[a]
Haihe River	4.91[a]	−12.53[a]	1.85	−0.18	−5.59
Yellow River	−0.62	−0.84	−3.38	0.34	−2.69
Yangtze River	−3.35	6.68[b]	−6.15[a]	2.56	−1.13
Huaihe River	−1.48	3.30	−7.60	1.86	−7.20
Southeastern rivers	−5.14	15.99[b]	0.55	7.37[a]	22.21
Southwestern rivers	5.72[a]	−2.70	0.59	−0.09	5.29
Pearl River	−4.27	0.94	−9.47	2.11	−5.71
All China	0.54	0.10	−2.76	1.26	1.10

[a] Significance at 95% level; [b] Significance at 90% level.

In summer, precipitation has markedly changed over the past half-century, showing large uptrends and downtrends with absolute values as large as >16 mm/10yr in most monsoon regions of China (Figure 5b). Specifically, significantly uptrends were observed in the Yangtze River and southeastern rivers, while significant downtrends were observed in Haihe River (Table 2).

Autumn precipitation showed a significant decrease from 1960 to 2013 in the whole of China, as well as in Songhuajiang River and Yangtze River (Table 2). Precipitation increases occurred only in limited areas of the Haihe River, northwestern and southwestern rivers (Figure 5c).

However, in winter, precipitation has increased significantly in China since 1960s, and nearly all basins experienced positive trends; significant in Songhuajiang River, northwestern rivers, and southeastern rivers (Table 2). The increase was more than 8 mm/10yr in the southeastern rivers, downstream areas of the Huaihe River and Yangtze River (Figure 5d).

Comparing changes in annual precipitation (Figure 2) to those in summer precipitation (Figure 5b), we can see that the large increase in summer rainfall in Southeastern China and large decrease in Northeastern and Southwestern China dominated the annual change during 1960–2013. At the basin scale, the strong impact of summer rainfall on annual precipitation change could be found in southeastern rivers and the Haihe River (Table 2). In the Yellow River basin, however, which experienced only small changes in summer rainfall, the annual decreasing trend represented precipitation decreases in the spring and autumn (Table 2). This was also observed in the Yangtze River, Huaihe River, and Pearl River, even though summer rainfall increased in these basins (Table 2). In northwestern rivers, the increase in annual precipitation was the result of increases during all seasons.

3.2.1. Seasonal Precipitation

As Figure 6a,b shows, there was less light and moderate precipitation during spring in regions from the mid-Yellow River to the Pearl River. Table 3 shows significant increasing trends in spring precipitation were observed in the Haihe River, northwestern rivers, and southwestern rivers (Table 3). In northwestern rivers, the increase in spring precipitation resulting from increase in light and moderate categories, while in southwestern rivers all categories led to the increase. The significant changes of light precipitation in basins, such as Songhuajiang River, northwestern rivers, Huaihe River, and southeastern rivers, and significant changes of moderate precipitation in basins, such as Liaohe River, Yangtze River, and southeastern rivers dominated changes in spring precipitation in these basins (Table 3). Moreover, in many basins, trends in moderate precipitation made a greater contribution to spring trends than trends in light precipitation.

Heavy and extreme precipitation were also observed in spring; these events occurred mainly in Southeastern China (Figure 6c,d). Specifically, heavy precipitation increased significantly in the Haihe River and southwestern rivers, decreased largely in the Yangtze River and Pearl River, but without significance (Table 3). Stations with significant extreme precipitation increase occurred in southeastern coastal regions (Figure 6d), and there was significant increase in Songhuajiang River, Haihe River, and southwestern rivers (Table 3).

Decreasing trends in light precipitation during the summer were observed in most basins, except for over the northwestern rivers, and moderate precipitation also decreased in many basins, such as Liaohe River, Haihe River, and Pearl River (Table 3). As Figure 6g shows, heavy precipitation increased significantly during the summer over the southeastern rivers, downstream areas of the Yangtze River and in some regions of the Huaihe River and Pearl River. Extreme precipitation occurred mainly in the summer (Figure 6h). The significant increase in extreme precipitation over the southeastern rivers and Yangtze River and significant decrease in Haihe River largely represented the changes in summer rainfall over these regions.

In autumn, both light and moderate precipitation exhibited significant decrease over China, and decreasing trends of light and moderate precipitation were observed in most basins during the study period (Table 3). Moreover Table 3 shows that in the Songhuajiang River, Yellow River, Yangtze River, Huaihe River, and Pearl River, light and moderate precipitation largely represented the overall decrease. Downward trends in heavy autumn precipitation in the Yangtze River, Huaihe River, and Pearl River also contributed to the autumn decrease. In the Haihe River; however, autumn precipitation increased substantially (1.85 mm/10yr), mainly resulting from an increase in moderate precipitation (1.03 mm/10yr).

Figure 6m–p shows winter precipitation trends. Stations that displayed significant trends in moderate and heavy precipitation were only in downstream areas of the Yangtze River and southeastern rivers (Figure 6n,o), while stations was rarely observed with changes in extreme precipitation across China during the past half-century (Figure 6p). Light precipitation increased significantly nearly in the whole country (Figure 6m) and, regionally, significant in Songhuajiang River, northwestern rivers, and Yellow River, which largely explained the increases in winter precipitation over these basins (Table 3). In the southeastern rivers, wetter winter conditions largely derived from increases in moderate and heavy precipitation, and in the Yangtze River and Pearl River, these conditions largely derived from increases in heavy precipitation (Table 3).

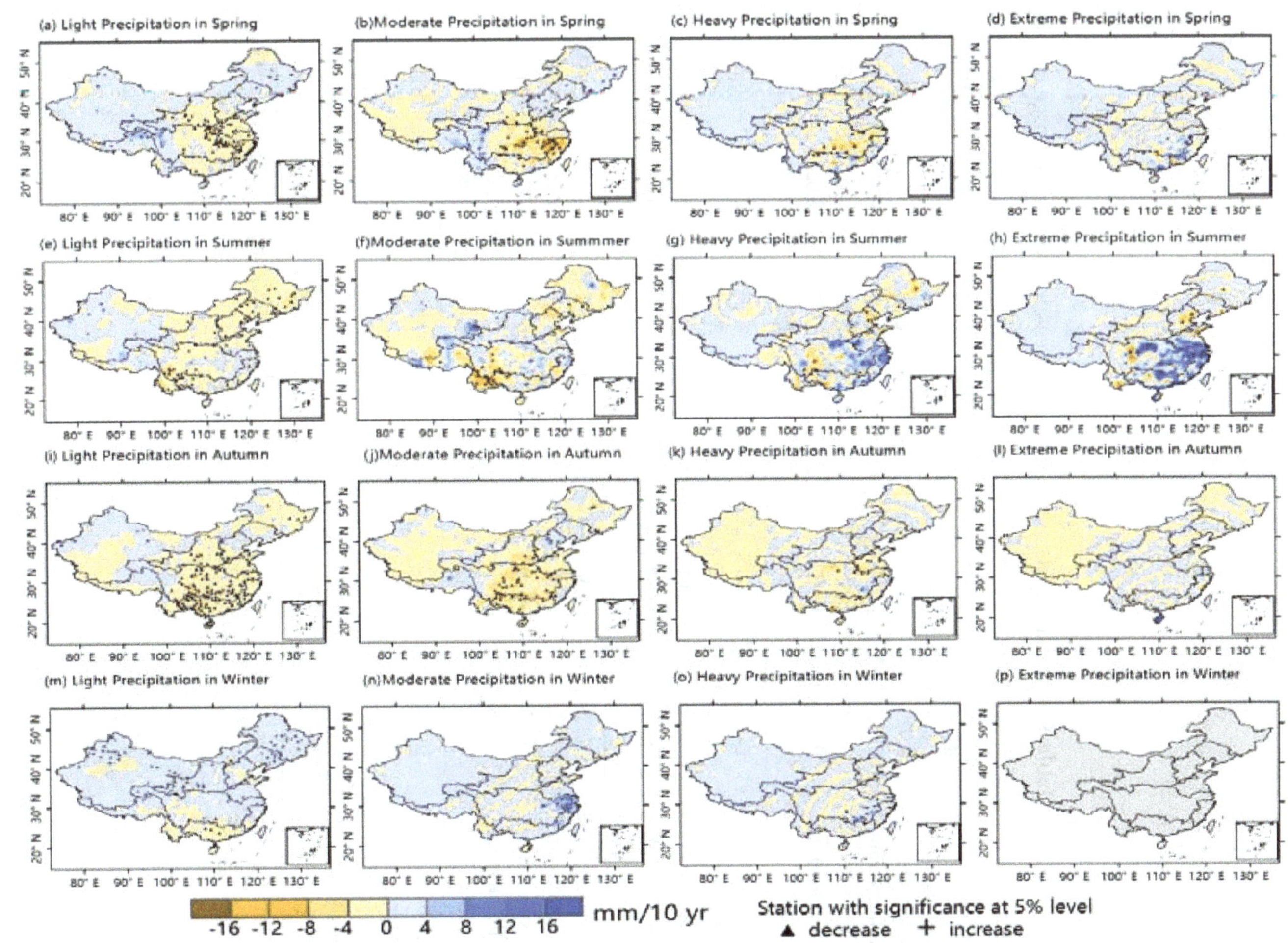

Figure 6. Seasonal trends in precipitation categories during 1960–2013. (**a**–**d**) Spring; (**e**–**h**) summer; (**i**–**l**) autumn; and (**m**–**p**) winter. For each season, the maps show light, moderate, heavy, and extreme precipitation from left to right.

Table 3. Seasonal trends of precipitation intensity over China's large river basins during 1960–2013 (mm/10yr).

Season	Intensity	Songhuajiang River	Liaohe River	Northwestern Rivers	Haihe River	Yellow River	Yangtze River	Huaihe River	Southeastern Rivers	Southwestern Rivers	Pearl River	All China
Spring	Light rain	1.02[b]	0.56	0.43[b]	0.28	−0.56	−0.69	−1.67[a]	−2.52[a]	1.21[a]	−0.82	0.11
	Moderate rain	0.84	1.29[a]	0.31[b]	1.93[a]	−0.11	−1.95[a]	−0.87	−5.89[a]	1.93[a]	−1.71	−0.15
	Heavy rain	0.03	0.35	0.13	1.15[a]	0.16	−1.53	−0.20	0.05	1.34[a]	−1.28	0.02
	Extreme rain	0.11[a]	0.08	0.15	0.28[a]	−0.02	0.40	0.95	2.66	0.89[a]	−0.66	0.36
	Total	2.03	2.21	0.97[b]	4.91[a]	−0.62	−3.35	−1.48	−5.14	5.72[a]	−4.27	0.54
Summer	Light rain	−2.21[a]	−2.28[a]	0.23	−1.58[a]	−0.56	−0.23	−0.74[b]	−0.23	−0.36	−1.22[b]	−0.57[b]
	Moderate rain	−2.12	−2.05[b]	0.52[b]	−2.71[b]	0.49	0.19	0.06	−0.27	−1.83	−2.13	−0.60
	Heavy rain	−1.22	−1.01	0.33	−2.08	−0.19	1.44	2.63	6.05[a]	−0.79	0.67	0.29
	Extreme rain	0.06	−2.49	0.36	−5.12[b]	−0.53	4.79[a]	2.95	8.88[b]	0.55	5.16	0.88[b]
	Total	−4.90	−7.75	1.36	−12.53[a]	−0.84	6.68[b]	3.30	15.99[b]	−2.70	0.94	0.10
Autumn	Light rain	−0.53	−0.48	0.20	−0.17	−1.46[b]	−1.95[a]	−1.81[b]	−1.42	−0.19	−3.55[a]	−0.86[a]
	Moderate rain	−0.97	−0.52	−0.03	1.03	−1.44	−3.04[a]	−1.88	−0.77	0.28	−3.54[a]	−1.09[a]
	Heavy rain	−0.31	−0.90	0.02	0.39	0.05	−1.01	−3.33[b]	0.58	0.23	−3.09	−0.38
	Extreme rain	−0.03	−0.19	0.05	0.37	−0.43[b]	−0.59	−0.40	1.90	0.18	0.29	−0.06
	Total	−2.16	−3.42	0.33	1.85	−3.38	−6.15[a]	−7.60	0.55	0.59	−9.47	−2.76[a]
Winter	Light rain	1.12[a]	0.36	0.53[a]	0.01	0.38[b]	0.46	1.00	0.17	−0.03	−1.21	0.38[b]
	Moderate rain	0.18[a]	0.03	0.05	−0.01	−0.01	0.63	1.11	3.75[b]	−0.10	1.00	0.27
	Heavy rain	0.05	0.00	0.03	−0.01	−0.01	1.15[a]	0.04	3.08[b]	0.24	1.93[a]	0.44[a]
	Extreme rain	0.00	0.00	0.02	−0.01	0.00	0.09	0.00	0.52[b]	0.01	0.55[b]	0.09
	Total	1.33	0.17	0.75	−0.18	0.34	2.56	1.86	7.37	−0.09	2.11	1.26

[a] Significance at 95% level; [b] Significance at 90% level.

3.2.2. Seasonal Distribution

As discussed above, the annual proportion of light precipitation decreased across nearly the whole country; the only regions where it increased were over the Songhuajiang River and Liaohe River, but without significance (Table 1). Seasonally, Table 4 shows the proportional changes in precipitation categories for each basin. For light precipitation, its decreasing proportion was observed across the country in seasons except winter, and was significant in spring and summer (Table 4), even when its amounts were observed to have increased in spring (Table 3). In winter, however, seven of the ten basins still displayed negative trends (Table 4).

In spring, the proportion of moderate, heavy and extreme precipitation all exhibited significant increase over China, as well as in northwestern rivers, Haihe River, and southwestern rivers (Table 4). However, in southeastern rivers, the proportion of moderate precipitation showed significant increase, and the distribution of precipitation shifted toward heavy and extreme categories (Table 4). Besides, the proportion of extreme precipitation increased significantly in Songhuajiang River and Huaihe River (Table 4).

Changes in proportions of summer precipitation categories are shown in Figure 7e–h; these were similar to those representing the whole year. In summer, the proportion of moderate, heavy and extreme precipitation significantly increased across China (Table 4). Light precipitation decreased in proportion nearly across all basins except Haihe River. In the Yangtze River, Huaihe River, southeastern rivers, and Pearl River, the proportion of moderate precipitation decreased significantly over 1960–2013, indicating precipitation became much more intense than basins such as northwestern rivers and Yellow River (Table 4). Contrary to other regions, the Liaohe River, Haihe River, and Yellow River experienced a negative trend in its proportion of extreme summer precipitation.

In winter, the proportion of heavy precipitation significantly increased over China, as well as in Yangtze River, southeastern rivers and Pearl River (Table 4). In Songhuajiang River, northwestern rivers, and Huaihe River, moderate precipitation exhibited significant increase in proportion (Table 4). These suggested the distribution of winter precipitation shifted toward the intense category in these basins. Changes in the distribution of autumn precipitation during the study period were inconspicuous across China. In general, however, decreasing trends in autumn precipitation were observed over nearly the whole of China, as discussed above.

Table 4. Seasonal trends in precipitation intensity proportions over China's large river basins during 1960–2013 (%/10yr).

Season	Intensity	Songhuajiang River	Liaohe River	Northwestern Rivers	Haihe River	Yellow River	Yangtze River	Huaihe River	Southeastern Rivers	Southwestern Rivers	Pearl River	All China
Spring	Light rain	−0.28	−1.93[a]	−0.51	−3.00[a]	−0.58	−0.13	−1.35[b]	−0.49[b]	−1.17[a]	0.14	−0.59[a]
	Moderate rain	0.46	1.05[b]	0.68[a]	1.49[a]	0.42	0.04	−0.05	-0.58_{a}	0.91[a]	−0.29	0.40[a]
	Heavy rain	−0.24	0.34	0.11	1.14[a]	0.39	−0.03	0.40	0.45[b]	0.23[b]	0.00	0.13[b]
	Extreme rain	0.04[a]	0.08	0.02	0.21[a]	0.02	0.27	0.87[a]	0.66[b]	0.17[b]	0.17	0.16[a]
Summer	Light rain	−0.14	−0.08	−1.17[a]	0.05	−0.32	−0.35[a]	−0.32[b]	−0.32[a]	−0.14	−0.26[b]	−0.56[a]
	Moderate rain	0.03	0.17	0.76[a]	0.19	0.41[b]	−0.29[a]	−0.31[b]	−0.61[a]	0.14	−0.42[a]	0.30[a]
	Heavy rain	0.04	0.24	0.14	0.22	0.13	0.08	0.13	0.28	0.05	0.05	0.16[a]
	Extreme rain	0.08	−0.28	0.09	−0.47	−0.11	0.54[a]	0.58	0.64	0.14[b]	0.65[b]	0.15[b]
Autumn	Light rain	0.29	0.60	−0.52	−0.97	−0.09	0.06	0.33	−0.77[b]	−0.67[b]	−0.26	−0.16
	Moderate rain	−0.38	0.14	0.23	0.98[b]	−0.15	−0.27	0.44	−0.12	0.48[b]	−0.06	0.05
	Heavy rain	−0.14	−0.18	0.02	0.43	0.59[b]	0.13	−1.07[a]	0.17	0.07	−0.34	0.06
	Extreme rain	−0.03	−0.06	0.05	0.22	−0.17	0.14	−0.04	0.66	0.06	0.42	0.09
Winter	Light rain	−0.36[a]	0.44	0.55	−0.03	−0.10	−0.24	−1.07	−1.88[a]	1.10[b]	−1.95[b]	−0.16
	Moderate rain	0.29_{a}	0.12	0.14[b]	0.03	0.11	−0.19	1.66[b]	0.58	0.10	0.18	0.14
	Heavy rain	0.03	0.00	0.05	−0.02	0.00	0.35[a]	0.10	1.21[b]	0.12	1.19[a]	0.20[a]
	Extreme rain	0.00	0.00	0.01	0.00	0.00	0.02	0.00	0.21	0.01	0.29[a]	0.03

[a] Significance at 95% level; [b] Significance at 90% level.

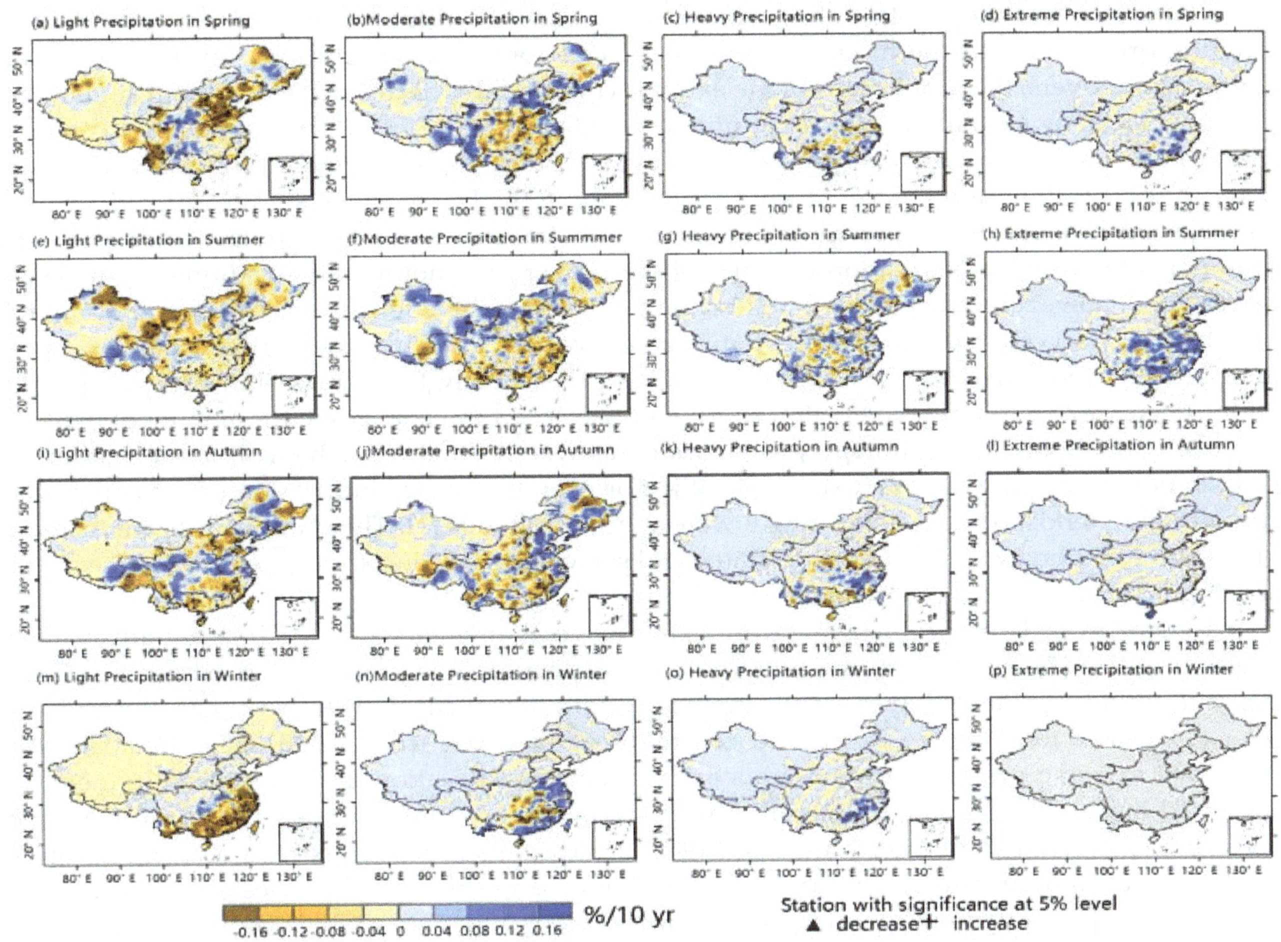

Figure 7. Seasonal trends in precipitation intensity proportions during 1960–2013. (**a–d**) Spring; (**e–h**) summer; (**i–l**) autumn; and (**m–p**) winter. For each season, the maps show light, moderate, heavy, and extreme precipitation from left to right.

4. Discussion and Conclusion

Long-term changes of precipitation are regarded as a practical subject for monitoring changes in water resources. In China, the result of our study showed that annual precipitation had increased slightly in the past half-century, which was consistent with some previous studies [7,42]. At the basin scale, significant increasing trends over the northwestern rivers was found, while annual precipitation decreased in large amplitude over the Liaohe River, Haihe River, Huaihe River, and Pearl River. Seasonally, Liu *et al.* found that precipitation had increased in winter and summer but decreased in spring and fall, over 1960–2000 [1]. Our study showed that autumn precipitation significantly decreased, while in spring, precipitation increased by 0.54 mm/10yr across China during the past half-century.

In hydrology, precipitation intensity is an important factor in rainfall-runoff processes [43]. Many studies have shown that over the past several decades, daily precipitation has become more intense in China [1,11,21,44]. In this study, changes in daily precipitation were analyzed by fixed categories. Results from trends in proportion showed the distribution of precipitation significantly shifted toward heavy and extreme precipitation over China. At the basin scale, precipitation intensity significantly increased in basins except Songhuajiang River, Liaohe River, Haihe River, and Yellow River. These increases are consistent the study of Zhang and Cong, which showed that precipitation intensity increased principally in the southern basins of China [11]. Seasonally, the shift in precipitation distribution from light to intense occurred in all seasons, significantly in spring and summer. However, this study showed, in northwestern rivers, precipitation also became more intense than before. Within the ten large basins, only the Haihe River experienced a significant decrease in extreme precipitation, while the proportions of the moderate and heavy categories increased there. In the Yellow and Songhuajiang Rivers, the proportions of the extreme category also decreased, but without significance.

Liu *et al.* showed that changes in annual precipitation could be attributed mostly to changes in the frequency and intensity of precipitation events in the top decile (90%) in China [1]. Using fixed thresholds, the results of this study showed only in the Haihe River, southeastern rivers, and southwestern rivers, extreme precipitation significantly influenced total precipitation. Furthermore, the impact of extreme precipitation was also obvious in spring, not merely in summer. The results also revealed, however, that changes in light and moderate precipitation also played an important role in the trends observed in the total precipitation in many basins, such as the northwestern rivers and Huaihe River, even in the Yangtze River and Pearl River. Seasonally, the increase in winter precipitation and the decrease in autumn precipitation also mainly derived from changes in the amount of light and moderate precipitation. In hydrology, precipitation in lower intensities also has considerable influence on basic flow, ground water, and so on [43,45]. For example, water infiltration of the soil surface during rainfall tends to be higher during lower-intensity rainfalls [1]. Therefore, significant increases in light precipitation over China during winter, and some basins during spring, could alleviate the dry conditions during the cold seasons in these regions.

China is prone to natural hazards from extreme weather events and flooding has always been a major problem [7]. Recently, numerous studies have documented precipitation extremes in China. Zhai *et al.* found significant increases in extreme precipitation events in Western China, the mid- and lower reaches of the Yangtze River, and coastal areas of China during 1951–2000 [42]. Xu *et al.* found that extreme precipitation amounts and extreme precipitation days significantly increased in the mid-and lower Yangtze River valley and Northwestern China [20].In those studies, extreme precipitation derived from percentile thresholds. As discussed above, extreme events by top 5% was just equal to moderate or light precipitation (defined in this study) in Western China. In this study, extreme precipitation is defined by fixed threshold (>50 mm/d), which could be more closely linked to floods. Results showed extreme precipitation in the Yangtze River, southeastern rivers, and southwestern rivers, increased significantly during 1960–2013, indicating a trend toward high risk of floods within these basins, as these basins are relatively developed regions of China with high population densities. What is more, the seasonal changes in extreme precipitation revealed in this study indicate that the risk of floods in some regions may be higher than suggested by previous studies; this should be highlighted in water resources management as follows. (1) Over the Yangtze River and southeastern rivers, a significant increase in extreme precipitation at the annual scale mainly derived from the summer, reflecting extreme events being concentrated in summer. (2) A significant increase in extreme precipitation was observed across the southwestern rivers and Songhuajiang River during spring, and Pearl River during winter, when vegetation cover is sparse and rainfall-runoff processes would be quick and heavy. (3) In the Haihe River, extreme precipitation during spring also significantly increased, although the yearly extreme precipitation showed a marked decrease. (4) However, in northwestern rivers, the increase in precipitation intensity was not related to an increase in the risk of floods over the past half-century, because it mainly resulted from an increase in moderate precipitation.

The amount and distribution of precipitation also play crucial roles in the occurrence of drought. As light precipitation generally occupied a fairly large proportion in rainfall events, a significant decrease in light precipitation over Yangtze, Huaihe, and Pearl Rivers might indicate an increase in dry-spells. Seasonally, autumn precipitation significantly decreased in the Songhuajiang and Yangtze Rivers, which might enhance the possibility of seasonal drought in these regions.

Acknowledgments: This work was supported by the Key Programs of the Chinese Academy of Sciences (KFZD-SW-301-5), the Important Science & Technology Specific Projects of Qinghai Province (2014-NK-A4-1).

Author Contributions: Aifeng Lv conceived the study. Bo Qu performed the data analysis and wrote the paper. Aifeng Lv, Shaofeng Jia and Wenbin Zhu read and edited the manuscript. All authors reviewed and approved the manuscript.

References

1. Liu, B.; Xu, M.; Henderson, M.; Qi, Y. Observed trends of precipitation amount, frequency, and intensity in China, 1960–2000. *J. Geophys. Res. Atmos.* **2005**, *110*. [CrossRef]
2. Rodrigo, F.S.; Trigo, R.M. Trends in daily rainfall in the Iberian Peninsula from 1951 to 2002. *Int. J. Climatol.* **2007**, *27*, 513–529. [CrossRef]
3. Field, C.B.; Van Aalst, M. Freshwater resource. In *Climate Change 2014: Impacts, Adaptation, and Vulnerability. Part A: Global and Sectoral Aspects*; Cambridge University Press: Cambridge, UK; New York, NY, USA, 2014; Volume 1.
4. Hulme, M.; Osborn, T.J.; Johns, T.C. Precipitation sensitivity to global warming: Comparison of observations with HadCM2 simulations. *Geophys Res. Lett.* **1998**, *25*, 3379–3382. [CrossRef]
5. Karl, T.R.; Knight, R.W. Secular Trends of Precipitation Amount, Frequency, and Intensity in the United States. *Bull. Am. Meteorol. Soc.* **1998**, *79*, 231–241. [CrossRef]
6. Brunetti, M.; Maugeri, M.; Monti, F.; Nanni, T. Changes in daily precipitation frequency and distribution in Italy over the last 120 years. *J. Geophys. Res. Atmos.* **2004**, *109*. [CrossRef]
7. Wu, Y.; Wu, S.Y.; Wen, J.; Xu, M.; Tan, J. Changing characteristics of precipitation in China during 1960–2012. *Int. J. Climatol.* **2015**, *36*, 1387–1402. [CrossRef]
8. Karl, T.R.; Knight, R.W.; Plummer, N. Trends in high-frequency climate variability in the twentieth century. *Nature* **1995**, *377*, 217–220. [CrossRef]
9. Zinke, P.J. Forest interception studies in the United States. In *International Symposium on Forest Hydrology*; Pergamon Press: New York, NY, USA, 1967; pp. 137–161.
10. Li, W.; Zhai, P.; Cai, J. Research on the Relationship of ENSO and the Frequency of Extreme Precipitation Events in China. *Adv. Clim. Chang. Res.* **2011**, *2*, 101–107. [CrossRef]
11. Zhang, X.; Cong, Z. Trends of precipitation intensity and frequency in hydrological regions of China from 1956 to 2005. *Glob. Planet Chang.* **2014**, *117*, 40–51. [CrossRef]
12. Chen, Y.; Chen, X.; Ren, G. Variation of extreme precipitation over large river basins in China. *Adv. Clim. Chang. Res.* **2011**, *2*, 108–114. [CrossRef]
13. Su, B.D.; Jiang, T.; Jin, W.B. Recent trends in observed temperature and precipitation extremes in the Yangtze River basin, China. *Theor. Appl. Climatol.* **2006**, *83*, 139–151. [CrossRef]
14. Zhang, Q.; Xu, C.; Gemmer, M.; Chen, Y.D.; Liu, C. Changing properties of precipitation concentration in the Pearl River basin, China. *Stoch. Environ. Res. Risk Assess.* **2009**, *23*, 377–385. [CrossRef]
15. Yang, T.; Shao, Q.; Hao, Z.; Chen, X.; Zhang, Z.; Xu, C.; Sun, L. Regional frequency analysis and spatio-temporal pattern characterization of rainfall extremes in the Pearl River Basin, China. *J. Hydrol.* **2010**, *380*, 386–405. [CrossRef]
16. Shi, W.; Yu, X.; Liao, W.; Wang, Y.; Jia, B. Spatial and temporal variability of daily precipitation concentration in the Lancang River basin, China. *J. Hydrol.* **2013**, *495*, 197–207. [CrossRef]
17. Hu, Y.; Maskey, S.; Uhlenbrook, S. Trends in temperature and rainfall extremes in the Yellow River source region, China. *Clim. Chang.* **2012**, *110*, 403–429. [CrossRef]
18. Vincent, L.A.; Zhang, X.; Bonsal, B.R.; Hogg, W.D. Homogenization of daily temperatures over Canada. *J. Clim.* **2002**, *15*, 1322–1334. [CrossRef]
19. Liu, W.; Fu, G.; Liu, C.; Charles, S.P. A comparison of three multi-site statistical downscaling models for daily rainfall in the North China Plain. *Theor. Appl. Climatol.* **2013**, *111*, 585–600. [CrossRef]
20. Xu, X.; Du, Y.; Tang, J.; Wang, Y. Variations of temperature and precipitation extremes in recent two decades over China. *Atmos. Res.* **2011**, *101*, 143–154. [CrossRef]
21. Gong, D.; Shi, P.; Wang, J. Daily precipitation changes in the semi-arid region over northern China. *J. Arid Environ.* **2004**, *59*, 771–784. [CrossRef]
22. Feng, S.; Hu, Q.; Qian, W. Quality control of daily meteorological data in China, 1951–2000: a new dataset. *Int. J. Climatol.* **2004**, *24*, 853–870. [CrossRef]
23. Mao-chang, C.; Hai, Z.; Shu-min, L.; Arpe, K.; Dumenil, L. Variability of daily precipitation in China (1980–1993): PCA and wavelet analysis of observation and ECMWF reanalysis data. *Chin. J. Oceanol. Limnol.* **2000**, *18*, 117–125. [CrossRef]
24. Svensson, C. Empirical orthogonal function analysis of daily rainfall in the upper reaches of the Huai River basin, China. *Theor. Appl. Climatol.* **1999**, *62*, 147–161. [CrossRef]

25. Qian, W.; Lin, X. Regional trends in recent precipitation indices in China. *Meteorol. Atmos. Phys.* **2005**, *90*, 193–207. [CrossRef]
26. Easterling, D.R.; Evans, J.L.; Groisman, P.Y.; Karl, T.R.; Kunkel, K.E.; Ambenje, P. Observed variability and trends in extreme climate events: A brief review. *Bull. Am. Meteorol. Soc.* **2000**, *81*, 417–425. [CrossRef]
27. Vicente Serrano, S.M.; Beguería, S.; López Moreno, J.I.; García Vera, M.A.; Stepanek, P. A complete daily precipitation database for northeast Spain: Reconstruction, quality control, and homogeneity. *Int. J. Climatol.* **2010**, *30*, 1146–1163. [CrossRef]
28. Brunetti, M.; Maugeri, M.; Monti, F.; Nanni, T. Temperature and precipitation variability in Italy in the last two centuries from homogenised instrumental time series. *Int. J. Climatol.* **2006**, *26*, 345–381. [CrossRef]
29. Wijngaard, J.B.; Klein Tank, A.; Können, G.P. Homogeneity of 20th century European daily temperature and precipitation series. *Int. J. Climatol.* **2003**, *23*, 679–692. [CrossRef]
30. Peterson, T.C.; Easterling, D.R.; Karl, T.R.; Groisman, P.; Nicholls, N.; Plummer, N.; Torok, S.; Auer, I.; Boehm, R.; Gullett, D. Homogeneity adjustments of *in situ* atmospheric climate data: A review. *Int. J. Climatol.* **1998**, *18*, 1493–1517. [CrossRef]
31. Vincent, L.A. A technique for the identification of inhomogeneities in Canadian temperature series. *J. Clim.* **1998**, *11*, 1094–1104. [CrossRef]
32. Alexandersson, H.; Moberg, A. Homogenization of Swedish temperature data. Part I: Homogeneity test for linear trends. *Int. J. Climatol.* **1997**, *17*, 25–34. [CrossRef]
33. Szentimrey, T. Statistical procedure for joint homogenization of climatic time series. In Proceedings of the 1st Seminar of Homogenisation of Surface Climatological Data, Budapest, Hungary, 6–12 October 1996; pp. 47–62.
34. Potter, K.W. Illustration of a new test for detecting a shift in mean in precipitation series. *Mon. Weather Rev.* **1981**, *109*, 2040–2045. [CrossRef]
35. Moberg, A.; Bergström, H. Homogenization of Swedish temperature data. Part III: The long temperature records from Uppsala and Stockholm. *Int. J. Climatol.* **1997**, *17*, 667–699. [CrossRef]
36. Vincent, L.A.; Mekis, E. Changes in daily and extreme temperature and precipitation indices for Canada over the twentieth century. *Atmosphere-Ocean* **2006**, *44*, 177–193. [CrossRef]
37. Kumar, S.; Merwade, V.; Kam, J.; Thurner, K. Streamflow trends in Indiana: Effects of long term persistence, precipitation and subsurface drains. *J. Hydrol.* **2009**, *374*, 171–183. [CrossRef]
38. Yue, S.; Pilon, P.; Phinney, B.; Cavadias, G. The influence of autocorrelation on the ability to detect trend in hydrological series. *Hydrol. Process.* **2002**, *16*, 1807–1829. [CrossRef]
39. Khattak, M.S.; Babel, M.S.; Sharif, M. Hydro-meteorological trends in the upper Indus River basin in Pakistan. *Clim. Res.* **2011**, *46*, 103–119. [CrossRef]
40. Sen, P.K. Estimates of the regression coefficient based on Kendall's tau. *J. Am. Stat. Assoc.* **1968**, *63*, 1379–1389. [CrossRef]
41. Bargaoui, Z.K.; Chebbi, A. Comparison of two kriging interpolation methods applied to spatiotemporal rainfall. *J. Hydrol.* **2009**, *365*, 56–73. [CrossRef]
42. Zhai, P.; Zhang, X.; Wan, H.; Pan, X. Trends in total precipitation and frequency of daily precipitation extremes over China. *J. Clim.* **2005**, *18*, 1096–1108. [CrossRef]
43. Wang, H.; Gao, J.E.; Zhang, M.; Li, X.; Zhang, S.; Jia, L. Effects of rainfall intensity on groundwater recharge based on simulated rainfall experiments and a groundwater flow model. *Catena* **2015**, *127*, 80–91. [CrossRef]
44. Wang, Y.; Zhou, L. Observed trends in extreme precipitation events in China during 1961–2001 and the associated changes in large-scale circulation. *Geophys. Res. Lett.* **2005**, *32*. [CrossRef]
45. Liu, H.; Lei, T.W.; Zhao, J.; Yuan, C.P.; Fan, Y.T.; Qu, L.Q. Effects of rainfall intensity and antecedent soil water content on soil infiltrability under rainfall conditions using the run off-on-out method. *J. Hydrol.* **2011**, *396*, 24–32. [CrossRef]

9

Spatial Patterns and Influence Factors of Conversion Coefficients between Two Typical Pan Evaporimeters in China

Yanzhong Li [1,2], Changming Liu [1] and Kang Liang [1,*]

[1] Key Laboratory of Water Cycle and Related Land Surface Processes, Institute of Geographic Sciences and Natural Resources Research, Chinese Academy of Sciences, Beijing 100101, China; liyz_egi@163.com (Y.L.); liucm@igsnrr.ac.cn (C.L.)
[2] University of Chinese Academy of Sciences, Beijing 100101, China
* Correspondence: liangk@igsnrr.ac.cn

Academic Editors: Tommaso Moramarco and Roberto Ranzi

Abstract: Pan measurement is a reliable and efficient method for indicating the evaporative demand of the atmosphere. There are several types of pan evaporimeters worldwide, and the estimation of the conversion coefficients (K_p) between them is necessary in hydrologic research. In China, E601B pans were installed at all meteorological stations beginning in 1998. They replaced the 20 cm pans (ϕ20). To fully use the records from the two pans and obtain long-term pan evaporation, the spatial patterns of K_p between ϕ20 and E601B and the factors that influence K_p are investigated based on records from 573 national meteorological stations from 1998 to 2001. In this study, The results show that higher K_p values are found in southwestern regions and lower values are found in northeastern regions during the warm seasons (from May to September), while K_p values are lower during warm seasons than during cold seasons (from October to April the following year). In addition, net radiation was found to be the dominant climate factor that affects variations in K_p, followed by relative humidity and the vapor pressure deficit. This study can improve the benefit of not only the selection of appropriate evaporimeters by meteorological departments, but also of the study of temporal variability and trends in the evaporative demand.

Keywords: conversion coefficients; pan evaporation; eight climate regions in China; 20 cm diameter pan; E601B (E601); evaporative demand

---

1. Introduction

Evaporation is a key hydrological process [1]. It is an important nexus between the water cycle and energy budget and can further impact regional and global climate [2–4]. The atmospheric evaporative demand can be evaluated based on potential evapotranspiration [5,6] or evaporimeters [7,8] and is regarded as the upper limit of evaporation [9–11]. The evaporative demand of the atmosphere is controlled by radiative and aerodynamic factors, and it can be calculated using meteorological variables, such as radiation, wind speed, air temperature, and humidity [7,10]. However, these meteorological factors are not always available, especially radiation. Instead, a simple and efficient observational measurement device, the pan evaporimeter, is often adopted to accurately quantify local atmospheric evaporation [8,12–14]. Many types of pan evaporimeters have been installed in different countries to measure the evaporative demand [1,15], including Class A evaporimeters in the US and Australia [16], the GGI-3000 in Russia [17], the MO tank in Britain [18], and the 20 cm pan (ϕ20) and

E601B in China [13,19]. These devices are used by meteorological and climate scientists, agricultural scientists, and hydrologists [12,20,21]. Although pan evaporation (E_{pan}) measured from various types of evaporimeters cannot completely represent the actual evaporation [22,23], it can provide in-depth insight into the trends and temporal variations in evaporation in the context of climate change [6,24] and anthropogenic interference [25,26].

Due differences between pan evaporimeters (for example, material type, geometric shape, and installation method), observed evaporation can vary greatly, even for evaporimeters in similar environments [27]. Pan evaporation values measured using different evaporimeters over the same time span, cannot be compared directly. Otherwise, considerable uncertainty would be introduced into the results. A conversion must be performed to make the values comparable. It is necessary to determine the conversion coefficient (K_p) between different measurements before using the values in studies of evaporation trends [28,29] or water resources [8]. A number of previous studies discussed K_p related to pan evaporation at the point scale [15,30], regional scale [31,32], and catchment scale [33]. For example, Hong et al. [30] investigated two types of evaporimeters, one buried in the soil and one exposed to the air, at Nansi Lake Station. The results showed that the buried devices, such as E601 and GGI-3000, had larger K_p values (ratio of the 20 m^2 evaporation tank to these evaporimeters, $K_p > 0.98$) compared to the K_p values ($K_p < 0.88$) of an exposed ones, such as Class A and ϕ20. Fu et al. [15] compared the K_p values of 15 types of evaporimeters to that of a 20 m^2 evaporation tank and found that the K_p values of ϕ20 ($K_p = 0.60$) and E601B ($K_p = 1.07$) were distinctly different at the annual scale. Liang et al. [32] discussed the K_p difference between E_{pan} and reference evaporation (ET_{ref}) in the West Songnen Plain of China and found that it varied significantly in space (0.48–0.68) and time at an annual scale. Xu et al. [33] also studied the K_p difference between E_{pan} and ET_{ref} in the Yangtze River basin. They observed higher K_p values in the central region of the basin, which has a relatively lower vapor pressure deficit, and K_p exhibited monthly variations in the three regions of the catchment. Due to the spatial and temporal variabilities in K_p, considerable uncertainties may be associated with using constant K_p values in large regions when conducting climate change research [28,34]. The spatial patterns of K_p and the dominant factors that control its variation must be determined. Revealing the spatial distribution of K_p and its driving factors is imperative and can improve hydrometeteorological studies.

China is an ideal location to study K_p patterns and the factors that influence K_p. Among the various evaporation pans used in China [15,19,35,36], the two most common are the ϕ20 and E601B pans (Figure 1 and Table 1). In 1998, the E601B pans were first installed at all meteorological stations across China, with the aim of replacing the ϕ20 pan. To maintain comparability between the two evaporimeters, simultaneous observations were collected from 1998 to 2001. By 2002, the ϕ20 pans were successfully replaced by E601B pans at all stations, and E601B became the standard evaporimeter for measuring evaporation in China. To effectively use the records from the two pan evaporation devices in long-term studies of evaporation trends (the ϕ20 pan records date to approximately 1951), the K_p value between them must be determined. In addition, China can be divided into eight climatic regions [13,37] due its large spatial extent. There are significant differences in climate between the regions [38]; for example, the temperature increases from north to south, and precipitation increases from northwest to southeast (Table 1). Ren et al. [35] compared the monthly and annual mean K_p values between the two evaporimeters in different provinces from 1998 to 2001. Additionally, Liu et al. [36] compared the K_p values between E601 (similar to E601B) and ϕ20 in several typical cities from 1986 to 1995. However, few studies have focused on K_p differences in different climatic conditions, which may considerably affect K_p. Mapping the spatial distribution of K_p between the two evaporimeters and determining the driving factors of K_p can allow the pan evaporation records in China to be fully used and provide an understanding of the underlying mechanisms of K_p variability.

Figure 1. Four types of evaporation evaporimeters: Class A from the USA (**a**); GGI-3000 from Russia (**b**); and ϕ20 (**c**) and E601B (**d**) from China.

Table 1. Detailed information regarding the four types of evaporimeters.

Evaporimeter Name	Size	Description
Class-A	Area: 12.56 ft^2 (diameter: 4 ft) Depth: 10 in	Supported by a wood frame and the bottom is 5 cm higher than ground, popular in USA.
GGI-3000	Area: 3000 cm^2 Depth: 60 cm (cylinder) C + 8.7 cm (circular cone)	Buried in the ground with rim about 7.5 cm above the ground level, popular in Russian.
ϕ20	Area: 314 cm^2 (diameter: 20 cm) Depth: 10 cm	Rim is 70 cm from the ground, in every meteorological station in China.
E601B	Same as GGI-3000	Buried like GGI-3000, fiberglass material, surrounded by water and soil circle. Installed in every meteorological station from 1998.

To address these issues, we compare the spatial distribution and temporal variability (monthly for one year) of K_p and investigate the factors that influence K_p between the E601B and ϕ20 pans in China. The main objectives of this paper are as follows: (1) investigate the monthly variation and spatial distribution of K_p for the two pan evaporations during the warm (May–September) and cold seasons (October–April); and (2) determine the contributions of several key climatic factors to the variation in K_p. The results of this study can improve the selection of the appropriate pan in different climate regions and provide information for hydrologic research, especially studies of evaporation trends. This study is structured as follows: in Section 2, the datasets and methods used in our study are described; in Section 3, the spatial distribution of K_p is mapped during the warm and cold seasons. Furthermore, the factors that influence K_p are also investigated in this section. The uncertainties and conclusions are shown in Sections 4 and 5.

This paper reports the initial stage of ongoing research work. The ongoing research and planned studies are as follows: (1) constructing a long-term series of pan evaporation records (from the

1950s–present) and investigating the trends in pan evaporation across the country in the context of climate change [5,13]; (2) performing experiments regarding the pan evaporation of ϕ20 and E601B and developing some novel approaches or formulations to explain the mechanisms of K_p [20,39]; and (3) investigating and quantifying the effects of various climatic factors on K_p using a modified PenPan model [27] and a partial differential method [6].

2. Data and Methods

2.1. Information and Measurements from Different Evaporimeters

Different types of evaporimeters have been used to measure the evaporative demand of the atmosphere [15]. Two typical pans are recommended by the World Meteorological Organization [40]: Class A from the USA (Figure 1a) and GGI-3000 from Russia (Figure 1b). However, these two evaporimeters are not widely installed in China, but the ϕ20 and E601B (Figure 1c,d) evaporimeters are commonly used. Additional information regarding the evaporimeters is listed in Table 1.

The ϕ20 pan with a screen to prevent bird drinking is made of metal and placed at a height of 70 cm (Figure 1c). This evaporimeter is weighed at 20:00 each day using a high-precision weighbridge, and it is then refilled with water to a depth of 20 mm. The daily evaporation rate can be calculated from the following equation:

$$E = \frac{W_1 - W_2}{31.4} + P \tag{1}$$

where E is the pan evaporation rate (mm·day^{-1}); W_1 and W_2 are the pan weights of the previous and current measurements, respectively (g·day^{-1}); P is the total precipitation (mm·day^{-1}, including rain and snow); and 31.4 is the weight of 1 mm of water in the pan (g·mm^{-1}). The E601B is made of fiberglass and has a relatively lower heat transfer to the surrounding area. The evaporation from this evaporimeter should be relatively close to the evaporation recorded from a moderately-sized water body (such as a 20 m^2 evaporation tank) [15,19]. The daily evaporation rate can be calculated using Equation (2):

$$E = P + (H_1 - H_2) \tag{2}$$

where E and P are the same as in Equation (1); and H_1 and H_2 are the water depths of E601B on the previous and current day, respectively. The water depths of H can be read directly from the indicator installed on the stylus holder (Figure 1d).

2.2. Meteorological and Evaporation Data

Daily meteorological data and evaporation data from the evaporimeters at 573 national meteorological stations were obtained from the China Meteorological Administration (CMA, Figure 2) for a four-year period (January 1998–December 2001). The data were quality controlled. Records that were missing for less than three consecutive days were interpolated based on the nearest data. For gaps of more than three days, the missing data were replaced using a simple linear regression based on the nearest stations. Finally, 573 of the 756 meteorological stations with continuous records were selected. Monthly data, which were summed from daily values, were used in the following sections.

To identify the effects of different climatic factors on the K_p values of the two pans, the entire region was divided into eight climatic regions [13,37]: northwest (NW), north center (NC), North China plain (NCP), northeast (NE), east (E), southeast (SE), southwest (SW), and Tibet plateau (TP). According to the aridity index (AI, Table 2), NW, NC, NCP, and NE are humid regions (AI > 1.0), while E, SE, and SW are non-humid regions (AI < 1.0). The climate characteristics of each region are listed in Table 2. Pan evaporation is an integrated process affected by various climate factors, such as the net radiation (R_n), wind speed (U_2), air temperature (T_{mean}), vapor pressure deficit (VPD), relative humidity (RH), and elevation (Elev) [19,21,41]. The U_2 at a height of 2 m was derived from a height of 10 m according to a logarithmic wind speed profile. R_n was calculated from the difference between the incoming net shortwave radiation (R_{ns}) and the outgoing net longwave radiation (R_{nl}) [21].

The coefficients, a_s and b_s, recommended by Allen et al. [21] to calculate the solar radiation (R_s) were not used. The optimized coefficients calibrated using the 116 solar radiation stations were adopted in our paper. These coefficients can significantly reduce the uncertainty associated with radiation [42].

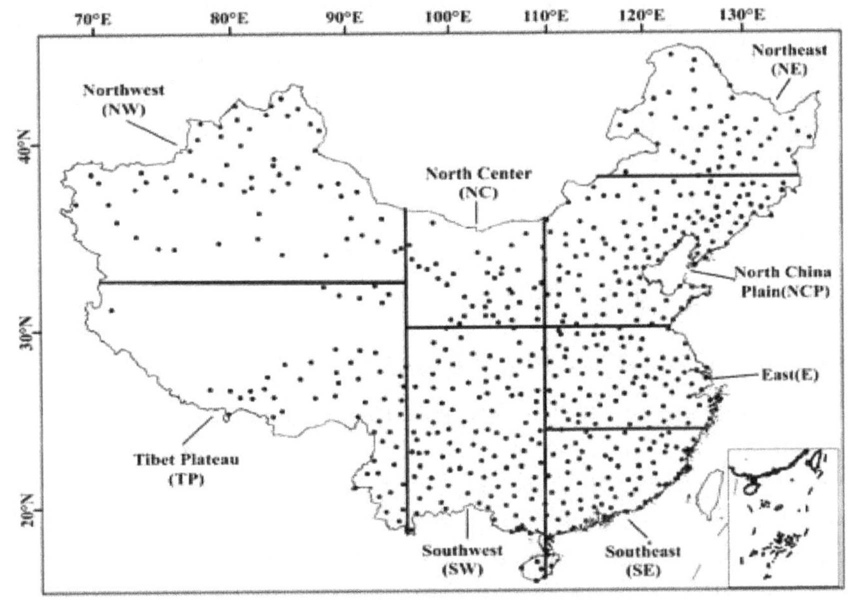

Figure 2. Locations of meteorological stations in China. The eight regional abbreviations are shown in the figure. This map was modified from Liu et al. [13].

Table 2. Climate factor characteristics in the eight climatic regions from 1998 to 2001.

Region	VP (kPa)	T (°C)	U_2 (m/s)	R_s (MJ/m^2/Day)	P_{re} (mm/a)	ET_{ref} (mm/a)	AI
NW (n = 57)	0.76	8.45	1.65	15.60	137.03	1078.22	7.87
NC (n = 39)	0.82	9.12	1.72	15.99	265.80	1088.08	4.09
NCP (n = 119)	1.04	9.12	1.91	13.95	494.26	938.91	1.90
NE (n = 48)	0.80	2.52	2.14	13.45	442.05	779.81	1.76
E (n = 83)	1.71	16.43	1.62	11.96	1164.64	902.78	0.78
SE (n = 67)	2.17	20.35	1.30	11.85	1728.42	948.11	0.55
SW (n = 103)	1.70	16.27	1.16	12.02	1179.91	886.25	0.75
TP (n = 57)	1.32	11.78	0.95	15.34	1035.75	941.24	0.91

Notes: VP is the vapor pressure; T is the air mean temperature; and U_2 is the wind speed; All of these parameters are taken at a height of 2 m above the ground. R_s is the solar radiation; P_{re} is the precipitation; ET_{ref} is the reference evapotranspiration calculated using the Penman-Monteith method [21,43]; and AI is the aridity index derived from the ratio of ET_{ref} to P_{re}.

Evaporation data for the E601B pan are not available during the winter in most northern regions in China because the water in the pan freezes. Thus, the year was divided into a warm season (May–September) and a cold season (October–April). This is reasonable because the warm season accounted for most of the annual evaporation (>60%), even in the southern areas; therefore, these months were the most important.

2.3. Calculations and Analysis Method for the K_p

K_p is defined as the ratio of the E601B evaporation to the ϕ20 pan evaporation:

$$K_p = \frac{E_{E601B}}{E_{20}} \quad (3)$$

where E_{E601B} is the monthly evaporation rate of the E601B pan; and E_{20} is the ϕ20 evaporation rate for the same month. Spatial interpolation using the Kriging method was performed in ESRI ArcGIS 10.0

software (Redlands, CA, USA) with the spatial analysis toolbox [6,44] to obtain the spatial distributions of monthly and annual K_p values.

A Pearson's correlation analysis, multiple stepwise regression analysis, and our knowledge of the potential physical driving factors (T_{mean}, RH, VPD, U_2, Elev, and R_n) were used to analyze the potential factors that influence K_p variation. The Pearson's correlation coefficients between K_p and selected climatic factors were calculated using SPSS Statistics 20 (SPSS Inc., Chicago, IL, USA). Stepwise multiple linear regressions were developed between K_p and the potential dominant factors to find the best predictors and the independent explanatory ability of each selected climatic factor based on the spatial variation in K_p.

3. Results and Analysis

3.1. Spatial Patterns of K_p for the Two Pan Evaporimeters

3.1.1. Spatial Distribution of Pan Evaporation during the Warm Season

The spatial distributions of ϕ20 and E601B evaporation during the warm season from 1998 to 2001 at the national scale are shown in Figure 3a,b, respectively. When the characteristics of the eight climatic regions were combined (Table 2), the differences between the two pan evaporations were spatially related to the variations in meteorological factors. For the warm season from 1998 to 2001, there were statistically significant differences ($p < 0.001$) between the ϕ20 and E601B pan evaporations, and the annual values were 1014 mm and 597 mm, respectively (Table 3). The spatial distributions of the pan evaporations of ϕ20 and E601B generally exhibited the same spatial pattern: a clear decreasing trend from northwest to southeast. The highest values were found in the NW region, while the lowest values were found in the TP, NW and NE regions. The ϕ20 pan evaporation varied from 739 mm to 5573 mm with an average of 1613 mm, and the E601B pan evaporation varied from 470 mm to 2831 mm with an average of 921 mm.

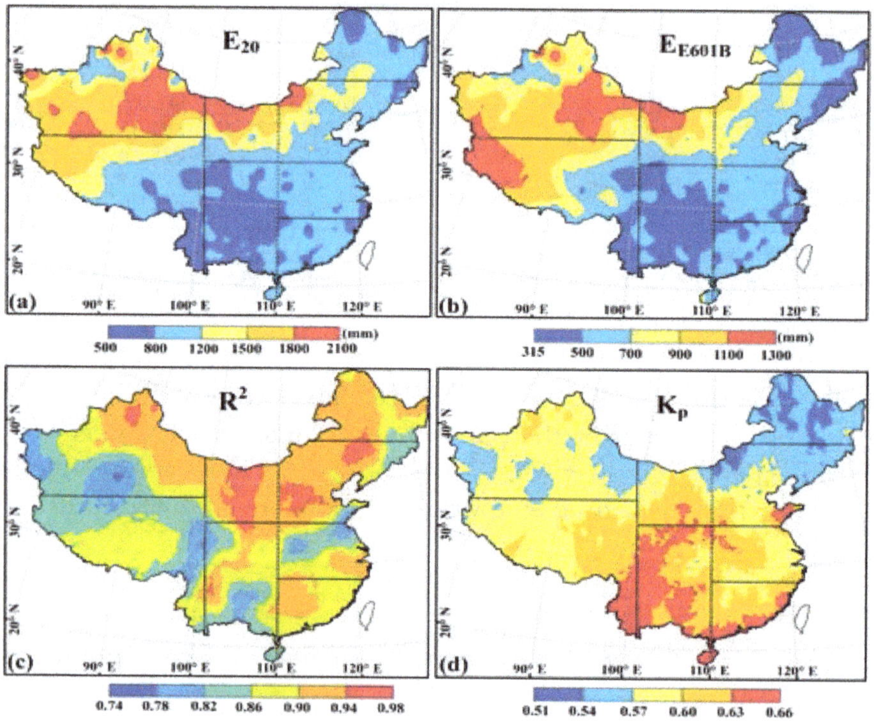

Figure 3. Spatial distributions of ϕ20 (**a**) and E601B (**b**) evaporation; the determination coefficient R^2 (**c**); and the conversion coefficient K_p (**d**) between the two pan evaporations during the warm season (May–September).

Table 3. Statistical information regarding E601B and ϕ20 pan evaporation (E_{E601B}, E_{20}), the determination coefficient (R^2), and the conversion coefficient (K_p) between the two pans in eight climatic regions during the warm season and three regions in the cold season. The minimum and maximum values are given in parentheses.

Seasons	Regions	E_{20}	E_{E601B}	R^2	K_p
Warm	NW	1613 (739, 5573)	921 (470, 2831)	0.89 (0.67, 0.97)	0.58 (0.50, 0.67)
	NC	1363 (832, 2382)	814 (477, 1436)	0.94 (0.86, 0.99)	0.61 (0.51, 0.70)
	NCP	1123 (648, 2166)	635 (386, 1107)	0.91 (0.52, 0.98)	0.57 (0.47, 0.92)
	NE	1004 (641, 1425)	543 (329, 803)	0.91 (0.68, 0.98)	0.55 (0.46, 0.62)
	E	872 (645, 1236)	524 (401, 792)	0.87 (0.67, 0.97)	0.60 (0.51, 0.75)
	SE	847 (468, 1033)	526 (304, 740)	0.89 (0.74, 0.97)	0.62 (0.53, 0.77)
	SW	752 (498, 1159)	472 (323, 708)	0.87 (0.57, 0.98)	0.63 (0.49, 0.89)
	TP	890 (521, 1766)	620 (328, 1288)	0.84 (0.48, 0.97)	0.62 (0.51, 0.74)
	National	1014 (468, 5573)	597 (304, 2831)	0.89 (0.48, 0.99)	0.60 (0.46, 0.92)
Cold	E	543 (402, 681)	345 (278, 419)	0.93 (0.83, 0.98)	0.66 (0.55, 0.80)
	SE	666 (403, 1224)	444 (250, 819)	0.90 (0.74, 0.97)	0.68 (0.58, 0.80)
	SW	692 (312, 1466)	440 (218, 889)	0.92 (0.76, 0.99)	0.67 (0.49, 0.80)
	Humid	664 (312, 1466)	427 (218, 921)	0.92 (0.74, 0.99)	0.66 (0.49, 0.81)

3.1.2. Spatial Distribution of the Correlation Coefficient during the Warm Season

To quantify the performance of the E601B pan in capturing the spatial distribution and temporal variation of the ϕ20 pan evaporation, the temporal determination coefficients (R^2) of the two pan evaporation series during the warm season were calculated at each station, and the spatial distributions of R^2 were mapped using the Kriging interpolation method [44]. The results showed that the two pan evaporations had high R^2 values in each region (Figure 3c) and varied significantly in one month (Figure 4). The mean value of R^2 in China can reach 0.89, indicating that the E601B values capture the variation in ϕ20 fairly well. This is not surprising because the two pans were influenced by the same meteorological conditions, such as solar radiation, wind speed, temperature, etc. Thus, it may be reasonable to convert ϕ20 evaporation values to E601B evaporation values by multiplying a constant.

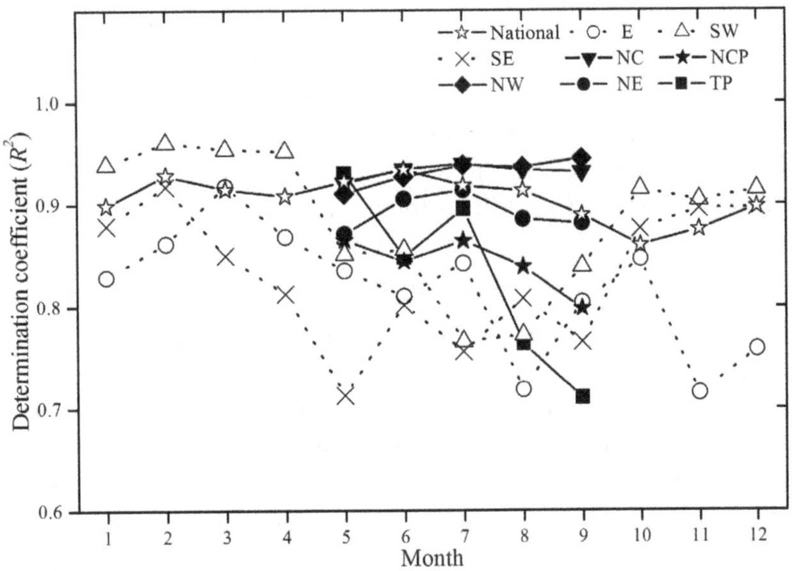

Figure 4. Monthly variation in the coefficient of determination (R^2) values of E601B and ϕ20 pan evaporation in the eight climate regions. The R^2 values in the three humid regions (E, SE, and SW) are illustrated by dotted lines, and the R^2 values in the four non-humid regions and TP region are illustrated by solid lines during the warm season (May–September). The national R^2 is illustrated by the solid black line with open star symbols.

Despite the overall high R^2 between ϕ20 and E601B in China, large differences were also found in the eight climatic regions. The R^2 values in the northern regions were generally higher than those in the southern regions, which indicated that the variabilities in the two evaporimeters became more uniform as evaporation increased. The average values of R^2 ranged from 0.89 to 0.94 in the northern regions. R^2 was highest in the NC region, and it varied between 0.86 and 0.99, with an average of approximately 0.94 (Table 3). By contrast, the average values of R^2 in the southern regions (E, SE, and SW) ranged from 0.87 to 0.89 with smaller deviations. The monthly R^2 values in the southern regions were high in the cold season and low in the warm season. The R^2 values in Northern China were high and stable in the warm season. The lowest average R^2 value (0.84) was found in the TP region, and it exhibited a decreasing trend from May to September (Figure 4). This R^2 value in TP may be caused by special climatic conditions. For example, the TP region is the highest plateau in the world with an average elevation of 4000 m, and it is also known as the "Third Pole" of the earth [45].

3.1.3. Spatial Distribution of K_p during the Warm Season

The spatial distribution of K_p during the warm season is shown in Figure 3d. K_p was calculated at each meteorological station independently, and the spatial distribution was obtained using the Kriging interpolation method [44]. The results show that the two pan evaporations exhibited the same spatial distribution (Figure 3a,b), while K_p exhibited significant spatial differences. Overall, K_p varied from 0.46 to 0.92 in China, with an average of approximately 0.60 and a standard deviation of 0.056 (Table 3). K_p was lower in the non-humid northern regions ($K_p < 0.60$) than in the humid southern regions ($K_p > 0.60$), which indicated that the bias in the evaporations of the two pan evaporations was smaller in the humid area during the warm season. Thus, the pan K_p between the two evaporimeters varied substantially, and this variation was larger at low latitudes and smaller at high latitudes. Researchers have documented that the additional heat absorbed by the pan wall has an important effect on K_p [46,47]. Additionally, the ϕ20 device can intercept more solar radiation at high latitudes (e.g., NE and NW) than at low latitudes (e.g., SW and SE) due to the solar zenith angle difference. The extra absorbed heat is subsequently transferred into the water through the pan wall, which increases the evaporation rate of the ϕ20 pan. The K_p pattern (Figure 3d) generally reflected this process. Additionally, pronounced differences between K_p values in the eight climatic regions were also detected. The smallest K_p values were found in the NE region, and the average K_p ranged from 0.46 to 0.62 with an average of 0.55. The largest values were observed in the SW and TP regions, both with averages of 0.63. The spatial distribution of K_p was similar to that noted by Chen, Gao [19], who compared ϕ20 evaporation to reference evapotranspiration in China. Therefore, the regional differences in K_p must be considered to obtain accurate evaporation data series when the evaporation values from the two pans are adopted to determine evaporation trends.

3.1.4. Monthly Variation in K_p during the Warm Season

In addition to the spatial differences, the K_p values of the two evaporimeters also exhibited significant temporal variability. The variation in K_p in eight regions during the warm season is shown in Figure 5. The results are as follows: (1) overall, the monthly average values of K_p in China increased during the warm season from approximately 0.58 in May to 0.63 in September, and the increase was especially rapid from June to August; (2) except in humid regions, K_p varied throughout the warm season with a similar increasing pattern (solid line); (3) in the humid regions, K_p decreased from May to June and then increased until August or September (dotted line), which confirmed the finding of Allen et al. [21] that K_p was high in humid environments; and (4) the lowest values of the entire warm season ($K_p = 0.54$) occurred in the NE region with mean monthly variation from 0.51 to 0.58. The largest values varied from the SW region in May to the TP region in June and July and to SW region again in August and September.

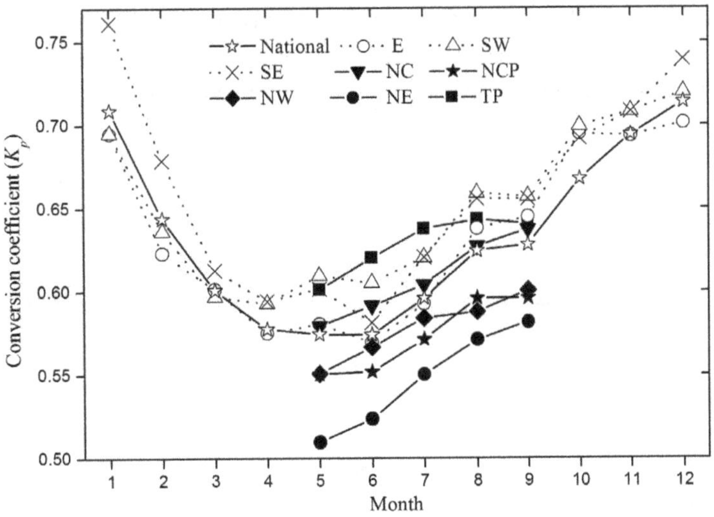

Figure 5. Monthly variation in the conversion coefficient (K_p) of E601B and ϕ20 pan evaporation in the eight climate regions. The K_p values in the three humid regions (E, SE, and SW) are illustrated by dotted lines, and the K_p values in the four non-humid regions and TP region are illustrated by solid lines during the warm season (May–September). The national K_p is illustrated by the solid black line with open star symbols.

3.1.5. Spatial Pattern and Temporal Variability of K_p in Humid Regions during the Cold Season

In humid regions (E, SE, and SW), the annual precipitation is larger than 1000 mm and the aridity index is less than 1.0 (Table 2). Due to the relatively high minimum temperature, there are no frozen periods throughout the whole year. Therefore, the evaporation records of E601B are still available during the cold season (October–April). The spatial pattern of evaporation (E_{20}, E_{E601B}) and the R^2 and K_p between the two pans during the cold seasons are shown in Figure 6. As was the case in the warm season (Figure 3), the evaporation from the two pans exhibited similar spatial patterns during the cold season in humid areas, increasing from the central to marginal regions (Figure 6a,b) and averaging 664 mm and 427 mm annually for the ϕ20 and E601B pans, respectively (Table 3). The R^2 values also increased from the central to marginal regions and were relatively high, with an average of 0.92. This trend indicated that the E601B evaporimeter could capture the variation in ϕ20 better during the cold season than during the warm season.

Although the two pan evaporations had similar spatial patterns, the spatial variations in K_p during the cold season were not constant. The highest values occurred in the northwest and southeast parts of the humid region, indicating larger differences between the two evaporations, while the lowest values occurred in the northeast and southwest areas, indicating smaller differences between the two evaporations. Overall, the K_p values ranged from 0.49 to 0.81 (with a mean of 0.66 $\pm$ 0.057) in the humid regions. These values were significantly different from those during the warm season ($p < 0.001$). Of the three regions, the K_p values in the E region were significantly lower than those in the others regions ($p < 0.01$), and there was no significant difference between the values in the SE and SW regions. The highest K_p values appeared in the SW region during the warm season and in the SE region during the cold season.

The K_p between the two pans varied significantly during the cold season (October–April of the following year): (1) the K_p increased from October to the maximum value ($K_p \approx 0.71$) in December or January of the following year, and then decreased until April ($K_p \approx 0.58$); and (2) in the three regions, K_p exhibited the same pattern of variations. The variation of K_p at the annual timescale showed an increase from approximately May to December and then a decrease until April. The monthly variation pattern was most similar to that found by Fu et al. (2009), whose study was conducted at an experimental evaporation station in the E region. The average value of K_p during the cold season (0.66 ± 0.057) was greater than that during the warm season (0.62 ± 0.047), indicating a smaller difference between the two pan evaporations during the cold season. The average value of K_p in all months in the humid regions was 0.64, with a standard deviation of 0.057.

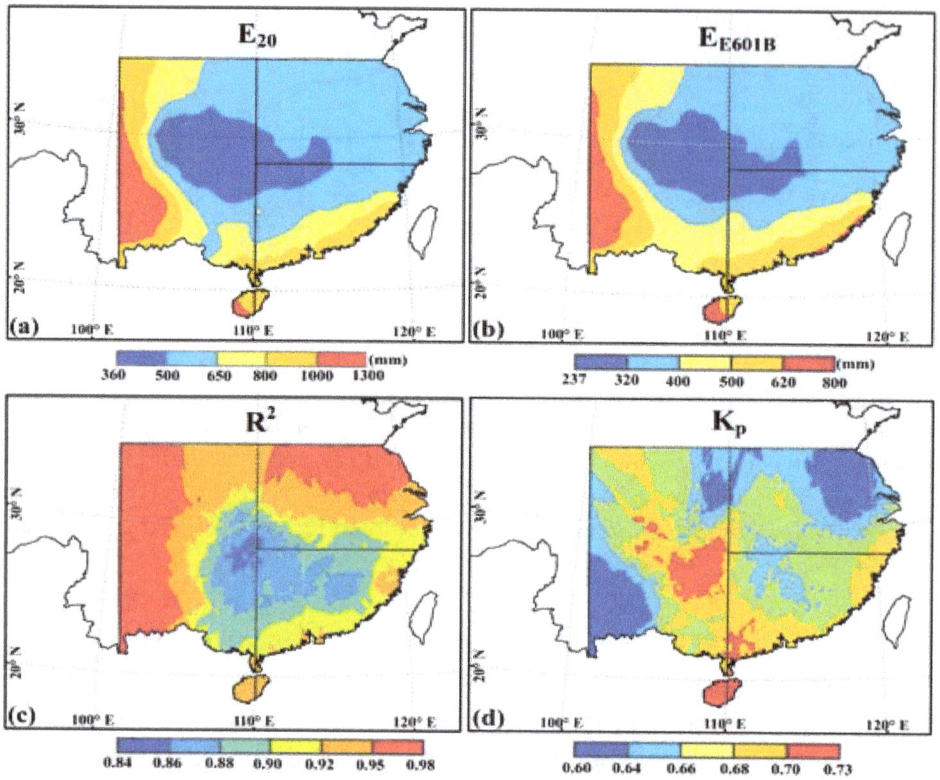

Figure 6. Spatial distributions of ϕ20 (**a**) and E601B (**b**) evaporation; the correlation coefficient R^2 (**c**); and the conversion coefficient K_p (**d**) between E601B and ϕ20 evaporation in the cold season (October–April).

As noted above, except for in the TP region, the monthly variability in K_p exhibited a similar pattern in both the warm and cold seasons, which indicated that some specific factors may control the variability in the K_p value. Therefore, meteorological factors were quantitatively explored to identify those that made the greatest contributions to K_p variability.

3.2. Potential Factors That Influence K_p

The Pearson correlation coefficients (r) and significance levels between K_p and several climatic factors are listed in Table 4. K_p was generally negatively correlated with R_n, VPD, U_2, Elev, and T_{mean}. The relationships between K_p and the climatic factors were similar to those noted by Xu et al. [33] in the Yangtze River basin, China. They investigated the K_p of ϕ20 and the Penman-Monteith reference evapotranspiration; and found that a high conversion coefficient was associated with a relatively high RH and low U_2. At the national scale, K_p was significantly negatively correlated ($p < 0.01$) with R_n ($r = -0.4$) and VPD ($r = -0.33$), and positively correlated with RH ($r = 0.33$) (Table 4). As climate change has occurred in recent decades [48], the pan evaporation paradox [12,13] associated with the variation in solar radiation has gained increasing attention at the global [49,50] and regional scales [51,52]. The change in net radiation (R_n) is expected to strongly affect K_p because of the high correlation between K_p and R_n; therefore, it should be considered in studies of evaporation trends that use different pan measurements. However, at the regional scale, the r values exhibited a significant geographic distribution. In five of the eight climatic regions, NCP, NE, E, SE, and SW, the K_p values had the highest correlation with R_n, followed by RH and VPD. In the other three regions, the K_p values were more associated with RH (especially in the NC and TP regions, with r values of 0.46 and 0.42, respectively), and the following factors were found in VPD and T_{mean} (Table 4). Therefore, further analysis must be performed to quantitatively separate the effects of different climate variables on K_p values.

Table 4. Pearson's correlation between the conversion coefficients (K_p) and various climate factors, and the independent explanatory powers of each variable based on K_p variation derived from the stepwise multiple linear regressions (%, in parentheses). T_{mean} is the mean air temperature, RH is the relative humidity, VPD is the vapor pressure deficit, U_2 is the wind speed at a height of 2 m, Elev is the elevation, and R_n is the net radiation. The dominant factors are highlighted in bold.

Variables and Regions	T_{mean}	RH	VPD	U_2	Elev	R_n	Combined
NW (n = 1528)	−0.04 (0.3)	**0.34** [a] **(11.2)**	−0.14 [a] (0.5)	−0.04 (0.8)	−0.12 [a] (1.4)	−0.12 [a] (1.2)	- (15.2)
NC (n = 984)	−0.21 [a]	**0.46** [a] **(21)**	−0.4 [a] (0.1)	−0.04 (4.5)	0.1 [b]	−0.36 [a] (5.8)	- (32)
NCP (n = 3065)	−0.03 (3.3)	0.21 [a] (0.8)	−0.19 [a]	−0.04 [b] (0.4)	−0.08 [a] (0.5)	**−0.24** [a] **(5.5)**	- (10.4)
NE (n = 973)	0.04 (8.6)	0.22 [a]	−0.16 [a]	−0.16 [a]	−0.04 (0.6)	**−0.3** [a] **(8.8)**	- (17.7)
E (n = 3858)	−0.3 [a] (3)	0.17 [a] (0.9)	−0.33 [a] (2.8)	−0.15 [a]	0.04 [b] (0.4)	**−0.43** [a] **(18.1)**	- (25.2)
SE (n = 3207)	−0.33 [a] (0.3)	−0.03 (1.1)	−0.23 [a] (0.3)	0.06 [b] (0.7)	0.01	**−0.35** [a] **(12.2)**	- (14.5)
SW (n = 4688)	−0.23 [a] (0.5)	0.25 [a] (3.9)	−0.32 [a] (2.2)	−0.11 [a]	−0.05 [a] (0.3)	**−0.35** [a] **(12.2)**	- (19.1)
TP (n = 1408)	0.3 [a]	**0.42** [a] **(17.7)**	−0.21 [a]	−0.06 [b] (1.9)	−0.21 [a] (2.8)	0.06 [b] (0.4)	- (22.6)
National (n = 19,848)	−0.17 [a] (0.2)	0.33 [a] (4.4)	−0.33 [a] (1.4)	−0.15 [a] (0.3)	−0.06 [a] (1.3)	**−0.4** [a] **(15.5)**	- (23.1)

Notes: [a] Significant at the 0.01 level; [b] significant at the 0.05 level; - is no value.

The percent contributions of each climate factor to K_p variation were analyzed using multiple stepwise regression with K_p as the dependent variable and the six climatic factors as the independent variables (Table 4). The climate factors together explained approximately one quarter of the K_p variation in China. Of the six climatic factors, R_n had the most explanatory power, with an independent explanation percentage of 15.5%, followed by RH and VPD (4.4% and 1.4%, respectively). The combined explanation percentage was 23.1%, which implied that more complicated mechanisms affect the variations in K_p. Similar to the r values, the highest explanatory percentages in five of the eight regions were found associated with R_n (the NCP, NE, E, SE, and SW regions) and ranged from 5.5% (NCP region) to 18.1% (E region), while RH in the other regions ranged from 11.2% (NW region) to 21% (NC region). The greatest combined explanation of the climatic factors was found in the NC region (32%), followed by the E (25.2%) and TP (22.6%) regions, and the weakest explanatory power was found in the NCP region (10.4%).

The two evaporimeters at each station experience the same macroclimate, so the microclimate differences between them can be magnified by the regional difference in climatic factors. For example, the volume and depth of the water in the pans are also important to K_p variability [16]. The difference in the water volume in the two pans caused the water temperature to increase at different rates. Due to the relatively low heat capacity, the water temperature of ϕ20 increases faster during the day and also decreases faster at night than that of the E601B pan. This phenomenon may be significant in regions with high temperature differences, such as the NE and NW regions (Table 3). By contrast, the E601B, compared to the ϕ20, has no effect because it was buried in the ground and surrounded by water (Figure 1). Similarly, Lim et al. [47], based on an energy balance experiment, found that the K_p between a Class A pan and steady state lake evaporation was mainly dependent on the additional radiation absorbed by the pan wall. The wall of the ϕ20 also had a large area based on the ratio of the wall to the water surface (Figure 1c). Thus, the wall absorbed additional radiation and transferred heat to the water, which increased the evaporation [27]. In addition, on nights without solar radiation, the appreciable storage of heat within the pan wall (Class A or ϕ20) may have caused additional evaporation [27], while E601B had a relatively small evaporation because it was buried in the soil. A negative correlation between K_p and R_n was found in this paper (Table 4), and this finding has also been confirmed by previous experimental data [15,41,53]. In addition, wind speed can decline rapidly with decreasing height due to higher surface roughness and friction near ground [21]. Therefore, a small ϕ20 evaporimeter located at a higher elevation and exposed to the faster wind speed should have more evaporation than the E601B evaporimeter, resulting in a small K_p. This condition can be amplified by regional differences in wind speed. For example, the average wind speed is high in the NE and NCP regions (Table 2); thus, the ϕ20 evaporation could be larger than that of E601, resulting in smaller K_p values than in other regions during the warm season (Table 3).

In conclusion, several factors, including water temperature, vapor pressure, wind speed, turbulence, heat transfer, and heat storage, can affect pan evaporation. The combined influence of various factors produces different spatial patterns and monthly variations in K_p. However, further analysis of the contribution of each factor to K_p requires rigorous experimental investigation of the physics of pan evaporation [47,53] or more reasonable evaporation models [14].

4. Uncertainties

Uncertainties of the spatial distribution and monthly variation of K_p existed in this study, as well as the explanatory powers of the climatic factors. First, although the spatial distribution of K_p is useful for evaporation studies, it should be noted that only larger R^2 values indicate a high confidence of converting the ϕ20 evaporation to E601B evaporation. Therefore, in combination with the spatial distribution of R^2 (Figures 3a and 4), it was reasonable to obtain E601B evaporation by multiplying K_p by the ϕ20 evaporation in the northern part of China, such as in the NC, NCP, and NE regions. However, there are uncertainties in the TP and NW regions due to the relatively sparse distribution of monitoring stations in such large areas (Figure 2); thus, K_p should be used cautiously in these regions. Second, the K_p value was calculated based on the assumption that K_p remains constant over time. This may not be true in some places, and its further testing of the inter-annual variability is required. In addition, the climatic factors that were selected to explain K_p may not be independent. For example, VPD is a function of RH and temperature [21].

Despite these uncertainties, this study presented the spatial distribution of K_p and investigated the dominant climatic factors. Our findings were similar to those of previous studies [19,33,41,53]. Therefore, more attention should be paid to the spatial and temporal variations in K_p before using evaporation data from different evaporimeters. Some researchers have documented that evaporation from Class A (Figure 1a), GGI-3000 (Figure 1b), and ϕ20 (Figure 1c) evaporimeters had relatively small K_p values compared those of 20 m^2 evaporation tanks [8,15,30]. By contrast, E601B evaporation had a high value of K_p and was close to the free water surface evaporation [15]. Researchers have compared various types of evaporimeters at the station scale in the E and NCP regions of China [15,30]. They found that the E601B, compared to other evaporimeters, had a larger conversion coefficient and smaller coefficient of variation with 20 m^2 evaporation tanks, indicating that the evaporation from E601 was much closer to evaporation from the free water surface. The Class A and ϕ20 pans, which are exposed to the air and can absorb additional radiation via the wall, had small conversion coefficients with 20 m^2 evaporation tanks [8,47]. Therefore, it is reasonable to substitute E601B for ϕ20 in China, not only because of its stability in different climate regions but also because of its relatively small difference relative to evaporation from large water bodies.

5. Conclusions and Suggestions

This paper analyzed the spatial distribution, correlation coefficient (R^2), and conversion coefficient (K_p) of evaporation from two typical pans in eight climate regions in China. The main conclusions are as follows:

(1) During the warm season, the spatial evaporation patterns of the two pans were similar and showed increasing trends from the southeastern to northwestern regions of China. The R^2 values were relatively high and ranged from 0.48 to 0.99 with an average of 0.89, which indicates that the E601B pans accurately captured the variation in ϕ20 evaporation. The K_p values showed significant spatial variability across China and varied from 0.46 to 0.92 with a mean of 0.60. The highest and lowest K_p values were found in the southwestern (SW and TP) and NE regions, respectively. Generally, K_p increased from May to September in all of the regions, especially during the summer (June, July and August). In the humid regions (the E, SE, and SW regions), the values of R^2 and K_p were higher during the cold season (means of 0.92 and 0.66, respectively) than during the warm season. The monthly K_p values at the annual scale had a unimodal distribution. They increased from May to December or January the following year, and then decreased until April.

(2) The Pearson correlations and the explanatory powers of the variables were calculated using the multiple stepwise regression method. R_n was the dominant climatic factor for the variation in K_p, exhibiting the best correlation ($r \approx 0.4$, $p < 0.01$) and highest independent explanatory power (approximately 15%) in five of the eight regions, followed by RH and VPD. The combined explanation percentage of all of the variables was 23.1%, and significant differences in explanation percentages were observed in the eight climate regions.

Although the contributions of climate factors to K_p variability were identified by the stepwise multiple linear regression method, the selected factors in this study had relatively weak explanatory powers. Further experiments must be performed to investigate the other potential mechanisms that affect K_p, such as the water vapor pressure at the surface and in the air, pan size, wind speed [53], and the energy balance around evaporimeters [47]. Based on this initial stage of this research, ongoing research and planned experiments, new approaches and evaporation equations [14,41] will be explored using the large and valuable meteorological records available in China. Quantification of the contributions of climatic factors to K_p and determination of the mechanisms of K_p variability can improve the understanding of long-term trends in pan evaporation across China.

Acknowledgments: This research was financially supported by Natural Science Foundation of China (No. 41330529, 41501032, 41571024). We will give many thanks to Wilfried Brutsaert and Guobin Fu for their good suggestions to the manuscript.

Author Contributions: This paper was designed and performed by all the authors. Yanzhong Li wrote the draft of the paper. Changming Liu provided the data and financial support. Kang Liang provided the idea and detailed directed the writing.

References

1. Brutsaert, W. *Evaporation into the Atmosphere: Theory, History and Applications*; Springer: Heidelberg, Germany, 1982; Volume 1.
2. Brutsaert, W.; Parlange, M. Hydrologic cycle explains the evaporation paradox. *Nature* **1998**, *396*, 30. [CrossRef]
3. Hetherington, A.M.; Woodward, F.I. The role of stomata in sensing and driving environmental change. *Nature* **2003**, *424*, 901–908. [CrossRef] [PubMed]
4. Jung, M.; Reichstein, M.; Ciais, P.; Seneviratne, S.I.; Sheffield, J.; Goulden, M.L.; Bonan, G.; Cescatti, A.; Chen, J.; De Jeu, R.; et al. Recent decline in the global land evapotranspiration trend due to limited moisture supply. *Nature* **2010**, *467*, 951–954. [CrossRef] [PubMed]
5. Zhang, Y.; Liu, C.; Tang, Y.; Yang, Y. Trends in pan evaporation and reference and actual evapotranspiration across the Tibetan Plateau. *J. Geophys. Res. Atmos.* **2007**, *112*, 1103–1118. [CrossRef]

6. Li, Y.; Liang, K.; Bai, P.; Feng, A.; Liu, L.; Dong, G. The spatiotemporal variation of reference evapotranspiration and the contribution of its climatic factors in the Loess Plateau, China. *Environ. Earth Sci.* **2016**, *75*, 1–14. [CrossRef]
7. Hobbins, M.; Wood, A.; Streubel, D.; Werner, K. What Drives the Variability of Evaporative Demand across the Conterminous United States? *J. Hydrometeorol.* **2012**, *13*, 1195–1214. [CrossRef]
8. Stanhill, G. Is the Class A evaporation pan still the most practical and accurate meteorological method for determining irrigation water requirements? *Agric. For. Meteorol.* **2002**, *112*, 233–236. [CrossRef]
9. Alkhafaf, S.; Wierenga, P.J.; Williams, B.C. Evaporative Flux from Irrigated Cotton as Related to Leaf Area Index, Soil Water, and Evaporative Demand. *Agron. J.* **1978**, *70*, 912–917. [CrossRef]
10. Azorin-Molina, C.; Vicente-Serrano, S.M.; Sanchez-Lorenzo, A.; McVicar, T.R.; Morán-Tejeda, E.; Revuelto, J.; El Kenawy, A.; Martín-Hernández, N.; Tomas-Burguera, M. Atmospheric evaporative demand observations, estimates and driving factors in Spain (1961–2011). *J. Hydrol.* **2015**, *523*, 262–277. [CrossRef]
11. Donohue, R.J.; McVicar, T.R.; Roderick, M.L. Assessing the ability of potential evaporation formulations to capture the dynamics in evaporative demand within a changing climate. *J. Hydrol.* **2010**, *386*, 186–197. [CrossRef]
12. Roderick, M.L.; Farquhar, G.D. The cause of decreased pan evaporation over the past 50 years. *Science* **2002**, *298*, 1410–1411. [PubMed]
13. Liu, X.; Luo, Y.; Zhang, D.; Zhang, M.; Liu, C. Recent changes in pan-evaporation dynamics in China. *Geophys. Res. Lett.* **2011**, *38*, 142–154. [CrossRef]
14. Singh, V.; Xu, C. Evaluation and generalization of 13 mass-transfer equations for determining free water evaporation. *Hydrol. Process.* **1997**, *11*, 311–323. [CrossRef]
15. Fu, G.; Liu, C.; Chen, S.; Hong, J. Investigating the conversion coefficients for free water surface evaporation of different evaporation pans. *Hydrol. Process.* **2004**, *18*, 2247–2262. [CrossRef]
16. Brouwer, C.; Heibloem, M. *Irrigation Water Management: Irrigation Water Needs. Training Manual*; Food and Agriculture Organization of the United Nations: Rome, Italy, 1986.
17. Golubev, V.S.; Lawrimore, J.H.; Groisman, P.Y.; Speranskaya, N.A.; Zhuravin, S.A.; Menne, M.J.; Peterson, T.C.; Malone, R.W. Evaporation changes over the contiguous United States and the former USSR. *Geophys. Res. Lett.* **2001**, *53*, 323–324. [CrossRef]
18. Symons, G.J. Evaporators and evaporation. *Br. Rainfall* **1867**, *7*, 9–10.
19. Chen, D.; Gao, G.; Xu, C.Y.; Guo, J.; Ren, G. Comparison of the Thornthwaite method and pan data with the standard Penman-Monteith estimates of reference evapotranspiration in China. *Clim. Res.* **2005**, *28*, 123–132. [CrossRef]
20. Brustaert, W. Evaluation of some practical methods of estimating evapotranspiration in arid climates at low latitudes. *Water Resour. Res.* **1965**, *2*, 187–191.
21. Allen, R.G.; Pereira, L.S.; Raes, D.; Smith, M. *Crop Evapotranspiration. Guidelines for Computing Crop Water Requirements*; FAO Irrigation and Drainage Paper 56; Food and Agriculture Organization of the United Nations: Rome, Italy, 1998.
22. Zuo, H.; Chen, B.; Wang, S.; Guo, Y.; Zuo, B.; Wu, L.; Gao, X. Observational study on complementary relationship between pan evaporation and actual evapotranspiration and its variation with pan type. *Agric. For. Meteorol.* **2016**, *222*, 1–9. [CrossRef]
23. Liu, W.; Wang, L.; Zhou, J.; Li, Y.; Sun, F.; Fu, G.; Li, X.; Sang, Y.F. A worldwide evaluation of basin-scale evapotranspiration estimates against the water balance method. *J. Hydrol.* **2016**, *538*, 82–95. [CrossRef]
24. Gao, G.; Xu, C.Y.; Chen, D.; Singh, V.P. Spatial and temporal characteristics of actual evapotranspiration over Haihe River basin in China. *Stoch. Environ. Res. Risk Assess.* **2012**, *26*, 655–669. [CrossRef]
25. Liu, X.; Zhang, D. Trend analysis of reference evapotranspiration in Northwest China: The roles of changing wind speed and surface air temperature. *Hydrol. Process.* **2013**, *27*, 3941–3948. [CrossRef]
26. Mao, D.; Wang, Z.; Li, L.; Song, K.; Jia, M. Quantitative assessment of human-induced impacts on marshes in Northeast China from 2000 to 2011. *Ecol. Eng.* **2014**, *68*, 97–104. [CrossRef]
27. Yang, H.; Yang, D. Climatic factors influencing changing pan evaporation across China from 1961 to 2001. *J. Hydrol.* **2012**, *414–415*, 184–193. [CrossRef]
28. Fu, G.; Charles, S.P.; Yu, J. A critical overview of pan evaporation trends over the last 50 years. *Clim. Chang.* **2009**, *97*, 193–214. [CrossRef]
29. Li, Y.; Liang, K.; Liu, C.; Liu, W.; Bai, P. Evaluation of different evapotranspiration products in the middle Yellow River Basin, China. *Hydrol. Res.* **2016**, *47*. [CrossRef]

30. Hong, J.; Fu, G.; Guo, Z. Experimental research on the water-surface evaporation of Nansi Lake in Shandong Province. *Geogr. Res.* **1996**, *15*, 42–49. (In Chinese)
31. Yang, Y.; Yang, L.; Wang, X.; Liu, J.; Qian, R. Variation Charicteristic and Influence Factors of Pan Evaporation in Xingtai of Hebei Province. *J. Arid Meteorol.* **2013**, *31*, 82–88. (In Chinese)
32. Liang, L.; Li, L.; Liu, Q. Spatio-temporal variations of reference crop evapotranspiration and pan evaporation in the West Songnen Plain of China. *Hydrol. Sci. J.* **2011**, *56*, 1300–1313. [CrossRef]
33. Xu, C.Y.; Gong, L.; Jiang, T.; Chen, D.; Singh, V.P. Analysis of spatial distribution and temporal trend of reference evapotranspiration and pan evaporation in Changjiang (Yangtze River) catchment. *J. Hydrol.* **2006**, *327*, 81–93. [CrossRef]
34. Brutsaert, W. Use of pan evaporation to estimate terrestrial evaporation trends: The case of the Tibetan Plateau. *Water Resour. Res.* **2013**, *49*, 3054–3058. [CrossRef]
35. Ren, Z.; Liu, M.; Zhang, W. Conversion coefficient of small evaporation pan into E-601B pan in China. *J. Appl. Meteorol. Sci.* **2002**, *13*, 508–512. (In Chinese)
36. Liu, X.N.; Wang, S.Q.; Wu, Z.X.; Wang, Y. Comparative analyses on two kinds of observed evaporation data in China. *Q. J. Appl. Meteorol.* **1998**, *9*, 321–328. (In Chinese)
37. Liu, B.; Xu, M.; Henderson, M.; Gong, W. A spatial analysis of pan evaporation trends in China, 1955–2000. *J. Geophys. Res. Atmos.* **2004**, *109*, 1255–1263. [CrossRef]
38. Zhang, Q.; Xu, C.Y.; Chen, Y.D.; Ren, L. Comparison of evapotranspiration variations between the Yellow River and Pearl River basin, China. *Stoch. Envrion. Res. Risk Assess.* **2011**, *25*, 139–150. [CrossRef]
39. Brutsaert, W. Equations for vapor flux as a fully turbulent diffusion process under diabatic conditions. *Hydrol. Sci. J.* **1965**, *10*, 11–21. [CrossRef]
40. World Meteorological Organization (WMO). *Guide to Meteorological Instruments and Methods of Observation*, 6th ed.; WMO Rep. 8; WMO: Geneva, Switzerland, 1996.
41. Yu, S.L.; Brutsaert, W. Evaporation from very shallow pans. *J. Appl. Meteorol.* **1967**, *6*, 265–271. [CrossRef]
42. Liu, C.; Zhang, D.; Liu, X.; Zhao, C. Spatial and temporal change in the potential evapotranspiration sensitivity to meteorological factors in China (1960–2007). *J. Geogr. Sci.* **2012**, *22*, 3–14. [CrossRef]
43. Zhang, D.; Liu, X.; Hong, H. Assessing the effect of climate change on reference evapotranspiration in China. *Stoch. Envrion. Res. Risk Assess.* **2013**, *27*, 1871–1881. [CrossRef]
44. Stein, M.L. Interpolation of spatial data: Some theory for kriging. *Technometrics* **2015**, *42*, 436–437.
45. Qiu, J. China: The third pole. *Nature* **2008**, *454*, 393–396. [CrossRef] [PubMed]
46. Linacre, E.T. Estimating U.S. Class A Pan Evaporation from Few Climate Data. *Water Int.* **1994**, *19*, 5–14. [CrossRef]
47. Lim, W.H.; Roderick, M.L.; Hobbins, M.T.; Wong, S.C.; Farquhar, G.D. The energy balance of a US Class A evaporation pan. *Agric. For. Meteorol.* **2013**, *182–183*, 314–331. [CrossRef]
48. Intergovernmental Panel on Climate Change (IPCC). *Summary for Policymakes: The Physical Science Basis*; Contribution of Working Group I to the IPCC Fifth Assessment Report Climate Change; IPCC: Geneva, Switzerland, 2013.
49. Wild, M.; Gilgen, H.; Roesch, A.; Ohmura, A.; Long, C.N.; Dutton, E.G.; Forgan, B.; Kallis, A.; Russak, V.; Tsvetkov, A. From dimming to brightening: Decadal changes in surface solar radiation. *Science* **2005**, *308*, 847–850. [CrossRef] [PubMed]
50. Gilgen, H.; Roesch, A.; Wild, M.; Ohmura, A. Decadal changes in shortwave irradiance at the surface in the period from 1960 to 2000 estimated from Global Energy Balance Archive Data. *J. Geophys. Res. Atmos.* **2009**, *114*. [CrossRef]
51. Hayasaka, T.; Kawamoto, K.; Shi, G.; Ohmura, A. Importance of aerosols in satellite-derived estimates of surface shortwave irradiance over China. *Geophys. Res. Lett.* **2006**, *33*, 178–196. [CrossRef]
52. You, Q.; Kang, S.; Flügel, W.A.; Sanchez-Lorenzo, A.; Yan, Y.; Huang, J.; Martin-Vide, J. From brightening to dimming in sunshine duration over the eastern and central Tibetan Plateau (1961–2005). *Theor. Appl. Climatol.* **2010**, *101*, 445–457. [CrossRef]
53. Brutsaert, W.; Yu, S.L. Mass transfer aspects of pan evaporation. *J. Appl. Meteorol.* **1968**, *7*, 563–566. [CrossRef]

10

An Integrated Approach for Site Selection of Snow Measurement Stations

Bahram Saghafian [1,*], Rahman Davtalab [2], Arezoo Rafieeinasab [3] and M. Reza Ghanbarpour [4]

1 Technical and Engineering Department, Science and Research Branch, Islamic Azad University, Tehran 1477893855, Iran
2 Department of Civil, Environmental and Construction Engineering, University of Central Florida, Orlando, FL 32818, USA; rdavtalab@saiengineers.com
3 Department of Civil Engineering, The University of Texas at Arlington, Box 19308, 416 Yates St. Suite 425, Arlington, TX 76019, USA; arezoo.rafieeinasab@gmail.com
4 Alberta Environment and Parks, Calgary, AB T2E 7L7, Canada; Reza.Ghanbarpour@gov.ab.ca
* Correspondence: b.saghafian@gmail.com

Academic Editor: Tommaso Moramarco

Abstract: Snowmelt provides a reliable water resource for meeting domestic, agricultural, industrial and hydropower demands. Consequently, estimating the available snow water equivalent is essential for water resource management of snowy regions. Due to the spatiotemporal variability of the snowfall pattern in mountainous areas and difficult access to high altitudes areas, snow measurement is one of the most challenging hydro-meteorological data collection efforts. Development of an optimum snow measurement network is a complex task that requires integration of meteorological, hydrological, physiographical and economic studies. In this study, site selection of snow measurement stations is carried out through an integrated process using observed snow course data and analysis of historical snow cover images from National Oceanic Atmospheric Administration Advanced Very High Resolution Radiometer (NOAA-AVHRR) at both regional and local scales. Several important meteorological and hydrological factors, such as monthly and annual rainfall distribution, spatial distribution of average frequency of snow observation (FSO) for two periods of snow falling and melting season, as well as priority contribution of sub-basins to annual snowmelt runoff are considered for selecting optimum station network. The FSO maps representing accumulation of snowfall during falling months and snowpack persistence during melting months are prepared in the GIS based on NOAA-AVHRR historical snow cover images. Basins are partitioned into 250 m elevation intervals such that within each interval, establishment of new stations or relocation/removing of the existing stations were proposed. The decision is made on the basis of the combination of meteorological, hydrological and satellite information. Economic aspects and road access constraints are also considered in determining the station type. Eventually, for the study area encompassing a number of large basins in southwest of Iran, several new stations and relocation of some existing stations are proposed.

Keywords: snow measurement network; site selection; satellite images; Iran

---

1. Introduction

Water demand increase in urban, agriculture and industry sectors all over the world continues to intensify the pressure on water resources. In the southwestern of Iran, the Karkheh, Dez, Karun and Marun river basins provide over a quarter of surface water resources in the country and even supply water for the central arid part of Iran via transboundary water transfers [1].

Nowadays, hydrological and water resources modeling play a key role in water resources management [2]. On the other hand, development and calibration of hydrologic and water resources models rely on adequate and accurate hydro-climatic observation data [3]. Reliable data may be acquired through appropriate design, installation and operation of an optimum network of hydro-climatic stations. Designing a proper measurement network results in the representative location and optimum number of stations so that interpolation techniques can lead to acceptable estimation of spatially distributed factors with adequate accuracy [4,5].

Among the hydro-climatic variables, snow is one of the most important factors in the hydrology of mountainous areas. Snow, as a renewable water resource, remains on the ground for weeks or months. Since the lag time of river flow from snowmelt is more than that of rainfall, management of this reliable resource is relatively easier. The amount of snowfall and snowmelt depends on both geographic (elevation, slope and aspect, latitude, etc.) and synoptic factors. Storing and melting snow is a continuous process especially in southwest of Iran with low latitude and relatively ephemeral snow. Therefore estimating accumulated snowpack and its spatiotemporal pattern is complex. Furthermore, due to limited access to high altitudes of mountainous areas, snow measurement is considered as one of the most difficult data collections in water resources. The position of snow gauges and the type of snow measurement instruments should be selected in a way to represent the spatial and temporal variations of snow characteristics. Selecting representative locations for automated snow-pillow (ASP) stations or manual snow-course (MSC) stations will ensure access to the reliable data required for snow storage estimation.

Nowadays, satellite images and remote sensing (RS) techniques are widely used for the estimation of snow cover areas. Derived snow cover areas from satellite images accompanied with field observations can provide a unique tool for estimation of snow storage, river flow forecasting and flood control required in water resource management.

ASP measurement networks have been developed in many countries. The snow measurement network in western U.S. states or SNOTEL [6], provincial or local networks in Canada, such as ASP measurement network in British Columbia [7], and the ASP measurement network in Pakistan [8] are good examples of available operating networks.

Nonetheless, there are insufficient studies on site selection of snow measurement networks at regional scale and a few existing studies are limited to local scale, focusing mainly on a single station [6]. According to the World Meteorological Organization [9], the number of snow measurement stations or snow courses and their positions depend on topography, the type and aspect of vegetation, the objectives of data collection, and economic considerations. WMO recommended one snow measurement station for about 2000 to 3000 km^2 in low homogeneous region and one for about 5000 km^2 for more homogeneous and plain regions [9]. For developing a snow measurement network, it is recommended to establish a temporary and dense network that measures the snow parameters for four to six years continuously in order to detect snow cover characteristics, related to the physiographical parameters. Then, according to the required accuracy, operation constraints and cost minimization, the network density may be reduced [4].

In this study, areas will be prioritized according to the need for snow measurement stations using field observations that are supported by satellite snow cover images. Furthermore, through assessment of the existing snow measurement network in each of the altitudinal range, installation of new stations, removal/maintaining/relocation of the existing stations MSC stations, or upgrade to ASP stations is proposed. The paper is organized as follows. Case study, data acquisition, and methodology, including the meteorological, hydrological, and remote sensing criteria used are described in Section 2. Results are presented in Section 3. Section 4 provides conclusions and recommendations.

2. Data and Methods

2.1. Study Area

The Zagros Mountains are one of the major snowy regions of Iran, which include the four large river basins of Karkheh, Dez, Karun and Marun. The study area encompasses the Karkheh basin, upstream of Karkheh Dam, with approximately 43,000 km^2 in area and 5.2 km^3 mean annual runoff; Dez Basin, upstream of Dez Dam with 17,300 km^2 in area and 8.4 km^3 mean annual runoff; Karun basin, upstream of Gotvand Dam with 32,000 km^2 in area and 12.5 km^3 mean annual runoff; and two tributaries of the Marun basin, first upstream of Shohadaye Behbahan Dam and second upstream of Jarreh Dam with 6800 km^2 in area and 2.3 km^3 mean annual runoff, in total. The area lies between the geographical latitude of 30°15′ N to 34°55′ N and 46°05′ E to 52°03′ E longitude. The mentioned river basins play a major role in food production and economy of more than 4 million people [10,11]. Figure 1 shows the location of these major river basins.

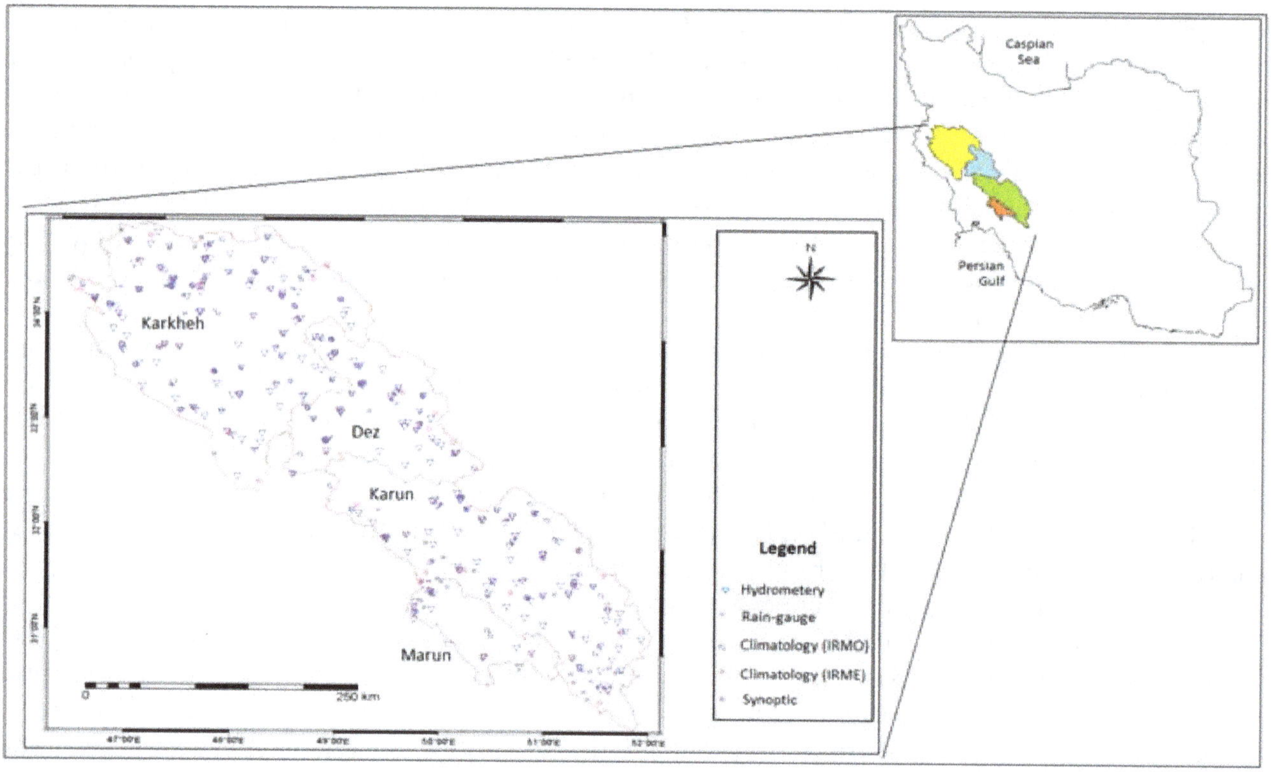

Figure 1. Location of the study basins and distribution of the meteorological and hydrometric stations.

2.2. Data Collection

The data used for this study comprise the monthly air temperature from 44 climatological stations (including the stations installed by Iran Ministry of Energy, IRME and Iran Meteorological Organization, IRMO), daily and monthly precipitation of 152 rain-gauge stations, daily temperature and 3-h precipitation depth and 6-h precipitation type (rainfall or snow) measured at 20 synoptic stations, snow water equivalent and snow depth at 45 available MSC stations accompanied by daily discharge at 40 hydrometric stations with long term data (Figure 1). For monthly temperature and rainfall, a 25-year data period is identified as a normal climate period (1974–1999). Missing data are reconstructed using linear correlation with the nearest neighboring station. The snow water equivalent may also be estimated through analysis of the amount and type of precipitation in the form of rainfall or snowfall, at synoptic stations (detail information is available in Saghafian and Davtalab, [12]). The only snow measurements available from MSC stations that are located in the Karun and Dez Basins. Data sampling in the mentioned stations is performed manually once a year. However, the sampling time varies between late February and late March. Therefore, the sampling period lasts about one month in any given year. Accordingly, the above-mentioned data are used for spatial analysis.

Quantity and quality of daily discharge data are assessed as well and the annual discharge series are obtained. The snow cover maps extracted from the National Oceanic Atmospheric Administration

Advanced Very High Resolution Radiometer (NOAA-AVHRR) series for the period of 1984 to 2003 have been used in this study. The maps are provided by Jamab [13] based on the methodology available in Porhemmat et al. [14]. The spatial resolution of the NOAA-AVHRR image is about 1.1 km, suitable for snow cover extraction over vast areas. There are two images for every 24 h, one of which corresponds to daytime. However, not all daytime images were suitable because of atmosphere noise and/or cloud cover. In all, 480 snow cover maps were available in Jamab's archive [13]. The snow cover maps are stored in binary format in GIS (1 denotes snow and 0 denotes non-snow pixel). The hypsometric maps, boundary of basins and sub-basins, digital elevation model (DEM), slope and aspect maps are derived from 1:250,000 topographic maps (Iran National Cartographic Center: http://www.ncc.org.ir) with 100 m contour interval. For instance, the contour map for Dez Basin is shown in Figure 2.

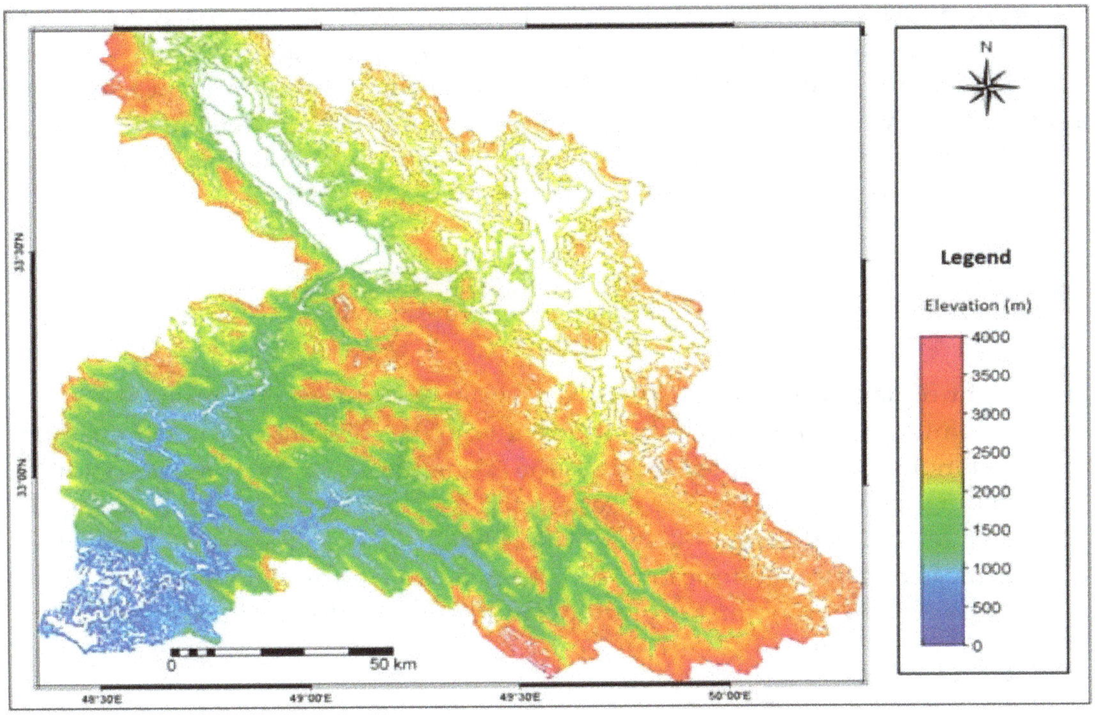

Figure 2. Elevation contour map of Dez Basin.

2.3. Methodology

The snow station site selection methodology is based on integration of meteorological, hydrological, and remote sensing studies. Meteorological analysis includes preparation of isothermal and isohyetal maps, determination of snowfall threshold temperature and precipitation estimation variance analysis [15]. Prioritization of the sub-basins has been done based on the contribution of snowmelt to the total annual runoff in each sub-basin. Analysis of remote sensing data includes the extraction of the FSO layers for two periods of snowfall accumulation and snowpack persistence. The details are given in the following sections of the paper.

2.3.1. Meteorological Criteria

Temperature is the most dominant meteorological factor in determination of precipitation type (snow and rain) and determines regions subject to snowmelt, snowfall, or both. Therefore, temperature and its spatial variation is a key factor in determining the proper location of snow measurement stations. In this study, the snowfall threshold temperature refers to the temperature under which the precipitation is assumed to be in the form of snow. Threshold temperature may vary from below 0 °C to 5 °C and may not be constant even in a single day [16]. Therefore, in many studies an average value is used as threshold temperature [2,17,18]. Daily precipitation data accompanied with the rain/snow observation at synoptic stations are used to estimate the threshold temperature [16].

In each station, through assigning the snowfall percentage in various temperature intervals, the temperature corresponding to 50% snowfall is determined as the threshold temperature. Based on the monthly and annual mean temperature for the study period at all stations, the spatial correlation relationships between temperature and elevation are found using DEMs. The areas below the threshold temperature are identified as potential snow storage areas using the isothermal maps. Hence, establishing snow gauges is considered in these areas.

In addition, the semi-variogram analysis [15] is conducted using observed snow water equivalent (SWE) data from snow measurement stations or MSC stations (average of SWE for the period of 1974 to 1999). The best variogram model is adopted to compute the spatial distribution of error variance. Estimation variance denoted by σ_e^2 demonstrates the variance of the error between observed and estimated values. In geostatistics, it can be computed from average values of semi-variograms using available supplementary functions [19]. In addition, for calculation of confidence levels, estimation variance is used in many other cases, such as designing an optimal measurement network, calculating the error reduction resulted from increasing the number of sampling, and assessing different sampling methods. Estimation variance can be defined with the following equation [20]:

$$\sigma_e^2 = 2\sum_{i=1}^{n} \lambda_i \times \gamma_{oi} - \gamma_o - \sum_{i=1}^{n}\sum_{j=1}^{n} \lambda_i \times \gamma_{ij} \quad (1)$$

where λ_i is weight of data at point i, γ_o is the variogram value for $h = 0$, (where h is the distance between paired points), γ_{oi} is the variogram value between the considered point and point i, γ_{ij} is the variogram value among i and j samples and σ_e^2 is the estimation variance at the ith point.

Further explanation about the error variance map is available in [19]. Error variance maps are used as the main reference in determining the new locations of a gauge or omitting redundant gauges. Therefore, snow measurement stations can be established where error variance is high. The reduction of spatial error variance corresponding to adding one station can also be determined.

2.3.2. Hydrological Criteria

In mountainous areas with a mixed snow-rain precipitation regime, the contribution of snowmelt to the annual runoff is an important factor. Because of the inherent uncertainty in the separation of flow components, a simple water balance model is used [21]. Precipitation over the melting season, annual runoff coefficient and snowmelt runoff determined for each hydrometric station are accounted in the water balance model, which can be expressed as follows:

$$R = (H_w + P) \times C \quad (2)$$

where R is the total runoff over the snowmelt season in millimeters, H_w is the depth of SWE at the beginning of the snowmelt season in millimeters, P is the precipitation over the snowmelt season (spring and summer) in millimeters and C is the average runoff coefficient that can be computed based on precipitation and discharge data. P is calculated based on monthly isohyetal maps. H_w represents the unknown of the model.

Having determined R at hydrometric stations, a regional model is developed to estimate R at ungauged locations. Basin characteristics such as area, average elevation, and the percentage of areas with elevation above 2000, 2500 and 3000 m are used as independent variables of the regional model [22]. As a result, a map of R values at sub-basin scale is produced such that sub-basins may be prioritized based on snowmelt runoff contribution. This map is used as one of the criteria in selection of snow measurement stations.

2.3.3. Remote Sensing Snow Observation Criteria

Time series of snow cover maps were available from Jamab [13]. These maps had been extracted from NOAA-AVHRR images for the period of 1984–2003. Detailed description of image geo-referencing, noise reduction, and snow cover extraction are available in Porhemmat et al. [14]. In brief, the methodology of snow detection by Porhemmat et al. [14] has two steps of split and merge technique on AVHRR satellite images, channels 2, 3 and 4. In the first step, two split-and-merge clustering/labeling procedures were used in order to separate clear land pixels from pixels contained by either snow or cloud as described by Simpson et al. [23]. In the second stage, the same procedure was used only on pixels containing cloud, snow or a mixture of both. At this stage, pixels containing snow were distinguished from those containing cloudy pixels. To improve and control this computerized procedure, the above methodology was combined with threshold techniques and filter analysis of AVHRR channels data, especially channel 2, 3, and 4. Due to the high contrast between snow covered and snow free surfaces, channel 1 facilitates the definition of snow pack boundaries. Also, channel 3 and channel 4 analyses were effectively used as the snow/cloud discrimination channels. Validation of the NOAA snow detection algorithm was performed against eye interpretation via on screen digitization of simultaneous NOAA and Landsat TM images.

The FSO derived from the mentioned snow cover maps is a key criterion for identifying the potential snowy areas. The following relationship for calculating FSO has been proposed by Saghafian and Davtalab [12]:

$$FSO(i) = \frac{\sum_{l=1}^{T} X(i,l)}{T} \times 100 \tag{3}$$

where *FSO(i)* is frequency of snow observations for the *i*-th pixel, *T* is the total number of snow cover maps over the study period and *X* is a binary number equal to one if it is snow and zero otherwise.

In this study FSO is calculated for two time periods of December–February and March–May. Since the study area receives snow from early December to late February [12], FSO during these months is considered as snowfall probability (FSO-SFP). Also, almost all snow storage of study area melts during March until end of May. Therefore, FSO during these months is considered as snowpack persistence probability (FSO-SPP). FSO-SPP is dependent on temperature, additional precipitation and topographic conditions of the land [24,25]. Snow may disappear quickly due to temperature increase and/or shallow depth of snow. Snowmelt and loss of snow storage occurs faster along southern aspects because of direct sunshine [24,25]. Since spring and summer are high water demand seasons in the study area, FSO-SPP or snowpack persistence is important for meeting the water demands.

2.3.4. Synthesis of Criteria in Site Selection

As previously mentioned, the study area is divided into 250 m elevation intervals. Establishment/relocation of new/existing snow measurement stations is analyzed within each interval. The area ratio of each elevation interval to the total basin area is considered as a weight index. The weighted average snowfall probability in each interval is also calculated, using FSO-SFP map. This criterion is used as one of the main factors in snow station site selection.

The areas with FSO-SFP higher than the weighted average snowfall probability are identified in each elevation interval that has no snow measurement station. If spatial distribution of the identified areas were like patchy snow cover, those small polygons were eliminated. Then the meteorological and hydrological criteria are taken into account so that the areas with no or minimal snowfall potential and/or sub-basins with low snowmelt runoff volume are eliminated in further analyses.

However, if a snow measurement station is already operational in an altitudinal range, the appropriateness of its present location is assessed based on the criteria such as FSO-SFP, FSO-SPP and field visit.

In addition, the spatial pattern of stations is investigated according to the estimation variance maps. Therefore, new stations are likely to be proposed in areas subject to high estimation variance. The FSO-SPP map is further used as a reference in very large areas of high variance to reduce the field visit area. Clearly, areas with FSO-SPP higher than the weighted average snowpack persistence probability are appropriate areas for establishing new stations, since they provide the snow storage data for simulating snowmelt runoff over dry seasons.

After determining suitable areas at regional or macro scale, topography, road access, and proximity with residential areas (to facilitate the operation and maintenance of stations) are the main factors for site selection of snow stations at local or micro scale. The final step for site selection was field study to determine the exact point for installing the stations based on WMO recommendations [26]. The following criteria has been considered during the field study for selecting the representative location for the snow course: leaving clear space within the radius of at least equal to the height of the nearest obstacle, absence of strong wind and snowdrift (as much as possible), safety of the station against avalanche, no significant changes in elevation in a radius of 50 m (if possible), no reverse slopes and hard surface conditions for ease of leveling, good drainage, homogeneous vegetation cover and land use in the vicinity, and having enough open space to avoid snow interception.

For sake of brevity, only Dez Basin results will be presented.

3. Results

3.1. Classification of Regions Based on Meteorological Criteria

Isothermal maps are developed using the digital elevation model (DEM) and the temperature gradient relationships. The average threshold temperature is 2.4 °C at most synoptic stations, which is assigned as the threshold for the entire region (more explanation is available in Saghafian et al. [16]). The area with average monthly temperature less than 2.4 °C are designated as the potential snowfall areas and areas with temperature below 0 °C are classified as the potential area for snow persistence. Therefore, the regions with temperature between 0 °C and 2.4 °C represent areas prone to coupled snowfall-snowmelt processes. Figure 3 illustrates the spatial distribution of potential snowfall, snowmelt, and snow persistence areas in February in Dez Basin.

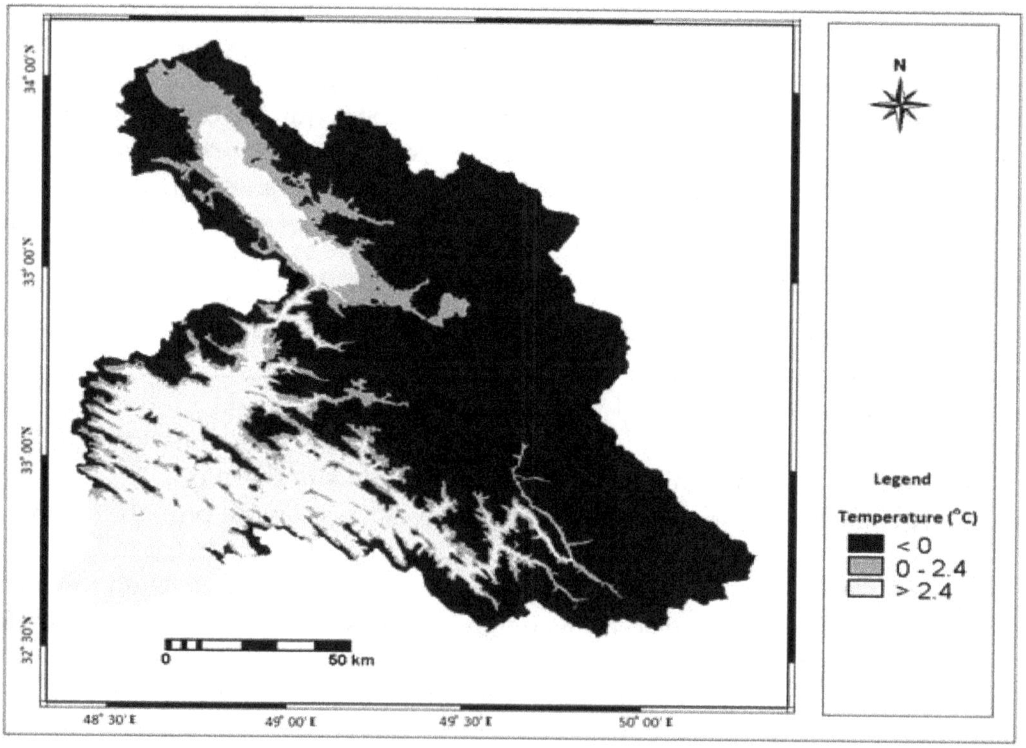

Figure 3. Potential areas for snow persistence ($T < 0$ °C), snowfall ($0 < T < 2.4$ °C) and snowmelt ($T > 2.4$ °C) in Dez Basin in February (for the period of 1974–1999).

Semi variogram analysis is carried out using the average of SWE data over the historical period, and the spatial distribution of estimation error variance is extracted using the best variogram model. Since the SWE data are available only for Dez and Karun basins, the performed estimation error variance analyses are limited to these basins. Figure 4 shows the error variance in the Dez Basin, areas with higher error are candidates for establishment of new stations.

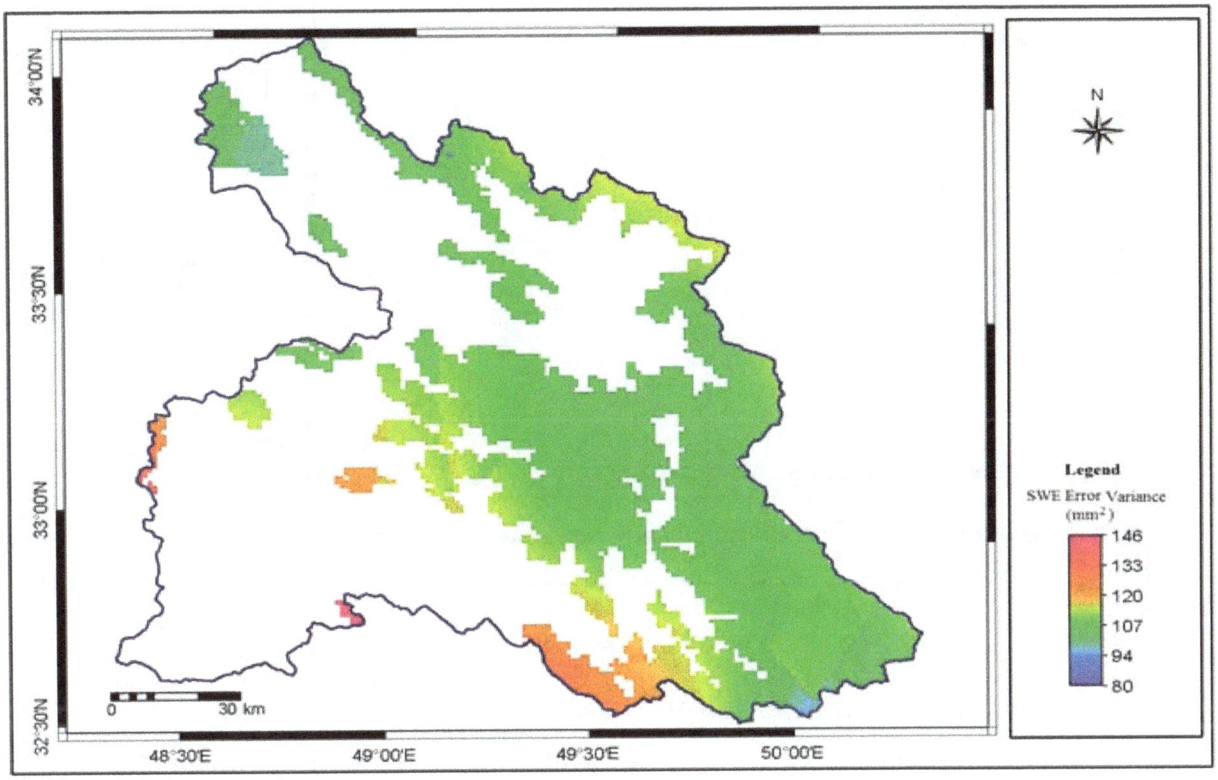

Figure 4. Error variance map of snow water equivalent for Dez Basin.

3.2. *Hydrological Criteria*

Snowmelt runoff volumes at 40 hydrometric stations are calculated based on the water balance model. Figure 5 shows the location of the selected stations in Dez Basin and all the hydrometric stations are shown in Figure 1. A regional regression model between snowmelt runoff volume and the percent of basin area higher than 2500 m is then built in order to provide estimates of snowmelt runoff at sub-basin outlets without hydrometric station (more information is available in Ghanbarpour et al. [22]). Snowmelt runoff volume and the corresponding priority in Dez Basin are given in Table 1 as an example.

Table 1. Snowmelt runoff volume and the corresponding priority in Dez Basin.

Sub-Basin	Snowmelt Runoff (mm)	Priority
Upstream of Tang Sezar	4.7	7
Upstream of Sepidan	12.4	6
Upstream of Sezar	44.7	2
Upstream of Tang Bakhtiari	42.5	3
Upstream of Zard Fahreh	84.2	1
Upstream of Daretakht	25.5	4
Upstream of Doroud	22.7	5

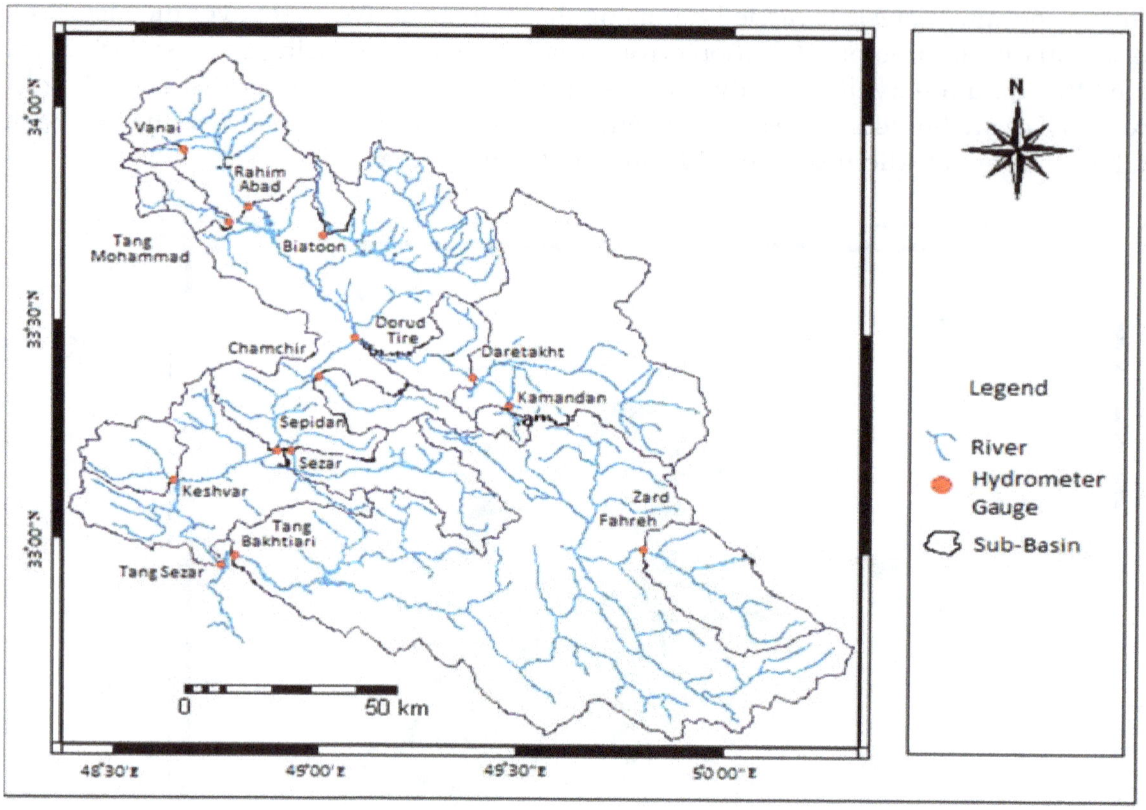

Figure 5. Selected hydrometric stations in Dez Basin.

3.3. *Snowfall and Snowpack Persistence Probability Criteria*

As previously indicated the FSO-SFP and FSO-SPP maps are derived based on Equation (3) over the snowfall and snowmelt seasons, respectively. These two layers represent the spatial distribution of the probability of the accumulation and persistence of snow that are key factors in site selection. Figures 6 and 7 show FSO-SFP and FSO-SPP maps in Dez Basin, respectively.

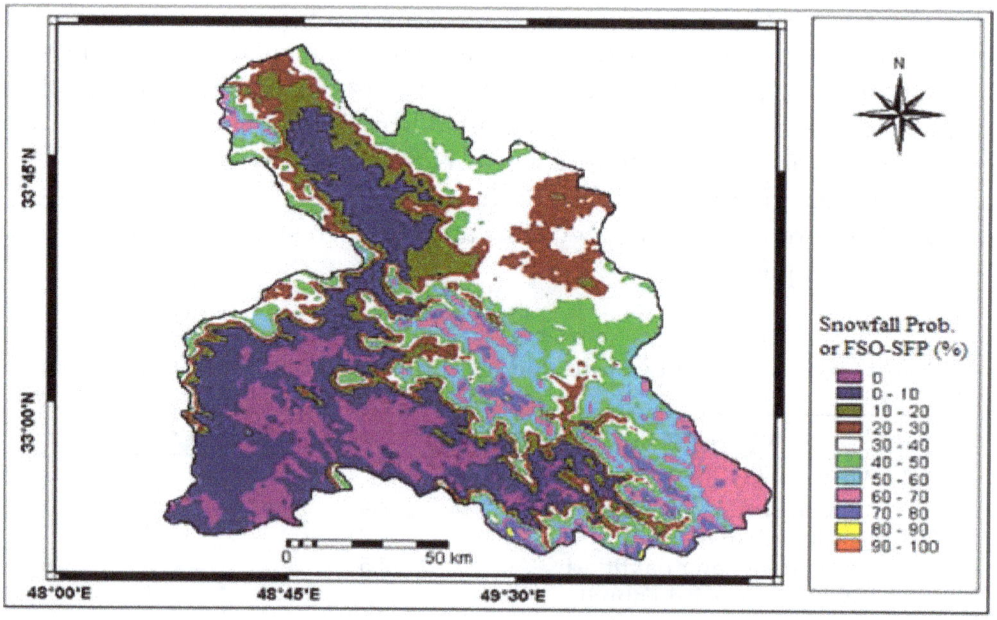

Figure 6. FSO-SFP Map in Dez Basin (for the period of 1984–2003).

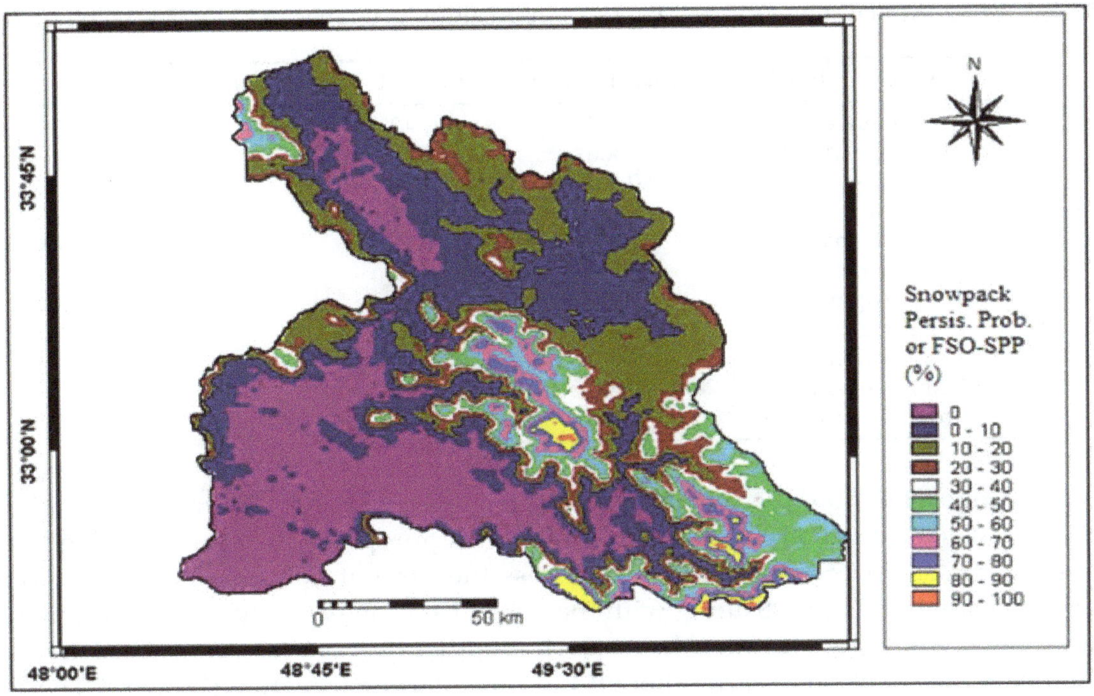

Figure 7. FSO-SPP Map in Dez Basin (for the period of 1984–2003).

3.4. Site Selection and Station Type

Site selection for snow stations is performed by integration of all available criteria including meteorological, hydrological and remotely-sensed variables. Site selection using the systematic and comprehensive approach described in this study will help decision makers to implement economically sound plans for optimizing a new set of stations or upgrading the existing network. This will be the case even where there is no historical snow data in the basin. In the study area, there is no historical snow data or MSC station available in Marun and Karkheh basins. Therefore, site selection of the snow stations is performed by integrating hydrological, meteorological and remotely-sensed variables. However, there are some operational MSC stations located in Karun and Dez Basins. Therefore, the refinement and optimizing of the existing network is performed based on the distribution of current stations and the error variance map, in addition to other variables mentioned above.

To make the proposed methodology more operationally practical, we choose to start the integration process from the altitudinal range of 250 m intervals and then generalize it to the basin scale. During the integrating phase, there is more emphasis on satellite images in terms of FSO-SFP and FSO-SPP criteria, due to the lack of homogeneous and long-term snow data. In the next step, meteorological and hydrological factors are also considered in order to compensate the impacts of snow data shortage. Since the site selection has been done in each individual elevation boundary, as an example, one of the altitudinal ranges of Dez Basin will be described in detail. This example shows step by step procedure that is followed in this research for all four major basins as follows.

The altitudinal range 2250–2500 m approximately covers 12% of the Dez Basin. The weighted average of FSO-SFP in this interval is about 40.8%. There are 11 MSC stations in this interval as listed in Table 2. According to this table, the distribution of stations over elevation is relatively good. Dalooni and Barfian stations have the highest and lowest snowfall probability or FSO-SFP in this altitudinal range, with 57% and 38%, respectively. In addition, Dalooni and Yazdgerd stations have the highest and lowest snowpack persistence probability or FSO-SPP, with 38% and 25%, respectively. Based on hydrologic criterion and road access, the ranking of each station is carried out for this altitudinal range. For example, in terms of road access, Ghaleh-Rostam and Yazdgerd stations are the most undesirable locations.

Table 2. Existing manual snow-course (MSC) stations in altitudinal range of 2250–2500 m in Dez Basin.

MSC Station	Elevation (m)	FSO-SFP (%)	FSO-SPP (%)
Vanaee	2250	42	30
Ghalavardeh2	2300	44	17
Vazmeh-dar	2300	48	32
Ghaleh-rostam	2300	32	10
Aziz-abad	2300	46	20
Barfian	2350	38	20
Yazdgerd	2500	44	25
Palang-dar2	2400	46	26
Dalooni	2500	57	38
Chahar-cheshmeh2	2400	46	18
Gardanehe-khakbad	2450	39	27

To make a final decision, the weighted average snowfall probability in this altitudinal range (rounded to 40%) is used as a criterion to determine the appropriate area for installing the stations. Thus, we set a threshold FSO-SFP of 40% to assess the present network. The selected zone for new stations is shown in Figure 8 along with the current station network. According to this figure, the existing network in this elevation has a desirable distribution, however, Ghaleh-rostam and Vazmeh-dar stations are close to each other and located at same elevation (2300 m). Therefore, one of the two stations should be removed. Vazmeh-dar station has more priority than Ghaleh-rostam station based on the FSO-SFP and FSO-SPP criteria, the ratio of winter precipitation to the annual precipitation, potential of snowfall, and road access. The two stations have similar contributions in snowmelt runoff according to the hydrological criteria. Therefore, Vazmeh-dar station is kept.

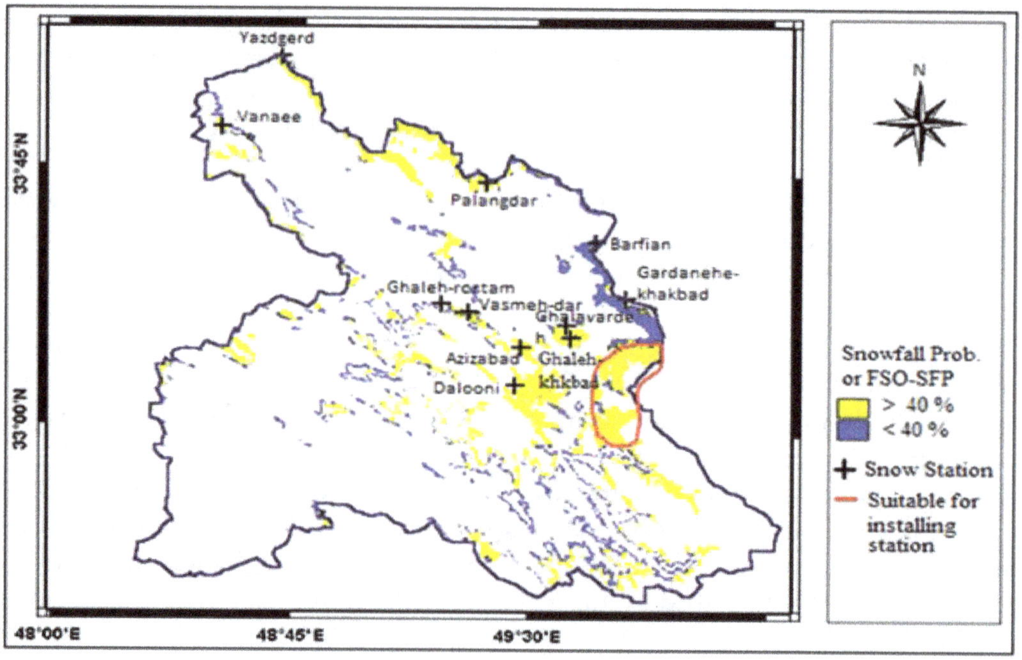

Figure 8. Selected zone for new snow stations and the existing station network in 2250–2500 m elevation interval in Dez Basin.

According to Figure 8 and the error variance map in Figure 4, spatial distribution of snow measurement stations in the 2250–2500 m elevation interval is generally appropriate while the error is relatively high in the west of the basin (shown by a thick polygon). Moreover, the average of snow persistence or FSO-SPP in this area exceeds 20% (Figure 7). By checking on topography and road access, one location is proposed for a new station in the mentioned area.

4. Conclusions

In this paper, a comprehensive approach for site selection of snow measurement stations in a mountainous area is proposed, based on an integration of meteorological, hydrological and remote sensing analysis. The approach is applied to a vast region encompassing four large river basins in the southwest of Iran. Snowfall and snowpack persistence probability maps are derived based on satellite images and used in a systematic site selection approach. Meteorological and hydrological variables such as basin average snowmelt and snowfall threshold temperature are included in the analysis to overcome lack of historical snow data availability.

The comprehensive site selection methodology for snow measurement stations not only ensures reliable snow data availability for the purpose of sustainable water resource management, but also provides an economically sound approach to optimize and maintain a cost-effective measurement network.

According to Moss [27], who emphasized economic considerations in network design, access (distance) to roads is considered as an indirect economic criterion for site selection of snow stations at local scale.

As a result of this study, six automatic snow-pillow stations (ASP) and four manual snow-course stations (MSC) in Karkheh basin; 17 ASP stations, 14 MSC stations and six control points in Dez Basin; 16 ASP stations, 10 MSC stations and 10 control points in Karun basin; and four ASP stations in Marun basin are proposed.

There is limited literature on design or evaluation of snow measurement networks in large spatial scale. The authors believe that the methodology described in this paper is a step forward for such studies, especially in data-poor regions where satellite information can play a significant role. However, there is room to improve on snow station network design in the future. One attractive alternative is to examine different scores for each input variable depending on their suitability, accuracy, and importance.

Acknowledgments: The authors acknowledge Jamab Consulting Engineering Co. for providing the snow cover map archive. Special thanks are also dedicated to the Khuzestan Water and Power Authority (KWPA) for partial support and providing companions in field trips.

Author Contributions: Bahram Saghafian and M. Reza Ghanbarpour proposed the original methodology for snow station site selection; Rahman Davtalab and Arezoo Rafieeinasab collected and analyzed the data and provided the results; Field trip were conducted by M. Reza Ghanbarpour and Rahman Davtalab; Bahram Saghafian, Arezoo Rafieeinasab, M. Reza Ghanbarpour and Rahman Davtalab jointly wrote the paper and addressed the comments.

References

1. Iran Ministry of Energy (IRME). *Karoun, Dez, Karkheh and Maroun River Basins Water Comprehensive Study—Second Study*; Iran Ministry of Energy (IRME): Tehran, Iran, 2012.
2. Gyawali, R.; Watkins, D.W. Continuous hydrologic modeling of snow-affected watersheds in the Great Lakes Basin using HEC-HMS. *J. Hydrol. Eng.* **2013**, *18*, 29–39. [CrossRef]
3. Larson, L.W.; Peck, E.L. Accuracy of precipitation measurements for hydrologic modeling. *Water Resour. Res.* **1974**, *10*, 857–863. [CrossRef]
4. World Meteorological Organization (WMO). *Snow Cover Measurements and Areal Assessment of Precipitation and Soil Moisture*; Operational Hydrological Report No. 35; World Meteorological Organization (WMO): Geneva, Switzerland, 1992.
5. Quanta Consulting Engineering Co. *Development and Modernization Plan of Iran Climatological Station*; Iran Meteorological Organization Publication No. 1; Quanta: București, Romania, 1976.
6. Molotch, N.P.; Bales, R.C. Scaling snow observations from the point to the grid element: Implications for observation network design. *Water Resour. Res.* **2005**, *41*. [CrossRef]

7. Brown, R.D.; Walker, A.E.; Goodison, B. Seasonal snow cover monitoring in Canada. In Proceedings of the 57th Eastern Snow Conference, Syracuse, NY, USA, 17–19 May 2000.
8. Bell, W.W.; Parmley, L.J.; Walk, H.; Afzal, H. Inflow forecasting for Pakistan's major reservoirs. *Int. Water Power Dam Constr.* **1994**, *46*, 21–26.
9. World Meteorological Organization (WMO). *Guide to Hydrological Practices No. 168*; World Meteorological Organization (WMO): Geneva, Switzerland, 1994.
10. Davtalab, R.; Madani, K.; Massah, A.; Farajzadeh, M. Evaluating the Effects of Climate Change on Water Reliability in Iran's Karkheh River Basin. In Proceedings of the World Environmental and Water Resources Congress 2014, Portland, OR, USA, 1–5 June 2014; pp. 2127–2135. [CrossRef]
11. Ashraf Vaghefi, S.; Mousavi, S.J.; Abbaspour, K.C.; Srinivasan, R.; Yang, H. Analyses of the impact of climate change on water resources components, drought and wheat yield in semiarid regions: Karkheh River Basin in Iran. *Hydrol. Process* **2014**, *28*, 2018–2032. [CrossRef]
12. Saghafian, B.; Davtalab, R. Mapping snow characteristics based on snow observation probability. *Int. J. Climatol.* **2007**, *27*, 1277–1286. [CrossRef]
13. Jamab Consulting Engineering Co. *Karoun, Dez, Karkheh and Maroun River Basins Water Comprehensive Study—First Study*; Jamab Consulting Engineering Co.: Tehran, Iran, 1991.
14. Porhemmat, J.; Saghafian, B.; Sedghi, H. An algorithm to mapping snow, cloud and land in NOAA AVHRR data, formulation, verification and evaluation. In Proceedings of the Fourth International Iran & Russia Conference, Shahrekord, Iran, 8–10 September 2004.
15. Goovaerts, P. *Geostatistics for Natural Resources Evaluation*; Oxford University Press, Inc.: New York, NY, USA, 1997.
16. Saghafian, B.; Davtalab, R.; Kafayati, M. Comparison of methods for determining the snowfall threshold temperature and potential area affected by snowfall in the Karkheh, Dez, Karun and Marun river basins. *Iran Water Resour. Res.* **2016**, *10*, 31–39.
17. Fleming, M.; Neary, V. Continuous hydrologic modeling study with the hydrologic modeling system. *J. Hydrol. Eng.* **2004**, *9*, 175–183. [CrossRef]
18. Garcia, A.; Sainz, A.; Revilla, J.A.; Alvarez, C.; Juanes, J.A.; Puente, A. Surface water resources assessment in scarcely gauged basins in the north of Spain. *J. Hydrol.* **2008**, *356*, 312–326. [CrossRef]
19. Saghafian, B.; Bondarabadi, S. Validity of Regional Rainfall Spatial Distribution Methods in Mountainous Areas. *J. Hydrol. Eng.* **2008**, *13*, 531–540. [CrossRef]
20. Severino, E.; Alpuim, T. Spatiotemporal models in the estimation of area precipitation. *Environmetrics* **2005**, *16*, 773–802. [CrossRef]
21. Rango, A.; Martinec, J. Snow accumulation derived from modified depletion curves of snow coverage. In Proceedings of the Exeter Symposium on Hydrological Aspects of Alpine and High Mountain Areas, Exeter, UK, 19–30 July 1982; IAHS Publ. No. 138; IAHS Press: Wallingford, UK, 1982; pp. 83–90.
22. Ghanbarpour, M.R.; Saghafian, B.; Saravi, M.M.; Abbaspour, K. Evaluation of spatial and temporal variability of snow cover in a large mountainous basin in Iran. *Nord. Hydrol.* **2007**, *38*, 45–58. [CrossRef]
23. Simpson, J.J.; Stitt, J.R.; Sienko, M. Improved Estimates of the Areal Extent of Snow Cover from AVHRR Data. *J. Hydrol.* **1998**, *204*, 1–23. [CrossRef]
24. Martinec, J. The degree day factor for snowmelt runoff forecasting. In Proceedings of the IUGG General Assembly of Helsinki, IAHS Commission of Surface Waters, Helsinki, Finland, 25 July–6 August 1960; IAHS Publ. No. 51; IAHS Press: Wallingford, UK, 1960; pp. 468–477.
25. Akyurek, Z.; Sorman, A.U. Monitoring snow-covered areas using NOAA-AVHRR data in the eastern part of Turkey. *Hydrol. Sci. J.* **2002**, *47*, 243–252. [CrossRef]
26. World Meteorological Organization (WMO). *The Planning of Meteorological Station Networks*; Technical Note No. 111; World Meteorological Organization (WMO): Geneva, Switzerland, 1970.
27. Moss, M.E. *Concepts and Techniques in Hydrological Network Design*; Operational Hydrology Report No. 19; World Meteorological Organization: Geneva, Switzerland, 1982.

11

From Surface Flow Velocity Measurements to Discharge Assessment by the Entropy Theory

Tommaso Moramarco *, Silvia Barbetta and Angelica Tarpanelli

Research Institute for Geo-Hydrological Protection, National Research Council, Via Madonna Alta 126, 06128 Perugia, Italy; s.barbetta@irpi.cnr.it (S.B.); a.tarpanelli@irpi.cnr.it (A.T.)
* Correspondence: t.moramarco@irpi.cnr.it

Academic Editor: Ataur Rahman

Abstract: A new methodology for estimating the discharge starting from the monitoring of surface flow velocity, u_{surf}, is proposed. The approach, based on the entropy theory, involves the actual location of maximum flow velocity, u_{max}, which may occur below the water surface (dip phenomena), affecting the shape of velocity profile. The method identifies the two-dimensional velocity distribution in the cross-sectional flow area, just sampling u_{surf} and applying an iterative procedure to estimate both the dip and u_{max}. Five gage sites, for which a large velocity dataset is available, are used as a case study. Results show that the method is accurate in simulating the depth-averaged velocities along the verticals and the mean flow velocity with an error, on average, lower than 12% and 6%, respectively. The comparison with the velocity index method for the estimation of the mean flow velocity using the measured u_{surf}, demonstrates that the method proposed here is more accurate mainly for rivers with a lower aspect ratio where secondary currents are expected. Moreover, the dip assessment is found more representative of the actual location of maximum flow velocity with respect to the one estimated by a different entropy approach. In terms of discharge, the errors do not exceed 3% for high floods, showing the good potentiality of the method to be used for the monitoring of these events.

Keywords: streamflow measurements; surface velocity; maximum velocity; entropy; ADCP; LSPIV; SVR

---

1. Introduction

The estimate of a reliable discharge is tightly connected to a robust velocity dataset where the mean flow velocity is inferred using velocity measurements carried out for high water levels. This condition is not so usual considering that at hydrometric sites the most common technique used for velocity sampling is based on the current meter from the cableway, and measuring velocity points during high floods in the lower portion of flow area is difficult and dangerous for operators [1]. The stage monitoring is straightforward and relatively inexpensive compared with the cost necessary to carry out flow velocity measurements, which are, however, limited to low flows and constrained by the accessibility of the site. Therefore, the discharge estimation is often obtained by extrapolating the relationship between stage and the discharge obtained for velocity measurements referring to low flows, and, as a consequence, a high uncertainty holds in the assessment [2]. For high flow, the mean flow velocity is hard to estimate, affecting de-facto the reliability of the discharge assessment for extreme events.

To date, ultrasonic devices, such as Acoustic Doppler Current Profilers (ADCP), can be adopted to address the velocity measure for high floods. Indeed, the introduction of ADCP on moving-vessels would allow operators to overcome the aforementioned issues, even if several limitations remain. During high floods, the magnitude of sediment transport may produce a reduction of the signal-noise ratio of acoustic sensors along with the capability of the signal to penetrate through the water [3]. In addition, high flow conditions, mean high velocity, and high turbulence can affect the water depth estimation because of the roll and pitch motions of the vessel [4] as well as vessel tracking [5]. The high velocity and high water levels represent big trouble for the operators considering that branches or trunks of trees can be carried by the flow and could still represent a risk for the instrument as well as the safety of the operators themselves.

On the other hand, the surface flow velocity can be easily monitored by using no-contact technologies as Surface Velocity Radar (SVR) [6,7] and/or Large Scale Particle Image Velocimetry (LSPIV) [8,9]. The radar technology may be affected during low flow when the backscatter of the beam can be blended by high noise, while LSPIV might be not suitable for measurements during scarce seeding, illumination conditions, and/or for any overnight floods [10]. However, the SVR and LSPIV, being no-contact techniques, are suitable for high flows monitoring, without encountering the issues connected to the floating surface material, sediment transport, and dangers for operators.

The information content given by SVR and LSPIV is the spatial distribution of surface flow velocity from which the discharge at a river site might be estimated. In this context, a focal point is how to turn the surface velocity, u_{surf}, measured at a location (vertical) into depth-averaged velocity, u_{vert}, considering the different hydraulic and geometric characteristics of river sites which affect the velocity profile shapes [11]. In the case of maximum flow velocity occurs on the water surface, the two-parameters power law velocity distribution developed by Dingman [12] could be applied to this end.

However, the aspect ratio W/D (with W the channel width and D the flow depth) unfolds a fundamental role on the characteristics of the velocity profile because the presence of the sidewalls may influence the velocity distribution, causing secondary currents that drive the maximum velocity below the water surface, so triggering the well-known dip phenomenon [13].

Overall, the surface velocities are transformed in depth-averaged velocity, assuming a monotonous velocity profile and multiplying the surface velocity, u_{surf}, by a velocity index, $k = u_{vert}/u_{surf}$, equal to 0.85 (e.g., [7] for SVR and [9] for LSPIV). However, the possible occurrence of the dip phenomenon could make the assumption 'weak' for create flows where a monotonous velocity distribution does not take place. Consequently, the velocity index value might be not representative of the two-dimensional spatial velocity distribution anymore, leading to failure of the assessment of discharge.

The entropic velocity distribution developed by [14] can be used to cope with this matter. This approach established a bridge between the probability domain, wherein a probability distribution of the velocity is surmised, and the physical space by deriving the cumulative probability distribution function in terms of curvilinear coordinates in the physical space [15,16]. Therefore, it would be possible to relate the surface velocity measured by SVR and/or LSPIV across the river site, and the maximum velocity occurring below the water surface corresponding to the vertical where the surface velocity is sampled, as shown by [17]. In this context, the identification of the dip, i.e., the location where u_{max} occurs below the water surface, is the main issue. The dip may significantly affect the estimation of the depth-averaged velocity from which, through the velocity area method, the mean flow can be assessed and, hence, the discharge.

This work is aimed to present a new methodology based on the entropy theory of estimating the discharge, starting from the surface flow velocity and taking the dip of the velocity profile into account. Moreover, remarks on the robustness of the velocity index using field data are discussed. The entropy method can be applied regardless of the technique adopted for the surface velocity measurement.

Five gage sites with different geometric and hydraulic characteristics are used as a case study.

The paper is organized as follows. Section 2 provides the theoretical background of the entropy method applied to measured surface flow velocities. Section 3 describes the case study along with the velocity dataset. Section 4 describes the results achieved for the five selected river gage sites and makes a comparison with the velocity index method. Finally, the last section features the conclusions drawn from this investigation.

2. Method

The method is based on the entropy probability density function of velocity, u, such as derived by [18] and simplified by [19] assuming that the entropy velocity profile at the y-axis, where the maximum flow velocity, u_{max}, occurs, holds for all verticals in flow area, yielding:

$$u(x_i, y) = \frac{u_{maxv}(x_i)}{M} \ln\left[1 + \left(e^M - 1\right) \frac{y}{D(x_i) - h(x_i)} \exp\left(1 - \frac{y}{D(x_i) - h(x_i)}\right)\right] \text{ for } i = 1 \ldots N_v \quad (1)$$

where $u_{maxv}(x_i)$ is the maximum velocity along the ith vertical, x_i is the position of the ith sampled vertical from the left bank, M is the entropic parameter, $h(x_i)$ is the dip, i.e., the depth of $u_{maxv}(x_i)$ below the water surface, $D(x_i)$ the vertical depth, y is the distance of the velocity point from the bed, and N_v is the number of verticals sampled across the river section. The entropic parameter M, which is a characteristic of the section [20], can be easily estimated through the pairs (u_m, u_{max}) of the available velocity dataset at a gauge site by using the linear entropic relation [14]:

$$u_m = \left(\frac{e^M}{e^M - 1} - \frac{1}{M}\right) u_{max} = \Phi(M) u_{max} \quad (2)$$

where u_m is the mean flow velocity and u_{max} is the maximum flow velocity in flow area.

In addition, Moramarco [21] showed that, at a gauged river site, $\Phi(M)$ is constant for any flow conditions and its estimation can be addressed for ungauged sites as well by leveraging the dependence of $\Phi(M)$ on hydraulic and geometric characteristics.

Even if u_{maxv} occurs below the water surface, it could be estimated for each vertical as a function of surface velocity, u_{surf}, such as proposed by [22]:

$$u_{maxv}(x_i) = \frac{u_{surf}(x_i,\ D(x_i))}{\frac{1}{M} ln\left[1 + (e^M - 1)\delta(x_i) e^{1-\delta(x_i)}\right]} \quad (3)$$

where $\delta(x_i) = \frac{D(x_i)}{D(x_i) - h(x_i)}$. Specifically, if $h(x_i) = 0$, it follows that $\delta(x_i) = 1$ and, hence, that $u_{maxv}(x_i) = u_{surf}(x_i,\ D(x_i))$. An application of Equation (3) in natural channels was done by [17], who calibrated a hydraulic model by leveraging the surface velocity measurements by SVR at different times.

2.1. Dip Estimate

Considering that the dip, $h(x_i)$, is generally unknown across the flow area, a method is proposed here for its estimation. The approach exploits the experimental outcomes obtained by [13] who investigated the mechanism of the dip phenomenon by leveraging a laboratory channel velocity dataset. Specifically, it was found that for different aspect ratios W/D (with W the channel width and D the flow depth), the location of $u_{maxv}(x_i)$ below the water surface, is mainly linked to the lateral position of the velocity profiles from the sidewalls. Based on this analysis, the dip phenomenon was more significant close to the banks and for large aspect ratio values, in accordance with the following relationship [13]:

$$\delta(x_i) = 1 + 1.3 e^{-x_i/D(x_i)} \quad (4)$$

It is worth noting that Equation (4) well represented the measured data obtained in different works dealing with smooth and rough natural channels and we refer to [13] for more details.

Placing Equation (4) into Equation (3) and both into Equation (1), the two-dimensional velocity distribution can be identified by leveraging the surface velocity measured by SVR or LSPIV. In this way, u_m and u_{max} can be estimated using the velocity-area method [23], and, hence, $\Phi(M)$ can be computed by Equation (2) [19]. However, the uncertainty due to the presence of secondary currents lead us to modify Equation (4) in:

$$\delta'_p(x_i) = a_p + \delta(x_i) \tag{5}$$

where a_p is a parameter representing the dip at the y-axis and which is updated through an iterative process, wherein at each iteration, p, a constant value (0.1 m) is added to the initial value $a_1(p = 1)$, assumed to be equal to 0.05 m. The iterative parameter refinement process ends by minimizing the objective function when the optimal value of $\Phi_{opt}(M_p)$ is achieved in accordance with:

$$\Phi(M_{obs}) - \Phi(M_p) = \epsilon \tag{6}$$

where ϵ is the threshold error, assumed equal to 0.01.

Steps of the procedure described above are detailed in the schematic diagram as shown in Figure 1. The method can be applied for gage sites in which a robust velocity dataset is available and $\Phi(M_{obs})$ can be easily estimated by the observed pairs (u_m, u_{max}). For ungauged sites, the M parameter can be even estimated by expressing its value in terms of hydraulic and geometric characteristics such as proposed by [21] and/or following [20], who identified a simple way to estimate M from the maximum surface velocity typically located near the middle of the channel [24].

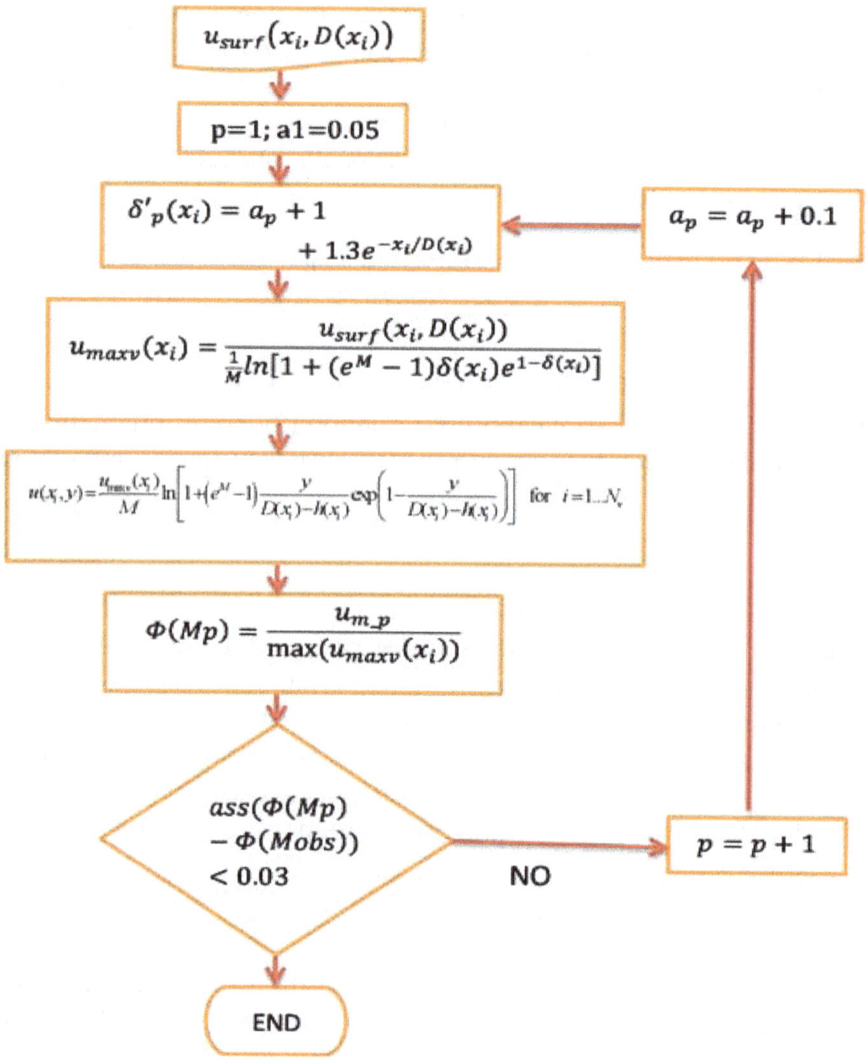

Figure 1. Schematic representation of dip estimation procedure using measured surface velocity, u_{surf} (for symbols see text).

For representing the dip in flow area, the relationship proposed by [25] may be also adopted:

$$\frac{h(x_i)}{D(x_i)} = 0.2ln\frac{58.3M\varnothing(M)}{e^M - 1} \tag{7}$$

Unlike the proposed method, Equation (7) does not need to consider an iterative procedure to estimate the dip along sampled verticals and would be simpler to apply. Specifically, Equation (7) provides $h(x_i)$ that can be used to estimate $\delta(x_i)$. This implies the interest in comparing the two approaches. However, it is worth noting that our target is not to estimate the effectiveness of these relationships in the dip assessment but rather to analyze whether, if given a distribution of dip in flow area, the proposed method is able to provide an accurate estimate of velocity, and considering that Equation (4) was tested with laboratory and field data, it is sufficient for our purpose. However, as stressed above, a comparison with Equation (7) is addressed as well and shown afterwards.

2.2. The Velocity Index

The velocity-area method is applied to estimate the mean flow velocity starting from the depth-averaged velocity, u_{vert}, estimated by velocity points along verticals sampled by using, e.g., current meter across the cross-sectional flow area. In this case, SVR is used for discharge monitoring, the measure of velocity is limited to the surface, and there is the need to turn the surface velocity u_{surf}, into u_{vert}. For that, an approach widely applied is to consider the velocity index, k, defined as:

$$k = u_{vert}/u_{surf} \tag{8}$$

As stressed by [7], k is set to 0.85 for many river gage site configurations (see also [6,9,10]). It has to be pointed out that the choice of k is of paramount importance, considering that Equation (8) refers to a monotonous velocity profile that, however, does not take the dip phenomenon into account. This issue could make the method 'weak' for estimating u_{vert}, in the case of secondary currents taking place.

3. Case Study

Considering that our main purpose is to evaluate the effectiveness of the proposed procedure, the analysis is addressed for the case of gage sites where the velocity measurements are limited only to the surface, e.g., by using SVR. This aspect is of considerable interest for the monitoring of high floods. Therefore, the proposed method is tested using a velocity dataset of five gauged sites. Four are located in the Tiber basin; Santa Lucia (900 km^2), Ponte Felcino (1970 km^2), and Ponte Nuovo (4100 km^2) along the Tiber River and Rosciano (1950 km^2) on the Chiascio River. One section is on the Po River, i.e., Pontelagoscuro (70,000 km^2).

The velocity dataset consists of velocity points data collected during velocity measurements carried out by conventional techniques, i.e., current meter from cableway. The number of velocity points sampled along the verticals is sufficient (more than eight for some verticals) to reconstruct the vertical velocity profile, even in presence of secondary currents. Table 1 shows for all the selected gaged sites the main flow characteristics and, as it can be seen, the measurements cover low and high flow conditions.

Table 1. Flow dataset characteristics: N_m = number of velocity measurements considered, N_v = total number of verticals (from the N_m measurements), $\Phi(M_{obs})$ = observed entropy parameter, Q = measured discharge, D = average flow depth, W = average channel width. The period of sampling is also shown.

River	Site	N_m	N_v	$\Phi(M_{obs})$	Q (m^3/s)	D (m)	W (m)	W/D	Period
Tiber	Santa Lucia	16	93	0.66	30–185	2.54	22	8.79	1987–2008
Tiber	Ponte Felcino	8	78	0.60	28–411	3.22	38	11.66	1990–2003
Tiber	Ponte Nuovo	12	77	0.66	10–540	3.98	50	12.30	1986–2005
Chiascio	Rosciano	8	67	0.60	20–378	2.74	35	12.90	1982–2002
Po	Pontelagoscuro	8	89	0.68	380–4000	6.32	270	43.63	1987–1992

4. Results and Discussion

The effectiveness of the proposed procedure in estimating the dip phenomena along with the velocity profile by Equation (1) is discussed here using the velocity dataset of the selected river gage sites. The errors in estimating the depth-averaged velocity along with the discharge are analyzed in-depth. Before that, however, the robustness of the velocity index is investigated using the velocity measurements carried out at gage sites, considering that these measurements contain velocity points well distributed along each vertical sampled in the cross-sectional flow area.

4.1. The Velocity Index Analysis

At a first step, the analysis of the velocity index, k, is addressed here. Unlike [7], who identified the variation of k with relative roughness and flow depth, we investigate how k can vary at each single site and if a dependence on the aspect ratio can be inferred. Considering the velocity dataset in the five gage sites, for each velocity measurement the depth-averaged velocity, u_{vert}, is estimated starting from the measured surface velocity, u_{surf}, i.e., $u_{vert}(x_i,D_i) = k\ u_{surf}(x_i,D_i)$, and compared with the depth-averaged velocity assessed by the sampled velocity points along each vertical.

The velocity at the water surface sampled by the current meter is used to mimic the surface velocity measured by SVR or LSPIV. It is worth noting that the analysis does not aim to show the capabilities of the above two technologies but only to verify the relationship between surface velocity and depth-averaged velocity, which is useful to estimate the mean flow velocity by using the velocity-area method. Figure 2 shows the box-plot of the velocity index values for the five gage sites as a function of the aspect ratio, W/D, in terms of the 5th and 95th percentile along with the median, minimum, and maximum values. As can be seen, the velocity index is quite scattered at each gage site, while the median value is inversely proportional with the aspect ratio, starting from 1.05 at Santa Lucia, the narrowest river section, and dropping to 0.86 at Pontelagoscuro, the widest site. Therefore, the average value of the k index observed for the five river gage sites deviates from the constant value of 0.85 used in the literature (e.g., [6,7]); for the Po River the mean value is close to 0.85. It is worth nothing that $k > 1$ means the dip phenomenon is significant in the velocity profile, causing a consistent curved shape approaching a parabolic one and leading to the depth-averaged velocity being greater than the surface velocity.

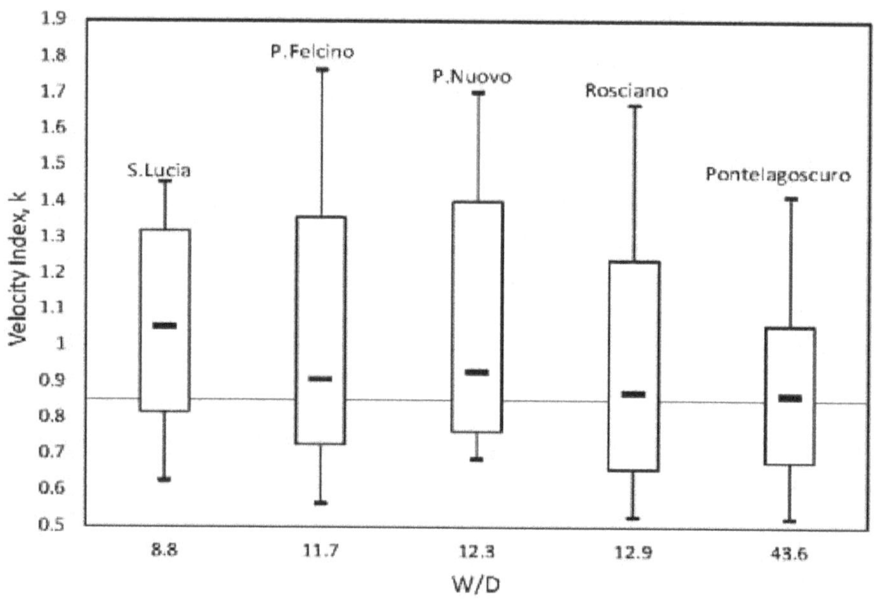

Figure 2. Box-plot of the values of the velocity index, k, as a function of the aspect ratios, W/D, for the five gage sites based on the velocity dataset. The line corresponding to the value of 0.85 is also shown along with the name of the gage sites.

Moreover, inspecting Figure 2, one infers that the velocity index is also influenced by the geometry of the river site, such as expressed by the aspect ratio, and while the median value tends to decrease for large aspect ratios (e.g., Pontelagoscuro site with W/D = 43.6), the distribution of the velocity index is always scattered, even in the widest channel. Indeed, as can be inferred from Figure 2, the mean value of k tends to be 0.85 for large values of the aspect ratio, while the variance seems to decrease only for the largest W/D occurring at the Pontelagoscuro site.

It has to be stressed that the limited sample of river gage sites does not allow a general rule for applying the velocity index to any river worldwide to be inferred. However, Figure 3 provides useful insights into understanding why $k = 0.85$ cannot be considered a parameter to be applied to rivers everywhere and for whatever flow condition.

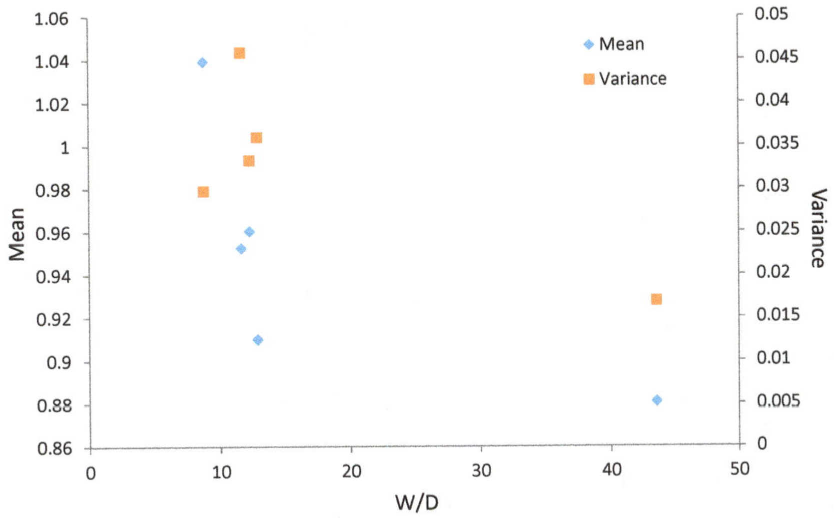

Figure 3. Mean and variance of the velocity index, k, for the selected gage sites as a function of the aspect ratio (W/D).

At a first attempt, Figure 3 could be useful for addressing the uncertainty in estimating the velocity index as a function of W/D. In this case, identifying the relationship expressing k_{mean} and $k_{variance}$ as a function of (W/D) could be of support in understanding the variability of the velocity index and, hence, to evaluate how much the depth-averaged velocity can vary at each vertical sampled in correspondence to the surface velocity measurement. However, more gage sites are necessary to identify a robust relationship that can be leveraged in addressing the uncertainty in the k estimation.

Therefore, the above results indicate that, by using new technologies as SVR and/or LSPIV, the application of a velocity index when dip phenomena occur may affect the accuracy of the mean flow velocity assessment and, hence, of the discharge.

4.2. Comparison of Depth-Averaged Velocities

Based on the previous analysis, it is evident that the assumption of a constant index velocity, i.e., 0.85, might lead to an incorrect estimate of the depth-averaged velocity and, hence, of the discharge. The variability of the index velocity for each vertical sampled in the cross-sectional flow area depends on the presence of dip phenomena due to effect of secondary currents caused by side walls, bridges, and wind effects, as well as the presence of floating material. These effects seem to tend to be smooth for large values of the aspect ratio. Therefore, in the case of streamflow measurements the use of the velocity index method should be applied with care and the reliability verified.

The entropy method proposed here allows the location of the maximum velocity below the water surface (dip) for each vertical sampled in the flow area to be estimated, starting from the measured surface velocity. The procedure illustrated in Figure 1 is applied for all measurements available at the selected gage sites and the results are compared with the ones obtained using the velocity index method with $k = 0.85$.

To this end, the error in magnitude for the depth-averaged velocity is computed for each gage sites and the results are shown in Figure 4 in terms of a box-plot referring to the median, 5th, and 95th percentile. As can be seen and expected, the differences are more noticeable for the narrower gage sites, with a median that does not exceed 6% and 11% for the entropy and the velocity index method, respectively; while for the Pontelagoscuro site, the differences are less evident, even though a slightly larger scatter is always present in the velocity index approach.

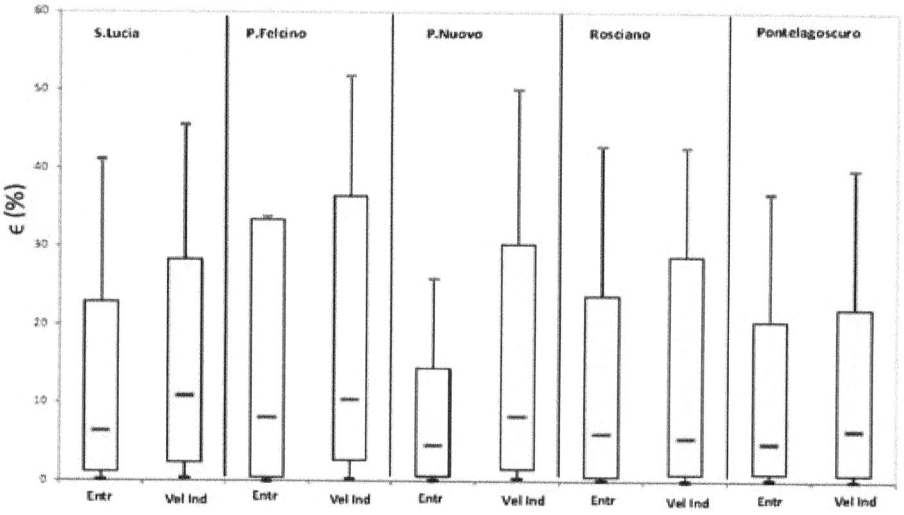

Figure 4. Box-plot of absolute percentage errors (ϵ) in estimating the depth-averaged velocity using the entropy approach (Entr) and the velocity index (Vel Ind).

The entropy approach has been also compared with Chiu and Tung's formulation [24], for which the dip is computed without an iterative procedure. In this case, the dip computed by Equation (7) is replaced in Equation (3), from which $u_{maxv}(x_i)$ is computed and used in Equation (1). By way of example, Figure 5 shows for the Rosciano and Pontelagoscuro gage sites the comparison between the errors in depth-averaged velocity computed using the dip assessment by [25] and the proposed iterative method. As can be seen, the iterative entropy approach performs better than the one based on Chiu's dip, and, as expected, the difference tends to decrease if the aspect ratio increases as for the Pontelagoscuro site. Similar performances obtained at the Rosciano site are for the three sites with lower aspect ratios (W/D). Table 2 shows the mean and standard deviation of the percentage errors in magnitude of estimating the depth-averaged velocity by using Equation (7) and the method proposed here for the dip assessment, showing again that the dip computed by the iterative procedure is more accurate than that based on Equation (7).

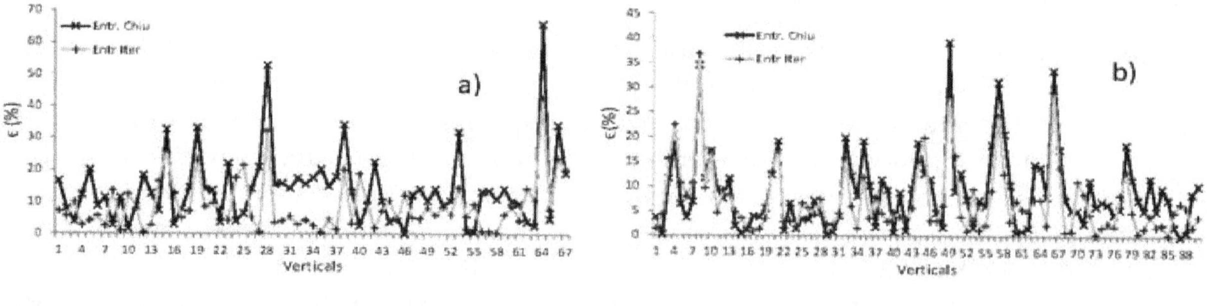

Figure 5. Absolute percentage error in estimating the depth-averaged velocity using the dip assessment by the iterative procedure proposed here (Entr Iter) and Chiu's formula (Entr Chiu) for (**a**) the Rosciano and (**b**) the Pontelagoscuro gage sites.

Table 2. Mean and standard deviation of the absolute percentage error in estimating the depth-averaged velocity using the dip assessment by the iterative procedure proposed here (Entr Iter) and Equation (7) (Entr Chiu) for the investigated gage sites.

Gage Site	Entr Iter		Entr Chiu	
	Mean (%)	Standard Deviation (%)	Mean (%)	Standard Deviation (%)
Santa Lucia	9.8	8.8	12.5	9.1
Ponte Felcino	11	9.6	19	11.1
Ponte Nuovo	8.6	7.9	15.4	10.2
Rosciano	8.9	8.1	15.1	11.4
Pontelagoscuro	7.2	6.9	8.9	7.8

4.3. Comparison of Mean Flow Velocity and Discharge

Once the depth-averaged velocities are computed for each measurement, the cross-sectional mean flow velocity can be estimated using the velocity-area method. Considering that the estimate of dip across the flow area by the proposed approach is more robust than that based on Equation (7), the analysis is here addressed for the proposed procedure and velocity index only.

Figure 6 shows the box-plot of the absolute percentage error in the computation of mean flow velocity for the whole velocity dataset for the five investigated gage sites. As can be seen, the entropy approach performs better than the velocity index one, as is inferable by comparing the median and 95th percentile of errors and the differences in terms of the median and scattering tend to smooth for larger aspect ratios, as for the Rosciano and Pontelagoscuro gage sites. The mean percentage error does not exceed 4.5% and 10% for the entropy and velocity index methods, respectively.

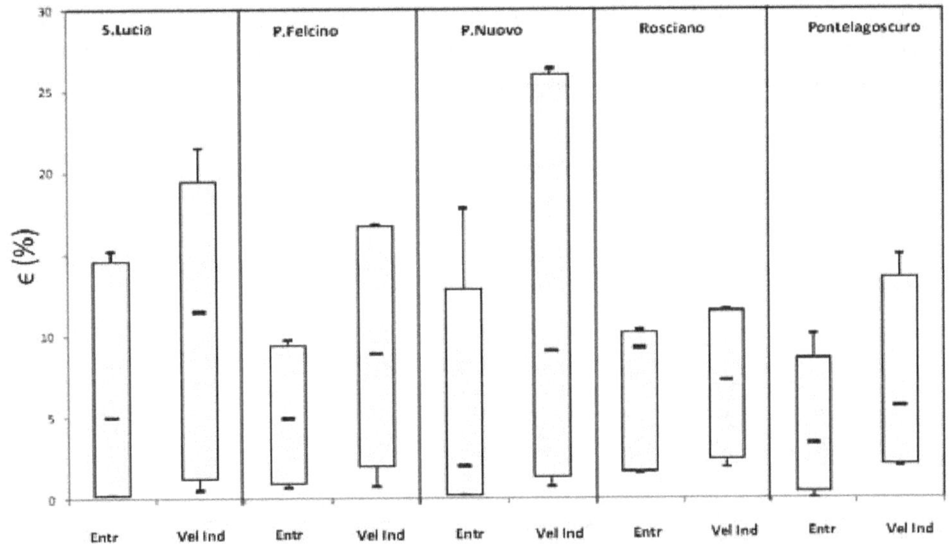

Figure 6. Box-plot of the absolute percentage error (ϵ) in estimating the cross-sectional mean flow velocity using the Entropy approach (Entr) and the velocity index (Vel Ind).

Moreover, considering the performances of the two methods the greatest differences are identified for higher mean flow velocity, as can be seen in Figure 7, where a comparison between the observed discharge and the one computed by the entropy and velocity index approaches is shown for the Santa Lucia, Ponte Nuovo, and Pontelagoscuro gage sites. Similar results are found for Ponte Felcino and Rosciano. Figure 7 demonstrates that, for the velocity index method ($k = 0.85$), the maximum errors shown in the graph in Figure 6 are referred to as the high discharges with an error greater than 26%. This proves that the velocity index method needs to be applied with care if high flood events are of interest. Conversely, for the proposed entropy approach the maximum error in percentage is obtained for the lower flows and does not exceed 14%, while for high floods the errors do not exceed

3%, thus showing the benefit in using the method for the monitoring of high floods starting from the sampling of surface flow velocity.

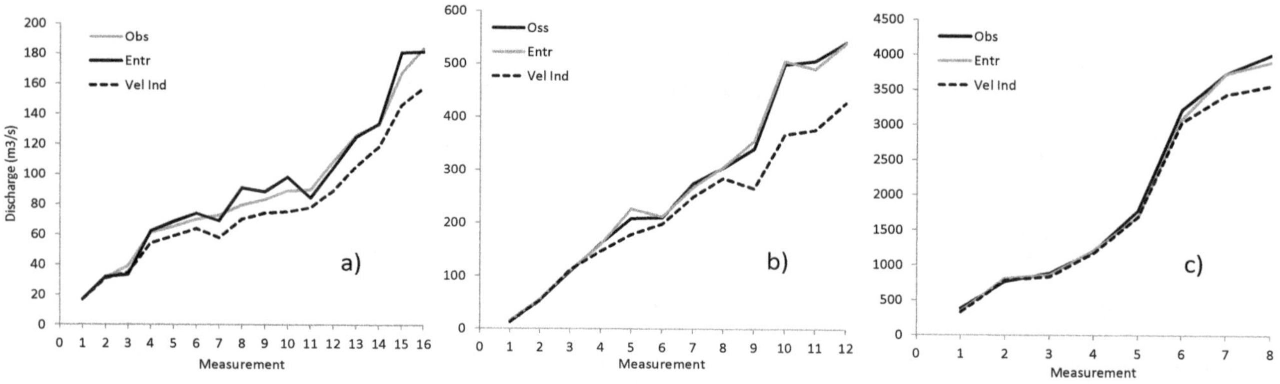

Figure 7. Comparison between observed discharge (Obs) and discharge computed by the Entropy approach (Entr) and the velocity index (Vel Ind) for (**a**) Santa Lucia; (**b**) Ponte Nuovo; and (**c**) Pontelagoscuro.

Finally, the correlation between u_{maxv} values observed and computed by the entropy method for each vertical is found good with the coefficient of determination, R^2, varying from 0.84 for the narrower sites (Santa Lucia and Ponte Felcino) to 0.94 for the wider ones. Moreover, by way of example, Figure 8 shows for the highest flood in the dataset of the Ponte Nuovo gauged site, the comparison between the velocity profile at four different verticals computed by the entropy method, and the velocity points sampled by current meter. It is worth noting that the velocity profiles are obtained just by sampling u_{surf}, and no information is used in terms of dip. It is noticeable how the velocity profile shape computed by Equation (1) is able to embed the observed points and, particularly, u_{maxv}. Similar performances are obtained for the velocity dataset of the other investigated gage sites.

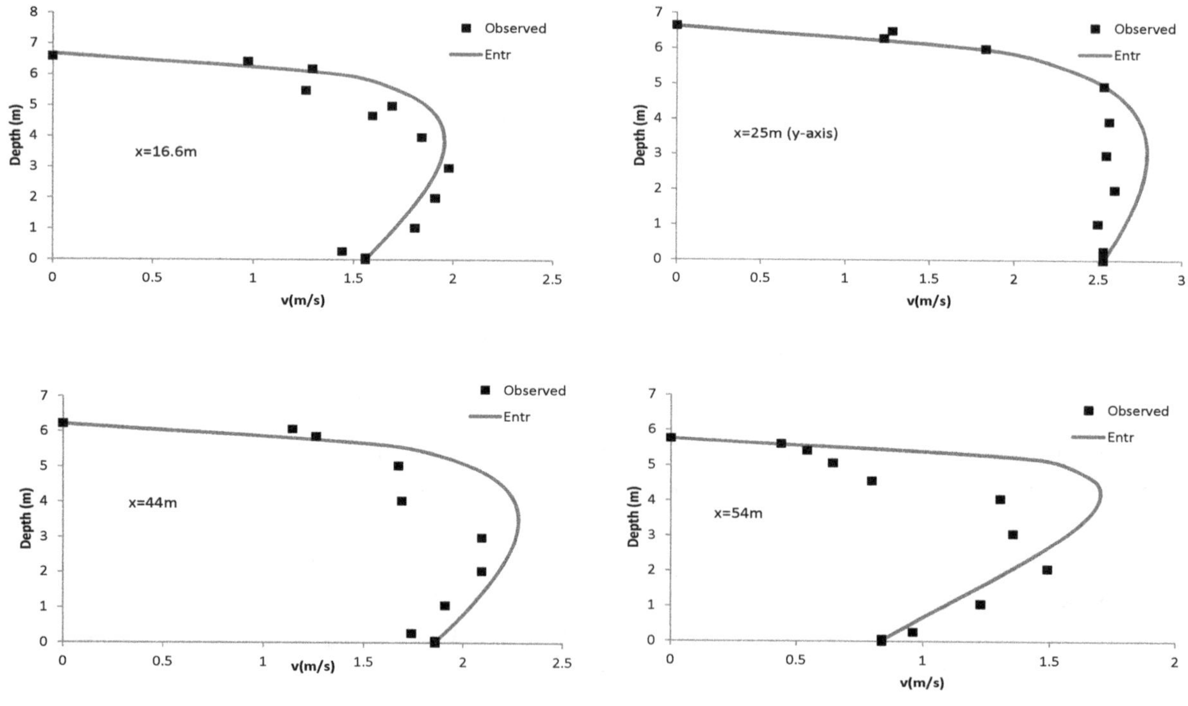

Figure 8. Velocity profiles estimated by entropy method (Equations (1) and (5)) plotted against velocity points sampled by current meter along four verticals during the highest flood of the velocity dataset at the Ponte Nuovo gage site. x is the distance from the left bank, while the y-axis represents the vertical where u_{max} occurs.

5. Conclusions

The trialing of emerging no-contact technologies, like SVR and LSPIV, for surface flow velocity measurement in rivers is of considerable interest and represents an effective alternative to conventional monitoring. These new technologies allow the issues related to the limited availability of hydraulic information due to low flow conditions and the difficulties and dangers to operators measuring velocity points in the lower portion of flow area using standard sampling techniques to be overcome.

In this context, this work addresses the crucial matter of turning the surface velocity into depth-averaged velocity, even if secondary current effects occur, for whatever natural channel geometry, providing some useful insights.

Based on the field data used, it has been shown that the proposed entropic algorithm well lends itself to being applied when advanced no-contact technology is used to estimate the discharge by monitoring the surface flow velocity across the river. The dip phenomenon, having a fundamental role in depth-averaged velocity assessment, is accurately identified by the entropy procedure for whatever flow conditions at the investigated gage sites.

The velocity index method, with k = 0.85, is of little use for rivers with a lower aspect ratio, with significant errors in estimating high discharges; while for wider rivers, its performance is found quite similar to that of the entropy method, considering that the dip phenomena are quite negligible. The relationship found between the mean and variance of the aspect ration of the k index needs to be further investigated by extending the velocity dataset and the case studies.

Acknowledgments: This work was partly funded by The Project of Interest NextData (MIUR-CNR). The authors wish to thank the Umbria Region Department of Environment, Planning, and Infrastructure, for providing Tiber River basin data. The authors want to thank Cristiano Corradini for the technical support. Finally, the Reviewers are gratefully acknowledged for their useful comments.

Author Contributions: All authors contributed extensively to the work. Tommaso Moramarco developed the methodology, conceived and performed the numerical experiments, and wrote the article. Silvia Barbetta and Angelica Tarpanelli contributed equally to the discussion of methodology, results analysis and the manuscript writing.

References

1. Moramarco, T.; Saltalippi, C.; Singh, V.P. Velocity profiles assessment in natural channel during high floods. *Hydrol. Res.* **2011**, *42*, 162–170. [CrossRef]
2. Di Baldassarre, G.; Montanari, A. Uncertainty in river discharge observations: A quantitative analysis. *Hydrol. Earth Syst. Sci.* **2009**, *13*, 913–921. [CrossRef]
3. Muller, G.; Bruce, T.; Kauppert, K. Particle image velocimetry: A simple technique for complex surface flows. In Proceedings of the International Conference on Fluvial Hydraulics, Balkema, Lisse, The Netherlands, 4–6 September 2002; pp. 1227–1234.
4. Simpson, M.R.; Oltman, R.N. *Discharge Measurement System Using an Acoustic Doppler Current Profiler with Applications to Large Rivers and Estuaries: U.S. Geological Survey Water-Supply Paper 2395*; U.S. Geological Survey: Reston, VA, USA, 1993; 32p.
5. Wagner, C.R.; Mueller, D.S. Comparison of bottom-track to global positioning system referenced discharges measured using an acoustic Doppler current profiler. *J. Hydrol.* **2011**, *401*, 250–258. [CrossRef]
6. Costa, J.E.; Spicer, K.R.; Cheng, R.T.; Haeni, F.P.; Melcher, N.B.; Thurman, E.M. Measuring stream discharge by non-contact methods: A proof-of-concept experiment. *Geophys. Res. Lett.* **2000**, *27*, 553–556. [CrossRef]
7. Welber, M.; Le Coz, J.; Laronne, J.B.; Zolezzi, G.; Zamler, D.; Dramais, G.; Hauet, A.; Salvaro, M. Field assessment of noncontact stream gauging using portable surface velocity radars (SVR). *Water Resour. Res.* **2016**, *52*, 1108–1126. [CrossRef]
8. Fujita, I.; Muste, M.; Kruger, A. Large-scale particle image velocimetry for flow analysis in hydraulic applications. *J. Hydraul. Res.* **1998**, *36*, 397–414. [CrossRef]
9. Tauro, F.; Porfiri, M.; Grimaldi, S. Orienting the camera and firing lasers to enhance large scale particle image velocimetry for streamflow monitoring. *Water Resour. Res.* **2014**, *50*, 7470–7483. [CrossRef]

10. Muste, M.; Fujita, I.; Hauet, A. Large-scale particle image velocimetry for measurements in riverine environments. *Water Resour. Res.* **2008**, *44*. [CrossRef]
11. Ferro, V. ADV measurements of velocity distributions in a gravel-bed flume. *Earth Surf. Proccess. Landf.* **2003**, *28*, 707–722. [CrossRef]
12. Dingman, S.L. Probability Distribution of Velocity in Natural Channel Cross Sections. *Water Resour. Res.* **1989**, *25*, 509–518. [CrossRef]
13. Yang, S.Q.; Tan, S.K.; Lim, S.Y. Velocity distribution and dip-phenomenon in smooth uniform open channel flows. *J. Hydraul. Eng.* **2004**, *130*, 1179–1186. [CrossRef]
14. Chiu, C.L. Application of Entropy Concept in Open-Channel Flow Study. *J. Hydraul. Eng.* **1991**, *117*, 615–627. [CrossRef]
15. Singh, V.P.; Fontana, N.; Marini, G. Derivation of 2D power-law velocity distribution using entropy theory. *Entropy* **2013**, *15*, 1221–1231. [CrossRef]
16. Fontana, N.; Marini, G.; De Paola, F. Experimental assessment of a 2-D entropy-based model for velocity distribution in open channel flow. *Entropy* **2013**, *15*, 988–998. [CrossRef]
17. Corato, G.; Moramarco, T.; Tucciarelli, T. Discharge estimation combining flow routing and occasional measurements of velocity. *Hydrol. Earth Syst. Sci.* **2011**, *15*, 2979–2994. [CrossRef]
18. Chiu, C.L. Velocity distribution in open channel flow. *J. Hydraul. Eng.* **1989**, *115*, 576–594. [CrossRef]
19. Moramarco, T.; Saltalippi, C.; Singh, V.P. Estimation of mean velocity in natural channel based on Chiu's velocity distribution equation. *J. Hydrol. Eng.* **2004**, *9*, 42–50. [CrossRef]
20. Farina, G.; Alvisi, S.; Franchini, M.; Moramarco, T. Three methods for estimating the entropy parameter M based on a decreasing number of velocity measurements in a river cross-section. *Entropy* **2014**, *16*, 2512–2529. [CrossRef]
21. Moramarco, T.; Singh, V. Formulation of the entropy parameter based on hydraulic and geometric characteristics of river cross sections. *J. Hydrol. Eng.* **2010**, *15*, 852–858. [CrossRef]
22. Fulton, J.; Ostrowski, J. Measuring real-time streamflow using emerging technologies: Radar, hydroacoustics, and the probability concepts. *J. Hydrol.* **2008**, *357*, 1–10. [CrossRef]
23. Herschy, R.W. *Streamflow Measurement*; Elsevier: London, UK, 1985.
24. Corato, G.; Melone, F.; Moramarco, T.; Singh, V.P. Uncertainty analysis of flow velocity estimation by a simplified entropy model. *Hydrol. Process.* **2014**, *28*, 581–590. [CrossRef]
25. Chiu, C.-L.; Tung, N.-C. Maximum velocity and regularities in open-channel flow. *J. Hydraul. Eng.* **2002**, *128*, 390–398. [CrossRef]

12

Rainfall Characteristics and Regionalization in Peninsular Malaysia based on a High Resolution Gridded Data Set

Chee Loong Wong [1,2,*], Juneng Liew [3], Zulkifli Yusop [1,4,*], Tarmizi Ismail [1,4], Raymond Venneker [5] and Stefan Uhlenbrook [6]

1 Faculty of Civil Engineering, Universiti Teknologi Malaysia, Johor Bahru 81310, Malaysia; tarmiziismail@utm.my
2 Department of Irrigation and Drainage, Jalan Sultan Salahuddin, Kuala Lumpur 50626, Malaysia
3 School of Environment and Natural Resource Sciences, Universiti Kebangsaan Malaysia, Bangi 43600, Malaysia; juneng@ukm.edu.my
4 Centre for Environmental Sustainability and Water Security, Research Institute for Sustainable Environment, Universiti Teknologi Malaysia, Johor Bahru 81310, Malaysia
5 UNESCO-IHE Institute for Water Education, Westvest 7, Delft 2611 AX, The Netherlands; r.venneker@unesco-ihe.org
6 World Water Assessment Programme (WWAP), UNESCO Villa La Colombella—Località di Colombella Alta, Perugia 06134, Italy; s.uhlenbrook@unesco.org
* Correspondence: wongcl_my@yahoo.com (C.L.W.); zulyusop@utm.my (Z.Y.)

Academic Editor: Tommaso Moramarco

Abstract: Daily gridded rainfall data over Peninsular Malaysia are delineated using an objective clustering algorithm, with the objective of classifying rainfall grids into groups of homogeneous regions based on the similarity of the rainfall annual cycles. It has been demonstrated that Peninsular Malaysia can be statistically delineated into eight distinct rainfall regions. This delineation is closely associated with the topographic and geographic characteristics. The variation of rainfall over the Peninsula is generally characterized by bimodal variations with two peaks, i.e., a primary peak occurring during the autumn transitional period and a secondary peak during the spring transitional period. The east coast zones, however, showed a single peak during the northeast monsoon (NEM). The influence of NEM is stronger compared to the southwest monsoon (SWM). Significantly increasing rainfall trends at 95% confidence level are not observed in all regions during the NEM, with exception of northwest zone (R1) and coastal band of west coast interior region (R3). During SWM, most areas have become drier over the last three decades. The study identifies higher variation of mean monthly rainfall over the east coast regions, but spatially, the rainfall is uniformly distributed. For the southwestern coast and west coast regions, a larger range of coefficients of variation is mostly obtained during the NEM, and to a smaller extent during the SWM. The inland region received least rainfall in February, but showed the largest spatial variation. The relationship between rainfall and the El Niño Southern Oscillation (ENSO) was examined based on the Multivariate ENSO Index (MEI). Although the concurrent relationships between rainfall in the different regions and ENSO are generally weak with negative correlations, the rainfall shows stronger positive correlation with preceding ENSO signals with a time lag of four to eight months.

Keywords: rainfall zonation; rainfall trends; rainfall variability; rainfall regionalization; Peninsular Malaysia

1. Introduction

Delineating climatic zones facilitates the investigation of the general climate characteristics of the region, and helps improve understanding of climate variability across a range of spatial and temporal scales. It plays an important role in gaining knowledge of water balance dynamics on various scales for water resources planning and management [1–3]. At low latitudes, where temperatures vary little within and over years, climate differences are largely determined by differences in rainfall patterns. Climatic boundaries are often plotted according to mean monthly rainfall, varying gradually except where the gradients steepen on mountain slopes or along seacoasts [4]. Distinction between zones is always arbitrary and relies on the cartographical convenience. For example, the climatic regions are grouped according to the same category of isohyet range, the climatic zones boundaries are not clear and are always subjectively determined [4]. Thus, an objective quantitative climatic classification is essential for discovering definite and distinctive climate zones boundaries using the climatic data.

The relationship between monsoon seasons and rainfall of Peninsular Malaysia was studied by Lim [5]. He delineated rainfall regions based on 29 rainfall stations and assessed pentad rainfall patterns that could be grouped roughly into five rainfall regions. These results became the reference for the researchers in investigating rainfall distribution across the country [6–9]. Bishop [10] who researched on integration of land use and climatic data, believed that this could produce good region classification. However, his efforts generated some doubts over certain regions, which were mainly attributed to the reliability of climatic data used in his research. He recommended that a more sophisticated statistical analysis procedure and better data sets be applied to overcome the uncertainty in defining the climatic regions.

With the defined climatic regions, the knowledge of past rainfall distribution and patterns can be assessed in order to qualify the nature of changes in time and space. In the past, most of the hydroclimatic studies focused on rainfall distribution and patterns of change over time. For example, Nieuwolt [11] adopted an agricultural rainfall index to quantify rainfall variability over time. Researchers further investigated the temporal and spatial characteristics of rainfall during 1990s, but often restricted their analyses to small catchments, e.g., an urbanized area [12] or a forested catchment [13].

Peninsular Malaysia has undergone development at a rapid pace over last decades. Thus, the overall picture of country water resources distribution has become important for future water resources planning and management. However, a comprehensive regional study of rainfall patterns is still very limited. The mean annual rainfall maps, derived from long-term monthly records (1950–1990) by the Economic Planning Unit [14,15], are only able to show the spatial distribution of average rainfall in the country instead of high temporal and spatial variability of rainfall patterns. Regional trends of rainfall distribution also have not received much attention from researchers.

The climate over Peninsular Malaysia is subjected to pronounced interannual variability which modulates hydrological variability, including floods and droughts [16–18]. The relation between Malaysian rainfall anomalies, sea surface temperature and El Niño-Southern Oscillation (ENSO) were studied by Tangang and Juneng [17] and Juneng and Tangang [19,20]. The global warming/temperature trends and variations were investigated by Tangang et al. [21] and Ng et al. [22]. However, these studies were based solely on station observations which have insufficient spatial coverage [23], biases associated with the gauge measurement process and homogeneity of rainfall time series [24], and problems of missing data [23]. The need for data with better accuracy and space-time resolution has been emphasized by many researchers [12,25–27].

The development of gridded rainfall data sets [27] can provide better spatio-temporal coverage across a region. It is also timely to reassess the climatic regions as defined by Lim [5] and Bishop [10], and investigate regional rainfall trends and distribution to update the understanding of the spatial and temporal variability in the country and support further water resource planning and management.

The aim of this study is two-fold. First, to delineate climatic regions using clustering algorithm based on the daily gridded rainfall data set for Peninsular Malaysia. Second, to explore the regional characteristics of rainfall distribution through trend analysis and analysis of spatial rainfall variability on annual and monsoon-seasonal basis for different climatic regions.

2. Materials and Methods

2.1. Study Area

Peninsular Malaysia is located in the tropics between 1° and 7° north and between 99° and 105° east. The total area of 132,000 km^2 is composed of highlands, floodplains and coastal zones. The Titiwangsa range forms the backbone of the Peninsula, running approximately south-southeast from southern Thailand over a distance of 480 km and separating the eastern part from the western part. Surrounding the central high regions are the coastal lowlands. In general, the Peninsula experiences a warm and humid tropical climate all year round, with uniform temperatures ranging from 25 °C to 32 °C. Rainfall is characterized by two rainy seasons associated with the southwest monsoon (SWM) from May to September and the northeast monsoon (NEM) from November to March [7,16,28].

2.2. Data Sources

The rainfall data used in this study is the gridded data set described by Wong et al. [27] for a period of 31 years from 1976 to 2006. The daily rainfall data originates from four data sources, i.e., the automatic station database of the Department of Irrigation and Drainage (DID), the Malaysian Meteorological Department (MMD), the Global Summary of the Day (GSOD) that archived at the National Climatic Data Center of the National Oceanic and Atmospheric Administration (NOAA/NCDC) was transmitted over the Global Telecommunication System of the World Meteorological Organisation (WMO-GTS), and additional Global Energy and Water Balance Experiment (GEWEX) research programme Asian Monsoon Experiment (GAME) data which also originate from MMD. The combination of all data sources brings an average of 123 daily point observations with maxima of 167 in years 2001 and 2003.

2.3. Methodology

Rainfall regions are delineated from annual rainfall cycles based on monthly climatological values using a clustering algorithm. The delineated regions were then analyzed statistically to assess the spatio-temporal rainfall distribution. Rainfall trends were investigated using Mann-Kendall nonparametric trend detection method. The spatial variability of annual and monsoon rainfall of all regions is calculated. The relationship and influence of El Ninõ-Southern Oscillation (ENSO) over the delineated regions is assessed by using the Multivariate ENSO Index (MEI) indicator.

2.3.1. Data Interpolation

The rainfall is interpolated directly from point observation values. The input data stations are mostly located at elevations below 300 m above sea level (m.a.s.l) with relatively few stations at higher altitudes between 1000 and 2000 m.a.s.l. Analysis of the available data did not reveal a clearly identifiable orographic relationship in daily rainfall between the station time series. The available rainfall data have been interpolated into daily grids with 0.05 degree resolution (approximately 5.5 km). The interpolation scheme is based on an adaption of Shepard's [29] angular distance weighting (ADW) procedure.

2.3.2. Climatic Regions Delineation

Zones of different annual rainfall cycles based on monthly climatological values were identified using a clustering algorithm. The identification of groups or clusters is based on the similarity between data properties. There are several algorithms developed for this purpose. Here, a non-hierarchical K-means clustering algorithm [30] coupled to an empirical orthogonal function (EOF) analysis was applied.

There are 4289 rainfall grid cells in the study areas. Hence, the size of monthly rainfall climatological data matrix is 4289 $\times$ 12. In order to reduce the dimensionality of the data and to filter the random noise, the EOF analysis [31] was applied to the monthly rainfall climatological matrix. The EOF analysis is also commonly known as principal component analysis (PCA) in statistical literature [32]. A Monte-Carlo randomization test suggests a cut-off point at the sixth principal component which describes a total of ~98% of the cumulative variance in the data matrix. The resulting data matrix (4289 × 6) was subjected to the non-hierarchical K-means clustering algorithm.

The K-means clustering algorithm minimizes the Euclidean distance between the data items and the cluster centroid by partitioning the data into k non-overlapping regions. Each of the data items is represented by six values corresponding to the six principal components selected earlier. The algorithm iteratively minimizes the total intra-cluster variance (V) described by the following:

$$V = \sum_{i=1}^{k} \sum_{x_j \in C_i} (x_j - \mu_i)^2 \tag{1}$$

where C_i shows the ith cluster ($i = 1, 2, \ldots, k$) and μ_i is the centroid of all the cluster points $x_j \in C_i$. The algorithm however requires the number of clusters, k, to be set a priori. Here, the algorithm was applied to the 4289 data items, setting the value of k from 3 to 10. The validity of the optimum number of cluster was then determined by the silhouette width [33] for each data items is calculated as:

$$S(i) = \frac{b(i) - a(i)}{\max\{a(i), b(i)\}} \tag{2}$$

where $a(i)$ is the averaged Euclidean distance of the i data item to all other data items in the same cluster, $b(i)$ is the minimum of the averaged distance of i data item to all data items in the other clusters. The averaged silhouette width, $\overline{S}$ evaluates the quality of the clustering solution by considering both compactness (distance between data items within the same cluster) and separation (distance between data items in two neighboring clusters). The closer $\overline{S}$ value is to 1, the better the grouping.

2.3.3. Trend Detection

The annual and monsoon seasonal averaged time series of rainfall of the climatic regions were analyzed using the Mann-Kendall nonparametric trend detection test [34,35]. The Mann-Kendall method has been widely used and tested to be an effective method to evaluate the presence of a statistically significant trend in climatological and hydrological time series [2,36–38]. In the trend test, the null hypothesis H_0 is that there is no trend in the series (the data are independent and identically distributed). The alternative hypothesis H_1 is that a trend exists in the data. The Kendall S-statistic is obtained from comparison between all possible (x, y) pairs of data and is given by:

$$S = \sum_{i=1}^{n-1} \sum_{j=i+1}^{n} \text{sgn}(x_i - x_j)\text{sgn}(y_i - y_j) \tag{3}$$

where, the sign function is defined (for any variable u) as

$$\text{sgn}\ (u) = \begin{cases} -1 \text{ if } u < 0 \\ 0 \text{ if } u = 0 \\ +1 \text{ if u} > 0 \end{cases} \tag{4}$$

Under the null hypothesis, the statistic S is approximately normally distributed with zero mean and variance given by

$$\text{Var}\ (S) = \frac{n\ (n-1)\ (2n+5) - \sum_{i=1}^{n} t_i i\ (i-1)\ (2i+5)}{18} \tag{5}$$

where t_i is the number of ties of extent i in either the x or y data. The summation term in the numerator is used only if the data series contains tied values. The standardized test statistic Z is obtained as

$$Z = \begin{cases} \frac{S+1}{\sqrt{\text{Var}(S)}} & \text{if } S < 0 \\ 0 & \text{if } S = 0 \\ \frac{S-1}{\sqrt{\text{Var}(S)}} & \text{if } S > 0 \end{cases} \tag{6}$$

The test statistic Z is used to measure the significance of a trend. In a two sided test, H_0 should be accepted if $|Z|$ is greater than $Z_{\alpha/2}$, where α represents the chosen significance level (e.g., 5% with $Z_{0.025} = 1.96$) then the null hypothesis is rejected implying that the trend is significant. A positive Z value indicates an upward trend, whereas a negative Z value indicates a downward trend. If a significant trend is present, the average rate of increase or decrease can be obtained from the slope of a simple linear regression.

2.3.4. Spatial Variation of Rainfall

Assessment of spatial variation is based on computation of the spatial variance within a particular delineated gridded region, as

$$\sigma^2 = \frac{1}{n}\sum_{i=1}^{n} (x_i - x)^2 \tag{7}$$

where $\overline{x}$ is the average areal rainfall and, x_i is the rainfall depth in each grid cell. The average areal rainfall and the spatial variance are computed for each region. The coefficient of variation of rainfall which is used to characterize the monthly spatial variability of rainfall is defined as the standard deviation divided by the average areal rainfall.

2.3.5. Rainfalls Correlation Analysis with El Ninõ-Southern Oscillation (ENSO)

The influence of ENSO over the delineated regions in Peninsular Malaysia is assessed by correlating the rainfall time series with the Multivariate ENSO Index (MEI). The MEI is based on six observed variables over the tropical Pacific Ocean, i.e., sea-level pressure, zonal and meridional components of the surface wind, sea surface temperature, surface air temperature, and total cloudiness fraction of the sky [39]. Negative values of the MEI represent the cold ENSO phase, i.e., La Niña, while positive MEI values represent the warm ENSO phase (El Niño). This index was chosen because it is less vulnerable to errors compared to station-based or single variable field based indices (e.g., Southern Oscillation Index, SOI) and has more robust key features in spatial variations and seasonal cycle [39].

To examine the relationship between the interannual rainfall and ENSO, each regional monthly rainfall index is correlated with the MEI time series concurrently. To further explore potential lagged influence of ENSO on NEM and SWM rainfall, the monsoon seasonal rainfall index is created to allow for comparison with the monthly MEI values preceding the seasons. The lag is defined by the time distance between the first month of monsoon season and monthly MEI value. For example, the correlation between the NEM rainfall and MEI at one month lag is the correlation coefficient value between the NEM, which is aggregated from November(0) to March(1)) rainfall and MEI value in October(0). The lagged correlation coefficient values were calculated with a lag time of 0–8 months (MEI leads the rainfall) as well as each individual monsoon month considered.

3. Results and Discussion

3.1. Characteristics of Delineated Climatic Regions

Table 1 shows the averaged silhouette width, $\overline{S}$ suggesting a best grouping of eight different regions for a maximum $\overline{S}$ value of 0.53. The eight delineated rainfall zones are shown in Figure 1. The classification patterns are generally consistent with Lim [5] that classified subjectively the Peninsular Malaysia rainfall into five different zones based on pentad data. The general similarity appears to be the existence of the northwest zone and the central interior rainfall zone over the central mountainous valley areas of Peninsular Malaysia. However, there exist some differences over the southern part of Peninsular Malaysia where current clustering algorithm further delineates the southwest zone of Lim [5] into two different rainfall zones. This rainfall characteristic is also observed by Dale [40] who reported the existence of a distinctive Port Dickson-Muar coastal rain belt in conjunction with other four larger rainfall zones. In addition, the east coast zone is further divided north-south into the northeastern zone and southeastern zone. The latter covers a large part of Johor state. Another difference between the current analysis and that of Lim [5] is the east-west separation of the west coast zone which is represented by region R6 (refer to Figure 1).

Table 1. Averaged silhouette width, $\overline{S}$ indicates better grouping if closer to one.

K, Non-Overlapping Regions	3	4	5	6	7	8	9	10
Averaged silhouette width, $\overline{S}$	0.35	0.41	0.43	0.46	0.43	0.53	0.50	0.48

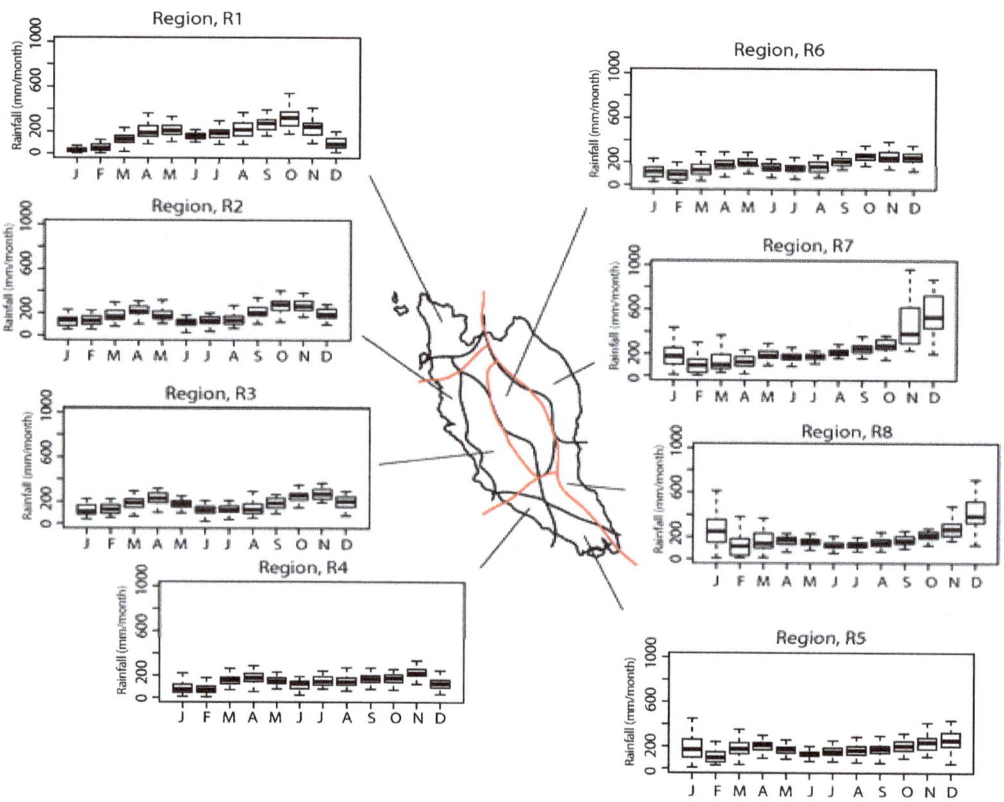

Figure 1. Box and whisker plot of areal average mean monthly rainfall (1976–2006) of the eight delineated rainfall regions in Malaysia. The solid line is the median, the height of the box is the difference between the third and first quartiles (IQR). The zones delineated by Lim [5] are shown in red lines.

An examination of the rainfall zonation patterns and the regional topographic structure suggests that the delineation is closely associated with the topographic characteristics. The temporal variation of rainfall over Peninsular Malaysia is generally characterized by a bimodal annual cycle, i.e., a primary peak occurring during the autumn transitional period and a secondary peak during the spring transitional period as shown in Figure 1. However, over the east coast regions (R7), the rainfall appears to have a single peak during the NEM. This region received 2940 mm/year of mean annual rainfall, 52% and 33% of which occurred during NEM and SWM periods, respectively, as shown in Table 2. The monsoons contribute 85% of the total annual rainfall in this region. During the NEM, the dry northeasterly cold surge winds become moist during the passage over the South China Sea. The interaction with the land along the east coast area creates deep convection clouds and rainfall during the NEM [41–43]. The difference between the northeast zone (R7) and southeast zone (R8) appears to be the higher monsoon rainfall during November–December over the northeastern coast region (see Figure 1). Based on numerical experiments, Juneng et al. [43] argued that the higher elevation in the interior parts of the north-eastern coast of Peninsular Malaysia (in R7) plays a crucial role in providing additional lifting and enhances convection and rainfall while interacting with the mesoscale circulations. Over the south of the eastern region (R8), the rather flat terrain allows the storm systems to reach further inland [43], hence resulting in higher inland rainfall.

Table 2. Rainfall contributions in Peninsular Malaysia for the period of 1976–2006.

Region	Mean Elevation (m) (min, max)	Annual Rainfall (mm/Year)	Standard Deviation	Northeast Monsoon (Nov.–Mar.) Rainfall		Southwest Monsoon (May–Sep.) Rainfall		Total Monsoon Rainfall	
				mm/Year	%	mm/Year	%	mm/Year	%
R1	272 (0, 860)	2118	440	567	26	1035	48	1602	75
R2	116 (0, 773)	2229	314	922	41	806	36	1728	77
R3	326 (16, 1475)	2154	347	921	42	765	35	1686	78
R4	74 (0, 189)	1807	464	705	39	743	41	1448	80
R5	50 (0, 183)	2206	604	1004	45	794	35	1798	81
R6	457 (68, 1280)	2174	329	868	39	868	39	1736	79
R7	266 (0, 1044)	2940	321	1530	52	993	33	2523	85
R8	128 (0, 352)	2383	550	1268	53	735	30	2004	84

Over the southwestern coast of Peninsular Malaysia, the bi-modal seasonal variation of the rainfall is significantly weaker. The terrain is relatively flat to undulating compared to the northern regions. The seasonal rainfall cycle shows that the southern area (R5) of the southwest zones has a rainfall peak in December and maintains a high amount of rainfall in January. Over the northern parts of the southwest zones (R4), rainfall peaks in November and decreases sharply in January the following year. The monsoon rainfall contributes 80% to the mean annual of 1807 mm/year (see Table 2). Note that this zone lies at the edge of the central mountain range (Titiwangsa mountain range) and the local meteorology may interact with the land-sea breeze to nurture storm intensity system over this part of Peninsular Malaysia [44,45].

The west coast region of Peninsular Malaysia, which is characterized by strong bi-modality, is delineated into the coastal band (R2) and the interior zone (R3). The seasonal variations of these two zones show similarity with two rainfall peaks during the monsoon transitional periods. The meteorology during this time of the year is dominated by strong diurnal variations [46] with high frequency of rainfall occurrence in the evening hours. However, the interior zone, which is characterized by rough terrain of the Titiwangsa mountain range, has a sharper drop in monthly rainfall after the annual peaks. The diurnal rainfall cycle over the Straits of Malacca is largely dominated by land-sea breeze forcing [44,45]. The blockage by the high north-south oriented terrain structures (Titiwangsa range) limits the in-land penetration of the sea breezes. This results in minor differences of rainfall climatology over this lower coastal band (R2) of the west coast region. The contribution of monsoon rainfall for both zones (R2 and R3) is relatively equal (see Table 2), but region R2 has higher annual rainfall of 2229 mm/year, compared to R3 with 2154 mm/year.

The northwest zone (R1) also shows a bimodal annual rainfall distribution. However, the primary peak during the autumn transitional period in October is higher compared to the secondary peak during the spring transitional period in April/May. The primary monthly rainfall peak drops dramatically and reaches at minimum level in January. Another significant difference that delineates this zone from the rest of the west coast region is the relatively high amount of rainfall received during the May–July months (>200 mm/month). This is likely due to the exposure of the northern region to the southwest monsoon wind influence. On the other hand, the rest of the west coast areas are shielded by the mountain ranges in Sumatra during this time of the year. Geographically, the northwest zone has a relatively flat topography separated from the rest of the regions by high and rough terrains in the northern parts of Peninsular Malaysia. The R6 region in the central interior part of the country received 2174 mm of mean annual rainfall, 79% of which occurred during the NEM and SWM (see Table 2). The reduction of rainfall amount in this region compared to the west coastal regions during NEM is due to the Titiwangsa mountain range, that appears to block the westward progression of the climatic system and therefore inhibits excessive rainfall over the central interior zone [43]. According to Nieuwolt [47], the rainfall produced in this region is mainly due to local convection caused by intense heating of land surface.

3.2. Rainfall Trends

Most climatic regions do not show any significant trends at the 95% confidence level as indicated in Table 3. The west coast interior region (R3) and south of eastern region (R8) were the only two regions with significant (at 90% confidence level) positive trends in the total annual rainfall. During SWM monsoon, most of the areas are drier over the years, in particular the northwest region (R1) which shows significant trends at the 95% confidence level, i.e., a −5.1 mm/season/year of rainfall declining trend. In contrast, significant increasing trends at the 95% confidence level have been found in the NEM rainfall for the northwest (R1) and west coast interior region (R3) of the Peninsula. During the NEM, the northwest (R1) and west coast interior regions (R3) show large increasing rainfall trends, i.e., 7.5 mm/season/year and 6.7 mm/season/year, respectively.

Table 3. Annual and monsoons rainfall trends (mm/year) of eight distinct climatic regions for the period of 1976–2006.

Region	Annual	NEM	SWM
R1	6.3	7.5	−5.1
R2	3.7	4.3	−1.5
R3	7.7	6.7	0.0
R4	2.1	3.1	−1.8
R5	7.1	4.9	−0.8
R6	5.8	5.0	−1.0
R7	13.3	11.4	−1.1
R8	9.1	5.8	−0.3

Notes: Light grey indicates significant rainfall (mm) trend at 90% confidence level; Dark grey indicates significant rainfall (mm) trend at 95% confidence level; NEM: northeast monsoon; SWM: southwest monsoon.

Focusing on monthly rainfall trend, Table 4 shows that all regions, with exception of the northern parts of the southwest zones (R4), have a significant increasing trend with at least 90% confidence level in January. It is noted that although most of the monthly rainfall throughout the NEM do not exhibit a significant trend at 95% confidence level, the monthly rainfall is at an increasing trend over years (with exception of regions R4 to R8 in November, R4 and R5 in February). On the other hand, most of the monthly rainfall for SWM (May to September) no trend was found, but the monthly rainfall is in decreasing condition over the years. It is interesting to note that although the monthly rainfall does not necessary show significant trends, the consistent sign of changes throughout the seasons resulted in significant trends (at 95% confidence level) when the seasonal means were analyzed.

Table 4. Monthly rainfall trends (mm/month) of eight distinct climatic regions.

Region	Jan.	Feb.	Mar.	Apr.	May	Jun.	Jul.	Aug.	Sep.	Oct.	Nov.	Dec.
R1	1.3	1.3	2.5	0.7	−3.5	0.8	−0.8	0.5	−2.0	2.6	0.2	2.8
R2	1.6	0.3	−0.2	0.1	−1.7	0.3	−0.3	0.5	−0.2	0.3	1.7	1.5
R3	2.2	0.2	1.2	0.8	−1.0	0.3	−0.1	0.5	0.4	−0.7	1.6	2.5
R4	2.0	−1.3	1.1	−0.2	−0.3	−0.2	−0.6	1.1	−2.0	0.1	−0.9	3.1
R5	5.6	−1.1	1.5	−0.2	−0.0	−0.9	0.0	0.8	−0.7	1.1	−0.4	1.5
R6	2.2	0.2	1.7	0.3	−1.4	0.9	−0.9	0.9	−0.6	0.1	−0.5	2.8
R7	5.0	2.4	3.5	−0.4	−0.6	1.1	−1.4	0.0	−0.2	1.0	−1.2	4.2
R8	5.8	0.1	1.1	−0.1	−0.5	−0.1	−0.2	0.9	−0.4	0.8	−0.3	2.0

Notes: Light grey indicates significant monthly rainfall (mm/month) trend at 90% confidence level; Dark grey indicates significant monthly rainfall (mm/month) trend at 95% confidence level.

3.3. Spatial Variability

Figure 2 shows the mean monthly spatial variability for the eight climatic regions. It is noted that the spatial rainfall variation is more uniform throughout the year in the east coast regions (R7 and R8) as shown in Figure 2g,f, although both regions have the largest variation in mean monthly rainfall as indicated in Figure 1. The southwestern coast (R4 and R5) and west coast (R2 and R3) regions receive smaller amounts of monthly rainfall, but the spatial variation differs, particularly during monsoons. For the inland region (R6), the largest spatial variability occurs in February, the month with the smallest rainfall amount of 94 mm/month (refer to Figure 1). The larger range of coefficients of variation is most visible for all regions during the NEM, when the largest amount of rainfall is received over the east coast region (see Table 1), and to a smaller extent during the SWM.

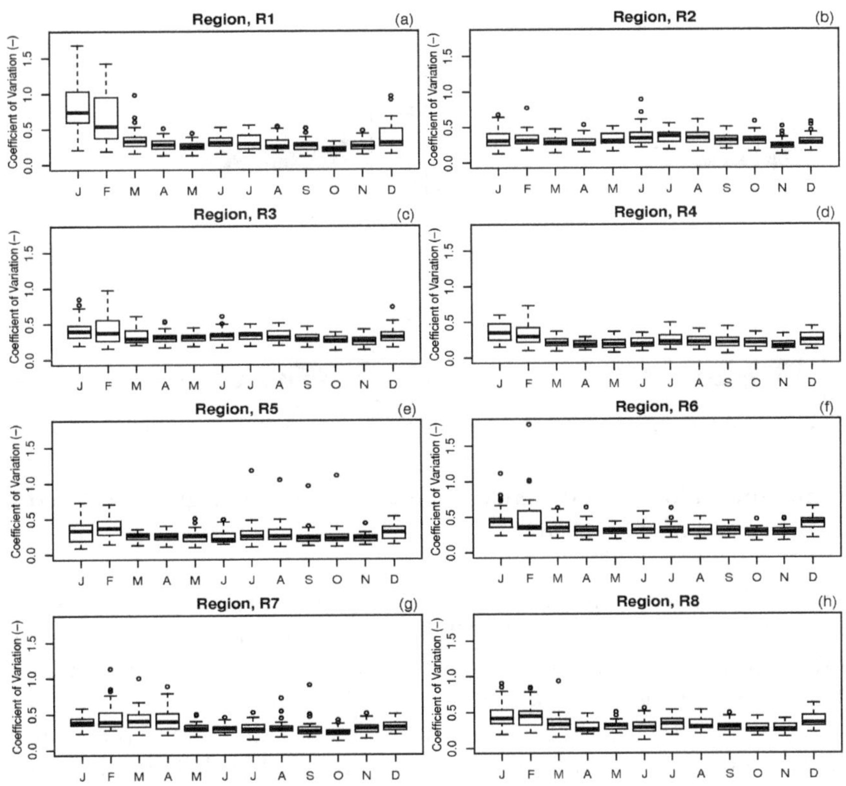

Figure 2. Box and whisker plot of mean monthly coefficients of spatial variations (1976–2006) of the eight delineated rainfall regions (Panel **a** to **h**). The solid line is the median, the height of the box is the difference between the third and first quartiles (IQR). Any data observation which lies 1.5 IQR lower than the first quartile or 1.5 IQR higher than the third quartile is considered an outlier in the statistical sense, indicated by open circles.

The northern region (R1) shows the largest interquartile range (0.19) and mean monthly coefficients of variation (0.38) compared to the other regions. In contrast, the northern part of the southwest region (R4) has a smaller interquartile range (0.13) and lowest mean monthly coefficients of variation (0.25), indicating that the areal rainfall is more uniformly distributed. The calculated median values of the coefficients of variation are close to the mean values for all regions, which suggest that the spatial rainfall variation is symmetrically distributed.

Figure 3 shows the coefficients of variation as a function of the spatial mean monthly rainfall for each climatic region. The relative coefficients of variation generally decreases exponentially with increasing mean monthly rainfall before reaching about 150 mm/month. The coefficients of variation remain fairly constant at 0.30 for mean monthly rainfall exceeding 150 mm/month. This implies the rainfall is more uniformly distributed across the region when the average rainfall amount is high. It is usually associated with heavy monsoon rainfall over the east coast of Peninsular Malaysia [43].

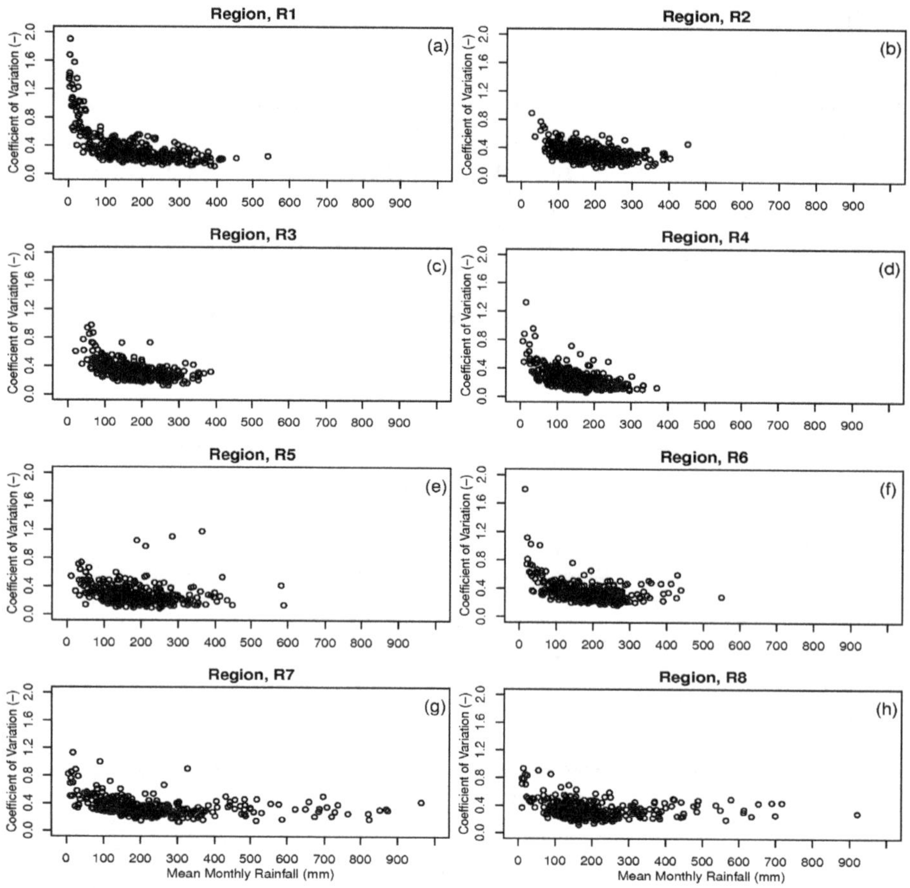

Figure 3. Mean monthly rainfall coefficients of variation over the eight delineated rainfall regions (Panel **a** to **h**) for the period of 1976–2006.

The topography and monsoon winds are likely the main factors controlling the magnitude of the spatial rainfall variation in the country. The Titiwangsa Range is a mountain range that forms the backbone of the Peninsula. During the northeast monsoon (NEM), stronger winds blow to the exposed areas, e.g., the east coast of Peninsular Malaysia [28,43,48], thus these areas receive substantial high amount of rainfall. Higher wind speeds promote more evaporation, which destabilizes the boundary layer and triggers deep convection, and hence, increases rainfall [49]. The inland and west coast areas, which are sheltered by mountain ranges, are relatively free from its influence. On the other hand, the southwest monsoon (SWM) wind speed is generally lighter. It tends to be wetter at the west coast of Peninsular Malaysia compared to the inland and east coast areas. The presence of mountain ranges separating the eastern and western parts of the Peninsula could be the best reason explaining the differences between the rainfall distributions of each region. The relatively flat landscape of the southern region (R5) and east coast regions (R7 and R8) result in reduced spatial rainfall variability.

3.4. Influence of ENSO

The evolution of the ENSO [19,50] may have implications on the rainfall distribution over the Peninsula. The 31-year long-term rainfall anomaly for Peninsular Malaysia was, as a whole, found to be rather weakly and insignificantly (at 95% level) correlated (−0.19) with MEI. In addition, the monthly rainfall of all eight regions generally showed weak concurrent relationships with MEI. Figure 4 shows the correlation coefficients between the rainfalls over the Peninsula and MEI on monthly basis. The most obvious monthly rainfall-MEI correlation (−0.58) was seen in April during the inter-monsoon season. The relationships between monthly rainfalls and MEI of each region were also relatively weak (see Figure 5). These weak relationships are found to be consistent with findings of Juneng and Tangang [19]. The impact of ENSO on rainfall anomalies over the maritime continent has shown considerable spatial variation throughout the different phases of ENSO. During the SWM months, ENSO signature is largely confined to the south of the equator, induced by the interaction between the regional anomalous circulation and the background flow. This anomalous circulation is maintained by the regional feedback processes between the regional seas and the atmosphere. Hence the impact does not extend far northward to the Peninsular Malaysia. During the NEM months, the ENSO signature is shifted northeastward due to the establishment of the northwestern Pacific anticyclone/cyclone anomalies as the Rossby wave response to the surface heating/cooling. Due to the location of this anomalous circulation in the northwestern Pacific, the impact is largely confined to the northern Borneo and southern Philippines, but does not extend far westward to the Malay Peninsula. Thus, the concurrent impact of ENSO is generally weaker over the Peninsular Malaysia.

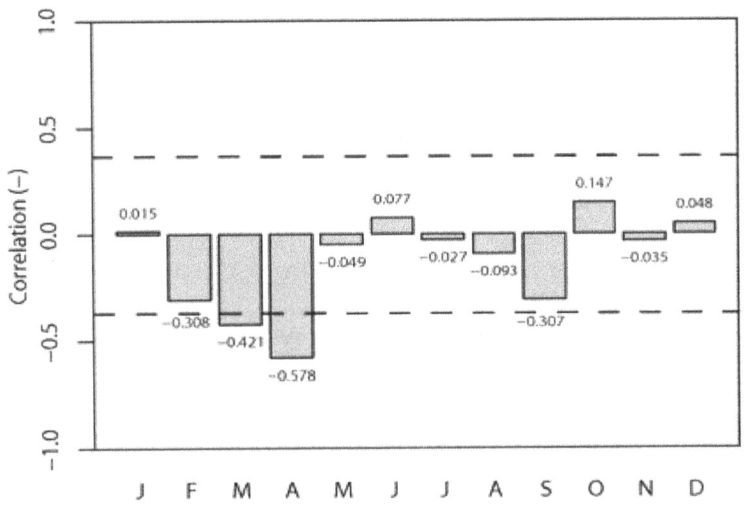

Figure 4. The correlation of concurrent monthly rainfall (mm/month) and Multivariate El Ninõ-Southern Oscillation (ENSO) Index (MEI) for Peninsular Malaysia. The dotted line indicates significance at 95% level.

A lagged correlation analysis to determine potential lag/lead relationship between monsoon rainfall and MEI is presented in Table 5, showing the lagged correlation between monthly NEM rainfall and MEI for the eight climatic regions. The months of NEM rainfall are mostly correlated negatively with MEI during the NEM months, i.e., from November to March, suggesting modulation of drier conditions during the warm phase of ENSO events during ENSO mature months. The northwest zone (R1), west coast interior zone (R3) and central interior part (R6) of Peninsular Malaysia are correlated negatively with MEI at 95% level of significance. It is interesting to note that the NEM rainfalls posed positive correlation at 95% level of significance with MEI, with a delay of between five and eight months. The correlation magnitude is generally larger compared to the concurrent coefficient values. This means any ENSO (La Niña or El Niño) events could affect Peninsular Malaysia with a time lag of between five and eight months. The lagged correlation between monthly rainfalls of SWM and MEI was found to have similar correlation results as shown in Table 6 where monthly rainfalls are mostly correlated negatively with the MEI indicator during SWM monsoon months, with few regions (as highlighted in dark grey colour in Table 6) posting negative correlation at 95% level of significance in August and September. Generally, the monthly rainfalls of SWM monsoon correlated positively with MEI at 95% level of significance, with a delay of 4–8 months. The analysis clearly shows that the influence of ENSO on rainfall in Peninsular Malaysia is delayed by several months before posing implications to rainfall distributions, especially during SWM. The stronger lagged impact of ENSO on Peninsular rainfall may be explained by the capacitor effect [51] of ENSO which sees the tropical Indian Ocean acting as a heat capacitor during the decaying of a warm ENSO event. It is noted that after the dissipation of El Nino in the spring over the equatorial Pacific, the sea surface temperature over the tropical Indian Ocean remains anomalously warm. This causes the tropospheric temperature to increase by a moist adiabatic adjustment in deep convection, emanating a baroclinic Kelvin wave into the Pacific [51]. Numerical experiment by Xie et al. [51] suggests that this generates low pressure on the equator and anomalous anticyclone over the western north Pacific. The sea surface temperature over the South China Sea is anomalously warm and this promotes deep convective activities over the region. Hence, together with the lower pressure, the increases the rainfall anomalously over the region as shown by the positive lagged correlations in Table 6. Conventionally, the warm ENSO phases are associated with dry rainfall anomalies over Malaysia [17] during the development of ENSO events and this relationship was used to forecast the seasonal rainfall anomalies in the country [20]. The lagged positive relationship found in this study is an important consideration in predicting seasonal rainfall anomalies over Peninsular Malaysia with a longer lead time.

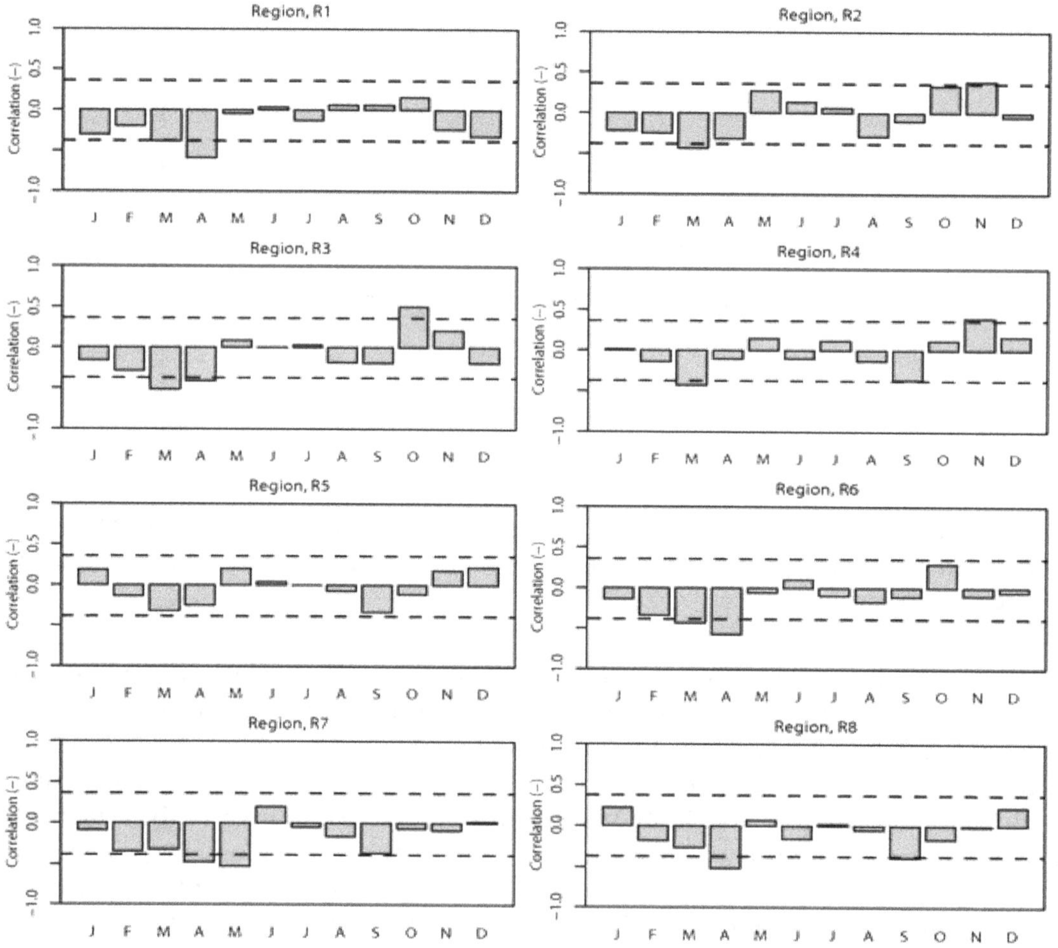

Figure 5. The correlation of concurrent monthly rainfall (mm/month) and Multivariate ENSO Index (MEI) of eight distinct climatic regions. The dotted line indicates significance at 95% level.

Table 5. Correlation coefficient values between the NEM rainfall and the MEI at various concurrent and leading months.

Region	Mar.	Apr.	May	Jun.	Jul.	Aug.	Sep.	Oct.	Nov.	Dec.	Jan.	Feb.	Mar.
R1	0.232	0.215	0.269	0.053	-0.118	-0.222	-0.267	-0.342	−0.435	−0.478	−0.489	−0.537	−0.536
R2	0.489	0.563	0.585	0.453	0.230	0.045	-0.051	-0.057	-0.153	-0.223	-0.243	-0.284	-0.287
R3	0.308	0.344	0.348	0.203	0.002	-0.143	-0.244	-0.236	-0.319	−0.393	−0.394	−0.444	−0.470
R4	0.293	0.392	0.408	0.414	0.280	0.146	0.030	0.055	-0.023	-0.087	-0.071	-0.114	-0.128
R5	0.396	0.439	0.395	0.363	0.259	0.166	0.100	0.147	0.059	0.045	0.056	0.012	-0.011
R6	0.542	0.527	0.559	0.300	0.048	-0.106	-0.212	-0.265	−0.383	−0.394	−0.401	−0.486	−0.478
R7	0.426	0.451	0.523	0.335	0.153	0.046	-0.036	-0.112	-0.247	-0.235	0.201	-0.258	-0.264
R8	0.538	0.582	0.589	0.529	0.341	0.222	0.145	0.136	0.022	0.040	0.052	-0.007	-0.016

Notes: Light grey indicates significant correlation coefficient at 90% confidence level; Dark grey indicates significant correlation coefficient at 95% confidence level.

Table 6. Correlation coefficient values between the SWM rainfall and the MEI at various concurrent and leading months.

Region	Sep.	Oct.	Nov.	Dec.	Jan.	Feb.	Mar.	Apr.	May	Jun.	Jul.	Aug.	Sep.
R1	0.395	0.411	0.381	0.398	0.405	0.355	0.330	0.280	0.255	0.218	0.086	−0.024	−0.090
R2	0.462	0.423	0.425	0.451	0.448	0.419	0.401	0.380	0.382	0.213	0.040	−0.117	−0.219
R3	0.373	0.323	0.285	0.341	0.332	0.265	0.250	0.253	0.249	0.061	−0.128	−0.249	−0.308
R4	0.527	0.450	0.371	0.389	0.371	0.328	0.322	0.288	0.201	−0.060	−0.205	−0.375	−0.464
R5	0.576	0.563	0.539	0.590	0.637	0.589	0.565	0.558	0.465	0.201	−0.085	−0.270	−0.321
R6	0.292	0.279	0.212	0.252	0.254	0.223	0.242	0.209	0.205	0.040	−0.156	−0.280	−0.326
R7	0.356	0.346	0.263	0.258	0.225	0.217	0.202	0.054	−0.035	−0.189	−0.335	−0.472	−0.499
R8	0.505	0.446	0.340	0.381	0.390	0.329	0.294	0.239	0.130	−0.094	−0.254	−0.378	−0.423

Notes: Light grey indicates significant correlation coefficient at 90% confidence level; Dark grey indicates significant correlation coefficient at 95% confidence level.

4. Conclusions

Statistically, Peninsular Malaysia can be differentiated into eight distinct rainfall regions. The results are generally in line with Lim [5] who classified subjectively the regional rainfall into five different zones based on pentad rainfall data. The differences between the current analysis and that of Lim [5] are (i) the east-west separation of the west coast zone; (ii) two different rainfall zones over south-west zone; and (ii) the east coast zone is divided into the northeastern zone covering parts of the states of Kelantan, Terengganu and Pahang, and southeastern zone covering large parts of Johor state.

The delineated rainfall zones are closely associated with the geography and topographic characteristics. The temporal variation of rainfall over the Peninsula is generally characterized by a bimodal distribution, i.e., a primary peak occurring during autumn transitional period and a secondary peak during the spring transitional period, except in the east coast zones with a single pronounced rainfall peak during the NEM. Generally, the NEM influence is relatively stronger and contributes more rainfall over all regions than the SWM. Significant increasing trends at the 95% confidence level have been found in the NEM rainfall for the northwest and west coast interior regions. The opposite is true during the SWM where most areas get drier due to decreases of rainfall over the year.

The east coast regions receive high monthly rainfall compared to other regions, but the rainfall is uniformly distributed spatially. For the southwestern coast and west coast regions, the spatial variation is more pronounced during monsoon periods, especially for NEM rainfalls. The larger range of the coefficients of variation is most visible during the NEM, and to a smaller extent during the SWM. The NEM also brings the largest amount of rainfall to the east coast region. It is noted that the inland regions, which received less rainfall in February, showed the largest spatial variation. The topography and monsoon winds are the main factors controlling the spatial rainfall variation in the country.

The relationship between monthly time series and MEI throughout the study period (31 years) showed a weakly negative correlation in the concurrent monthly rainfall and MEI for all regions. The influence of ENSO showed a correlation lag in the range of a 4–8 month period. Such delays may have implications for the rainfall distribution over the country, especially during NEM and SWM.

The results presented in this study demonstrate the advantages of using high quality gridded data sets in understanding hydrological regimes over Peninsular Malaysia. With a consistent spatio-temporal pattern of the data set, various precipitation indices can be properly defined, offering new insights to different hydrological regimes over the regions in the future. In addition, the enhanced understanding of the spatial and temporal rainfall variability is useful for more effective planning and management of water resources.

Acknowledgments: The authors thank the Department of Irrigation and Drainage Malaysia (DID), Malaysia Meteorological Department (MMD), and NOAA/NCDC for providing the data. Special thanks are also dedicated to Universiti Teknologi Malaysia for the financial support from Vote number RJ130000.7809.4F910.

Author Contributions: Chee Loong Wong and Juneng Liew designed the analytical framework and analyzed the data; Chee Loong Wong wrote the paper; Zulkifli Yusop, Tarmizi Ismail, Raymond Venneker and Stefan Uhlenbrook provided valuable comments on the papers.

References

1. Sato, Y.; Ma, X.; Xu, J.; Matsuoka, M.; Zheng, H. Analysis of long-term water balance in the source area of the Yellow river basin. *Hydrol. Process.* **2008**, *22*, 1618–1629. [CrossRef]
2. Yang, D.; Li, C.; Hu, H.; Lei, Z.; Yang, S.; Kususa, T.; Koike, T.; Musiake, K. Analysis of water resources variability in the Yellow river of china during the last half century using historical data. *Water Resour. Res.* **2004**, *40*, W06502. [CrossRef]
3. Abbaspour, K.C.; Rouholahnejad, E.; Vaghefi, S.; Srinivasan, R.; Yang, H.; Kløve, B. A continental-scale hydrology and water quality model for europe: Calibration and uncertainty of a high-resolution large-scale SWAT model. *J. Hydrol.* **2015**, *524*, 733–752. [CrossRef]

4. Thornthwaite, C.W. An approach toward a rational classification of climate. *Geogr. Rev.* **1948**, *38*, 55–94. [CrossRef]
5. Lim, J.T. Rainfall minimum in peninsular Malaysia during the northeast monsoon. *Mon. Weather Rev.* **1976**, *104*, 96–99. [CrossRef]
6. Deni, S.M.; Jemain, A.A.; Ibrahim, K. Fitting optimum order of markov chain models for daily rainfall occurrences in peninsular Malaysia. *Theor. Appl. Climatol.* **2009**, *97*, 109–121. [CrossRef]
7. Suhaila, J.; Jemain, A.A. Investigating the impacts of adjoining wet days on the distribution of daily rainfall amounts in peninsular Malaysia. *J. Hydrol.* **2009**, *368*, 17–25. [CrossRef]
8. Suhaila, J.; Yusop, Z. Spatial and temporal variabilities of rainfall data using functional data analysis. *Theor. Appl. Climatol.* **2016**. [CrossRef]
9. Deni, S.M.; Suhaila, J.; Wan, Z.W.Z.; Jemain, A.A. Spatial trends of dry spells over peninsular Malaysia during monsoon seasons. *Theor. Appl. Climatol.* **2009**, *99*, 357–371. [CrossRef]
10. Bishop, I.D. Provisional climatic regions in peninsular Malaysia. *Pertanika* **1984**, *7*, 19–24.
11. Nieuwolt, S. Tropical rainfall variability—The agroclimatic impact. *Agric. Environ.* **1982**, *7*, 135–148. [CrossRef]
12. Desa, M.N.; Niemczynowicz, J. Temporal and spatial characteristics in Kuala Lumpur, Malaysia. *Atmos. Res.* **1996**, *42*, 263–277. [CrossRef]
13. Noguchi, S.; Nik, A.R. Rainfall characteristics of tropical rain forest and temperate forest: Comparison between bukit tarek in peninsular Malaysia and hitachi ohta in Japan. *J. Trop. For. Sci.* **1996**, *9*, 206–220.
14. Economic Planning Unit. *Masterplan for the Development of Water Resources in Peninsular Malaysia 2000–2050*; Economic Planning Unit, Prime Minister's Department: Kuala Lumpur, Malaysia, 1999.
15. Department of Irrigation and Drainage. *Review of the National Water Resources Study (2000–2050) and Formulation of National Water Resources Policy*; Department of Irrigation and Drainage Malaysia: Kuala Lumpur, Malaysia, 2011.
16. Tangang, F.T. Low frequency and quasi-biennial oscillations in the Malaysian precipitation anomaly. *Int. J. Climatol.* **2001**, *21*, 1199–1210. [CrossRef]
17. Tangang, F.T.; Juneng, L. Mechanism of Malaysian rainfall anomalies. *J. Clim.* **2004**, *17*, 3616–3622. [CrossRef]
18. Chan, N.W. Impacts of disasters and disaster risk management in Malaysia: The case of floods. In *Resilience and Recovery in Asian Disasters: Community Ties, Market Mechanisms, and Governance*; Aldrich, D.P., Oum, S., Sawada, Y., Eds.; Springer: Tokyo, Japan, 2015; pp. 239–265.
19. Juneng, L.; Tangang, F.T. Evolution of enso-related rainfall anomalies in southeast Asia region and its relationship with atmosphere-ocean variations in indo-pacific sector. *Clim. Dyn.* **2005**, *25*, 337–350. [CrossRef]
20. Juneng, L.; Tangang, F.T. Level and source of predictability of seasonal rainfall anomalies in Malaysia using canonical correlation analysis. *Int. J. Climatol.* **2008**, *28*, 1255–1267. [CrossRef]
21. Tangang, F.T.; Juneng, L.; Ahmad, S. Trend and interannual variability of temperature in Malaysia: 1961–2002. *Theor. Appl. Climatol.* **2006**, *89*, 127–141. [CrossRef]
22. Ng, M.W.; Camerlengo, A.; Khairi, A.A.W. A study of global warming in Malaysia. *Jurnal Teknol.* **2005**, *42*, 1–10.
23. Moten, S. Multiple time scales in rainfall variability. *J. Earth Syst. Sci.* **1993**, *102*, 249–263.
24. Groisman, P.Y.; Legates, D.R. Documenting and detecting long-term precipitation trends: Where are we and what should be done. *Clim. Chang.* **1995**, *31*, 601–622. [CrossRef]
25. Yatagai, A.; Alpert, P.; Xie, P. Development of a daily gridded precipitation data set for the middle east. *Adv. Geosci.* **2008**, *12*, 1–6. [CrossRef]
26. New, M.; Todd, M.; Hulme, M.; Jones, P. Precipitation measurements and trends in the twentieth century. *Int. J. Climatol.* **2001**, *21*, 1899–1922. [CrossRef]
27. Wong, C.L.; Venneker, R.; Jamil, A.B.M.; Uhlenbrook, S. Development of a gridded daily hydrometeorological data set for peninsular Malaysia. *Hydrol. Process.* **2011**, *25*, 1009–1020. [CrossRef]
28. Camerlengo, A.; Demmler, M.I. Wind-driven circulation of peninsular Malaysia's eastern continental shelf. *Sci. Mar.* **1997**, *61*, 203–211.
29. Shepard, D. A Two-dimensional Interpolation Function for Irregularly-spaced Data. In Proceedings of the 1968 23rd ACM National Conference, New York, NY, USA, 27–29 August 1968; ACM: New York, NY, USA, 1968; pp. 517–524.

30. MacQueen, J.B. Some Methods for Classification and Analysis of Multivariate Observations. In Proceedings of the Fifth Berkeley Symposium on Mathematical Statistics and Probability, Oakland, CA, USA, 21 June–18 July 1965; University of California Press: Berkeley, CA, USA, 1967; Volume 1, pp. 281–297.
31. Von Storch, H.; Zwiers, F.W. *Statistical Analysis in Climate Research*; Cambridge University Press: Cambridge, UK, 1999.
32. Joliffe, I.T. *Principal Component Analysis*; Springer: New York, NY, USA, 1986; p. 271.
33. Rousseeuw, P.J. Silhouettes: A graphical aid to the interpretation and validation of cluster analysis. *J. Comput. Appl. Math.* **1987**, *20*, 53–65. [CrossRef]
34. Kendall, M.G. *Rank Correlation Methods*; Hafner: New York, NY, USA, 1948.
35. Mann, H.B. Non-parametric test against trend. *Econometrika* **1945**, *13*, 245–259. [CrossRef]
36. Zhang, X.; Harvey, K.D.; Hogg, W.D.; Yuzyk, T.R. Trends in canadian streamflow. *Water Resour. Res.* **2001**, *37*, 987–998. [CrossRef]
37. Liu, Q.; Yang, Z.; Cui, B. Spatial and temporal variability of annual precipitation during 1961–2006 in yellow river basin, China. *J. Hydrol.* **2006**, *361*, 330–338. [CrossRef]
38. Burn, D.H.; Elnur, M.A.H. Detection of hydrologic trends and variability. *J. Hydrol.* **2002**, *255*, 107–122. [CrossRef]
39. Wolter, K.; Timlin, M.S. El niño/southern oscillation behaviour since 1871 as diagnosed in an extended multivariate enso index (mei.Ext). *Int. J. Climatol.* **2011**, *31*, 1074–1087. [CrossRef]
40. Dale, W.L. The rainfall of Malaya, Part I. *J. Trop. Geogr.* **1959**, *13*, 23–37.
41. Chang, C.P.; Harr, P.A.; Chen, H.J. Sypnotic disturbancess over the equatorial south china sea and western maritime continent during boreal winter. *Mon. Weather Rev.* **2005**, *113*, 489–503. [CrossRef]
42. Salimum, E.; Tangang, F.T.; Juneng, L. Simulation of heavy precipitation episode over eastern peninsular Malaysia using mm5: Sensitivity to cumulus parameterization schemes. *Meteorol. Atmos. Phys.* **2010**, *107*, 33–49. [CrossRef]
43. Juneng, L.; Tangang, F.T.; Reason, C.J.C. Numerical case study of an extreme rainfall event during 9–11 December 2004 over the east coast of peninsular Malaysia. *Meteorol. Atmos. Phys.* **2007**, *98*, 81–98. [CrossRef]
44. Joseph, B.; Bhatt, B.C.; Koh, T.Y.; Chen, S. Sea breeze simulation over the malay peninsular in an intermonsoon period. *J. Geophys. Res.* **2008**, *113*, D20122. [CrossRef]
45. Sow, K.S.; Juneng, L.; Tangang, F.T.; Hussin, A.G.; Mahmud, M. Numerical simulation of a severe late afternoon thunderstorm over peninsular Malaysia. *Asmos. Res.* **2011**, *99*, 248–262. [CrossRef]
46. Varikoden, H.; Samah, A.A.; Babu, C.A. Spatial and temporal characteristics of rain intensity in the peninsular malaysia using trmm rain rate. *J. Hydrol.* **2010**, *387*, 312–319. [CrossRef]
47. Nieuwolt, S. Diurnal rainfall variation in Malaya. *Ann. Assoc. Am. Geogr.* **1968**, *58*, 313–326. [CrossRef]
48. Lim, E.S.; Das, U.; Pan, C.J.; Abdullah, K.; Wong, C.J. Investigating variability of outgoing longwave radiation over peninsular Malaysia using wavelet transform. *J. Clim.* **2013**, *26*, 3415–3428. [CrossRef]
49. Back, L.E.; Bretherton, C.S. The relationship between wind speed and precipitation in the pacific itcz. *J. Clim.* **2005**, *18*, 4317–4328. [CrossRef]
50. Cheang, B.K. Interannual variability of monsoons in malaysia and its relationship with ENSO. *J. Earth Syst. Sci.* **1993**, *102*, 219–239.
51. Xie, S.-P.; Hu, K.; Hafner, J.; Tokinaga, H.; Du, Y.; Huang, G.; Sampe, T. Indian Ocean capacitor effect on Indo-western Pacific climate during the summer following El Niño. *J. Clim.* **2009**, *22*, 730–747. [CrossRef]

13

Local- and Plot-Scale Measurements of Soil Moisture: Time and Spatially Resolved Field Techniques in Plain, Hill and Mountain Sites

Giulia Raffelli [1], Maurizio Previati [1,*], Davide Canone [1], Davide Gisolo [1], Ivan Bevilacqua [1], Giorgio Capello [2], Marcella Biddoccu [2], Eugenio Cavallo [2], Rita Deiana [3], Giorgio Cassiani [4] and Stefano Ferraris [1]

1 Interuniversity Department of Regional and Urban Studies and Planning (DIST), Polytechnic and University of Torino, viale Mattioli, 39, 10125 Torino, Italy; giulia.raffelli@gmail.com (G.R.); davide.canone@unito.it (D.C.); davide.gisolo92@gmail.com (D.G.); Ivan.bevilacqua@gmail.com (I.B.); stefano.ferraris@unito.it (S.F.)

2 Institute for Agricultural and Earthmoving Machines, Italian National Research Council, Strada delle Cacce 73, 10135 Torino, Italy; g.capello@ima.to.cnr.it (G.C.); m.biddoccu@ima.to.cnr.it (M.B.); e.cavallo@imamoter.cnr.it (E.C.)

3 Department of Cultural Heritage (dBC), University of Padova, 35127 Padova, Italy; rita.deiana@unipd.it

4 Department of Geosciences, University of Padova, 35127 Padova, Italy; giorgio.cassiani@unipd.it

* Correspondence: maurizio.previati@unito.it

Abstract: Soil moisture measurement is essential to validate hydrological models and satellite data. In this work we provide an overview of different local and plot scale soil moisture measurement techniques applied in three different conditions in terms of altitude, land use, and soil type, namely a plain, a mountain meadow and a hilly vineyard. The main goal is to provide a synoptic view of techniques supported by practical case studies to show that in such different conditions it is possible to estimate a time and spatially resolved soil moisture by the same combination of instruments: contact-based methods (i.e., Time Domain Reflectometry—TDR, and two low frequency probes) for the time resolved, and hydro-geophysical minimally-invasive methods (i.e., Electromagnetic Induction—EMI, Ground Penetrating Radar—GPR, and the Electrical Resistivity Tomography—ERT) for the spatially resolved. Both long-term soil moisture measurements and spatially resolved measurement campaigns are discussed. Technical and operational measures are detailed to allow critical factors to be identified.

Keywords: soil moisture measurements; TDR; FDR Sensors; EMI; ERT; GPR; case studies

---

1. Introduction

1.1. *Soil Moisture and Soil–Water Relations*

Soil moisture can be defined as the water in the uppermost part of a field soil, and it is strictly affected by soil physical properties, such as texture, organic matter content and stone presence, but also by land cover (vegetation), land use, topography and rainfall [1–3]. Soil moisture represents a key state variable to understand surface hydrological processes (such as drainage, evaporation and plant uptake) and it controls water and energy exchanges between the land surface and the atmosphere [4,5] contributing also, as key factor, in soil–snow interactions and snow gliding [6]. Through direct evaporation and plant transpiration, soil moisture regulates the partitioning of the incoming solar energy at the land surface into the outgoing sensible, latent, and ground heat fluxes. Furthermore, soil moisture plays a significant role in the organization of natural ecosystems and biodiversity [4], also with consequences for human disease (e.g., malaria transmission [7]). This is especially important in water-scarce environments where more frequent drought events occur [8].

Considering agricultural and irrigation management practices, since soil moisture is extremely variable in space and time, it is very important to accurately evaluate its spatiotemporal dynamics in the root zone at the field scale. In fact, soil water content strongly influences crop production, crop health, and soil salinization. In addition, understanding the factors controlling soil moisture variability allows for an improvement in irrigation management strategies with respect to crop production and optimal use of water resources [5,9,10].

During the last decades new approaches and techniques for monitoring, modelling and using soil moisture data have been developed e.g., [11–13]. With reference to soil moisture monitoring, several studies have been focused on surveying test sites having very different morphological conditions. Lin et al. [1] did year-round monitoring at different sites in the Shale Hills Catchment in central Pennsylvania and found a temporal stability of the soil moisture spatial pattern, which was governed by soil types and landforms. Tromp-van Meerveld and McDonnell [2] measured soil moisture content at different depths between the soil surface and the soil-bedrock and in different locations across the trenched hillslope in the Panola Mountain Research Watershed, GA, USA. They observed that spatial differences in soil depth and total water available at the end of the wet season and during the summer period appeared to be responsible for the spatial differences in basal area and species distribution between upslope and mid-slope sections of the hillslope. Teuling et al. [3] investigated the role of interannual climate variability on soil moisture spatial dynamics for a field site in Louvain-la-Neuve, Belgium, considering three different years (intermediate conditions—1999; wet conditions—2000; extremely dry conditions—2003). They observed that climate variability induces non-uniqueness and two distinct hysteresis modes in the yearly relation between the spatial mean soil moisture and its variability.

In this work we show the soil moisture estimation at local and plot scale in three different field conditions: a plain permanent meadow, a mountain permanent meadow and a hilly vineyard (all located in Northwestern Italy). The main goal is to provide a synoptic view of techniques supported by practical case studies to show that in very different conditions it is possible to estimate a time and spatially resolved soil moisture by the same combination of instruments: contact-based methods for the time resolved, and minimally invasive hydro-geophysical methods for the spatially resolved. Both long-time measurement series and single-measurement campaigns performed simultaneously with different techniques are presented.

The paper is organized as follows. In the following part a general overview of the field measurement technique is presented. In Section 2 we describe some of the available measurement techniques (also employed for the purpose of this paper) to evaluate soil water content both at the local and at the plot scale, focusing on how they work from a theoretical point of view. In Section 3 we present three different case studies (together with the obtained results), showing how we applied the methods. In Section 4 we discuss the results. In Section 5 we draw conclusions and provide an outlook on current opportunities and challenges.

1.2. Soil Moisture and Field Measurement Techniques: A General Overview

According to Robinson et al. [11], soil moisture in situ measurement methods can be divided into contact-based and contact-free methods. Contact-based methods imply direct contact with the soil. They are mostly applied at the local scale, and include both the destructive sampling (e.g., gravimetric methods—also defined as properly "direct" method—[14]), and indirect methods that account for the effects exerted by water content in the soil on its dielectric properties such as: Time Domain Reflectometry—TDR e.g., [15–17]; Frequency Domain Reflectometry—FDR e.g., [11]; capacitance sensors e.g., [18,19]; and Time Domain Transmission (TDT) sensors, e.g., [20]. The sensor networks (wireless in particular) are generally constituted by capacitance, FDR, and TDT sensors, e.g., [5,21,22]; however, more complex but reliable solutions are represented by TDR multiplexing systems, e.g., [23,24], similar to the one presented in the plain permanent meadow case study. These techniques give spatially and resolved temporally highly measurements at the local scale (e.g., vertical soil moisture profile) and the spatiotemporal dynamics of soil water content at the field scale [4,25]. Contact-free methods include hydrogeophysical methods and remote sensing. Hydrogeophysical methods provide multi-point measurements using electromagnetic soil moisture sensors, with a spatial resolution down to several meters [5]. The most used hydrogeophysical methods are: (1) Ground Penetrating Radar (GPR) e.g., [26–32], in which the propagation velocity of high frequency (1 MHz to 1 GHz) electromagnetic waves is used to determine the soil dielectric permittivity and thus soil moisture content. It can be used with different setups, such as crosshole, air-launched and surface-based [5,26]; (2) Electromagnetic Induction (EMI) e.g., [33–35], and Electrical Resistivity Tomography (ERT) e.g., [32,35–40], which measure the apparent electrical conductivity (EC_a) of the soil, and will be treated more in detail in the following chapters. Note that strictly speaking ERT is rather a non-invasive method, as galvanic contact with the ground must be ensured. Remote sensing methods include: (a) passive microwave remote sensing e.g., [5,41–43], in which radiometers measure the thermal radiance emitted from the earth surface of a bare or cropped soil, using frequencies between 1 and 12 GHz (L- to X-band), obtaining the dielectric permittivity [5]; (b) airborne and spaceborne active microwave remote sensing (i.e., radar), such as TerraSAR-X, European Remote Sensing 2 (ERS-2), Advanced Synthetic Aperture Radar (ENVISAT/ASAR), Radarsat 2, and Phased Array type L-band Synthetic Aperture Radar (ALOS/PALSAR), which estimate surface soil moisture by the interpretation of the backscattering coefficient (expressed in decibels and depending on the dielectric permittivity of soil and surface roughness) [44,45]; (c) a cosmic-ray probe [46,47], which counts secondary fast neutrons near the soil surface created by primary cosmic-ray particles in the atmosphere and in the soil. The cosmic-ray probe was applied in different environmental and agricultural settings, e.g., [48–50]. Despite effortz, e.g., [51–54], to downscale the remote sensing soil moisture products (from several km to the local scale), the promising results obtained in scale root-zone soil moisture, e.g., [55–58], and the challenging opportunities offered by the future missions under launch in the coming decades, e.g., [59], remote sensing methods have typically been used to detect surface and near-surface soil moisture (topsoil—0–0.05 m) with coarse spatial and temporal resolution. Extensive reviews of these topics are offered by Peng et al. [60] and Mohanty et al. [61]. Within this context, a cosmic-ray probe is actually the only one used at the field scale, because it has the ability to measure integral soil moisture content with an acceptable temporal resolution [5]. However, it is essential to highlight that (i) the cosmic ray footprint radius is some hundreds of meters depending on the humidity, soil moisture and vegetation; (ii) the detector is extraordinarily sensitive for the first few meters close to the detector; and (iii) the penetration depth (on the order of a few decimetres) decreases exponentially with the distance to the sensor, e.g., [46,62–64].

In the following sections we will describe in more detail the soil measurement methods we adopted in our experimental sites. Namely: TDR and two different low-frequency types of soil moisture sensors (ECH$_2$O—5TM (Decagon Devices Inc., Pullman, WA, USA) and CS616 (Campbell Scientific Inc., Logan, UT, USA)) directly connected to data loggers, which give local measurements of soil water content in time and space; and EMI, GPR, and ERT, which represent plot-scale measurement techniques.

2. Soil Moisture Field Measurement Methods: A Focus on the Techniques Used

2.1. Time Domain Reflectometry (TDR)

As it is non-destructive and less time-consuming with respect to gravimetric measurements, Time Domain Reflectometry (TDR) is a widely used method for soil moisture measurement. Thanks to the accurate results achievable, together with its applicability to a large range of soils and settings, TDR progressively strengthened its position, replacing in the end the gravimetric technique [13]. In this work, a Tektronix 1502 C TDR cable tester coupled with two wire rods probes has been used.

TDR estimates the apparent dielectric permittivity of the soil by measuring the travel time a step voltage pulse takes to propagate along the probe and back. The probe must be placed in the soil either vertically or horizontally at the selected measuring depth. The dielectric permittivity of the system that surrounds the probe depends, in turn, on the soil moisture [65–67]. Calibration requirements are minimal (in many cases soil-specific calibration is not needed), but soil-specific calibration is possible for applications that demand high accuracy.

Once a soil permittivity value has been calculated starting from the travel time measurements, the soil moisture can be obtained by employing either empirical relationships [15] or quasi-physically based equations [68,69]. Considering that soils can have rather different electrical properties, the choice of the proper relationship/equation is connected with the physico-chemical characteristics of the surveyed material.

Different types of TDR field probes were studied in the literature, e.g., [70–77]. Robinson et al. [17] indicated the probes with two parallel rods as those to be preferred in field conditions, since they investigate a bigger volume in a more homogeneous way. Probes with two parallel rods can also be inserted more easily into the soil. Nissen et al. [78] suggested that a balun transformer could be omitted due to the results they obtained in their study.

2.2. *Soil Moisture Sensors Directly Connected to Data Loggers*

An increasing number of soil moisture sensors directly connected to data loggers are deployed to measure volumetric soil water content (θ) for agricultural, ecological, and geotechnical applications. While time-domain reflectometry (TDR) and transmissometry (TDT) operate at GHz range frequencies, the above mentioned sensors generally operate between 20 and 300 MHz [79]. With respect to time domain techniques, lower-frequency sensors are less expensive and, despite TDR still being more accurate, modern electronics coupled with a better understanding of the theory progressively improve low-frequency sensors' performance [80]. In this work, we present two contact-based sensor types: ECH_2O—5TM capacitance/frequency domain sensors (Decagon) and CS 616 water content reflectometers (Campbell Scientific). The frequency of 70 MHz at which both sensors are operating minimizes textural effects (and salinity) [81–83].

ECH_2O—5TM sensors employ a capacitance technique to determine soil moisture. In particular, these sensors determine the dielectric permittivity of the medium by measuring the charge time of a capacitor, using the soil as a dielectric medium. CS616 measures a period value. The period in air is approximately 14.7 microseconds, and the period in saturated soil with porosity equal to 0.4 is approximately 31 microseconds. The output is a ± 0.7 volt square wave with the frequency dependent on water content. This frequency is scaled down in the water content reflectometer circuit output to a frequency easily measured by a data logger. The probe output frequency or period is empirically related to water content using a calibration equation. Concerning the geometry, 5TM probes have three flat 52 mm long prongs spaced 5 mm apart, while CS616s have two 300 mm long rods with a diameter of 3.2 mm, spaced 32 mm apart. Measurement rods of CS616s are therefore approximately six times longer and six times wider apart than those of 5TMs. With reference to the porous medium measurement volume, considering that it is restricted to the direct surroundings material of the prongs/rods, the 5TMs investigate a soil volume much smaller than the CS616 [82,84]. Particular attention must be paid to the probe–soil contact.

2.3. *Electromagnetic Induction (EMI)*

As already mentioned, soil is a three-phase (solid–water–air) system, in which the main conducting phase is the aqueous solution. Due to this assumption, apparent Electrical Conductivity (EC_a) measurements can be used to evaluate soil water content [85]. Friedman [85] also states that there are three main categories of factors that can affect soil EC_a: (i) factors describing the bulk soil and the respective volumetric fractions occupied by the three phases and possible secondary structural configurations (aggregation): porosity, water content and structure; (ii) factors quantifying solid particle, which are relatively time-invariable: particle shape and orientation, particle size distribution, cation exchange capacity (CEC) and wettability; (iii) factors representing soil solution attributes, changing quickly in response to alterations in management and environmental conditions (e.g., environmental factors): ionic strength, cation composition and temperature. Electrical resistivity thus can change in space and time as an effect of (a) lithology (time-independent); (b) moisture content; and (c) pore water salinity. This is accounted for in all constitutive models describing the dependence of bulk electrical resistivity on soil parameters and state variables, starting with Archie's law [86]. The possible misleading effects must be accounted for, as shown in a number of recent literature reports, e.g., [87]. The same problem applies of course to EMI and ERT (see next section) as both measure electrical resistivity/conductivity.

Since it is a fast, non-invasive technique easy to use in field conditions, Electromagnetic Induction (EMI) is often used to map soil properties [88], but also soil moisture changes [35]. EMI follows the principle that within a time-varying electromagnetic field, any conductive object carries a current. Each instrument has two coils, a transmitter and a receiver, placed at either a fixed or variable spacing. Vereecken et al. [5] report typical fixed coil separations between 0.3 and 4.0 m, and a frequency operational range of between 5 kHz and 50 kHz.

EMI induces an electrical current into the soil; the depth of penetration is influenced by the separation of the coils and by the frequency of the current. Since EC_a is affected by soil's properties, the signal reaches a specific depth, also related to the uniformity of the soil. If the soil is very conductive close to the surface, then the signal will be locally dissipated without going deeper. Therefore, the depth of investigation of EMI equipment depends on four principal factors: (1) the resistivity of the investigated soil; (2) the coil separation of the instrument; (3) the coil orientation (vertical or horizontal); and (4) the waveform frequency. By varying the conditions of (2), (3) and (4), multiple depths of investigation can be achieved. The depth of investigation can be obtained as a function of the waveform frequency, the magnetic permeability and the electrical conductivity.

Advanced inversion techniques hold the potential of being applied effectively on small-scale EMI data, such as described above, e.g., [89,90]. While this is not yet common practice for soil moisture monitoring, some notable applications already exist, e.g., [91]. A recent review of EMI methods for hydrological and environmental applications is given by Boaga [92].

A number of EMI instruments are available in the market. For this study, we used a commercial GF Instruments CMD1 system (GF Instruments, Brno, Czech Republic) as well as a commercial GSSI Profiler EMP-400 (Geophysical Survey Systems Inc., Nashua, NH, USA). No inversion was carried out in this study, and maps show apparent electrical conductivity as traditionally estimated using a small induction number approach [93].

2.4. *Electrical Resistivity Tomography (ERT)*

The ERT is a non-destructive, cost-effective, indirect method that can also acquire data concerning soil moisture data within the subsurface, e.g., [5,40,94,95]. ERT uses a set of electrodes: two that inject electrical current into the soil, and two others that monitor the resulting voltage difference. Given a certain number of deployed electrodes, a very large number of measurement combinations is possible, giving rise to the measured dataset. Different configurations are possible, with pros and cons [94]. The measured voltage differences versus injected currents are summarized in terms of the relative ratio, i.e., the measured resistances (one for each measurement quadripole). A good practice consists of collecting both direct and reciprocal resistances (i.e., swapping current and voltage electrode pairs). Theoretically the two configurations shall give the same resistance value, so differences can be taken as an estimate of measurement error, to be used later for a more educated inversion procedure, e.g., [96]. Data are then inverted to give an estimate of the electrical resistivity spatial distribution that causes the observed resistances. Inversion requires the minimization of the overall difference between measured and simulated resistances, the latter being modelled using a physico-mathematical simulator that reproduces a DC current in a heterogeneous medium. This least-squares approach must generally be supplemented by some form of regularization [94], i.e., some a priori information about the spatial variability of electrical resistivity. In the most common case, the smoothest spatial distribution is sought, according to the so-called Occam's inversion approach that makes use of error estimates derived, e.g., from reciprocal analysis [94–97]. The resolution of the resulting resistivity distribution estimates decreases with depth, but is also dependent on the resistivity values themselves. The resulting distribution is generally reconstructed in 2D, but 3D imaging is possible if dense electrode distributions are available in 2D at the ground surface or in boreholes [5].

In order to reconstruct soil moisture patterns from ERT data, petro-physical relations such as Archie's law are used [40,98]. ERT measurements are widely employed to monitor vadose zone water dynamics, e.g., [32,94,99,100]. Since ERT allows providing images of spatial scale soil water content variability along either 2D transects or 3D soil volumes (considering high spatial resolution and daily temporal resolution) [5], it has been widely used in the studies of root water uptake by plants and temporal and spatial variations of soil–moisture interactions, e.g., [5,35,101–103]. However, there is evidence that this inversion procedure can produce mass balance errors due to a rapid decrease of ERT resolution with depth [40]. This problem may be solved through a coupled hydro-geophysical approach [40,101]. However, if only qualitative assessment of spatial distribution and time variations of soil moisture content are sufficient, an uncoupled approach can be used [32] based on the following steps: (1) inversion of geophysical field data gives the spatial distribution of electrical resistivity; (2) application of a petro-physical relationship to obtain an estimation of moisture content distribution from the electrical resistivity.

Many ERT instruments are commercially available. In this study we used a commercial IRIS Instruments SYSCAL Pro 10 resistivimeter with 72 channels (IRIS Instruments, Orleans, France). Inversion was carried out using the freely accessible inversion codes developed by Andrew Binley [32,104–106].

2.5. Ground Penetrating Radar (GPR)

GPR is a non-invasive and non-destructive geophysical method that uses radar pulses to image the subsurface. Given that the velocity of an electromagnetic wave in a soil is connected to its bulk dielectric constant (which is, in turn, related to the soil moisture), over the last three decades GPR has also been widely applied to estimate soil moisture in the unsaturated zone. A transmitter emits pulses (high-frequency radio waves typically in the range between 10 MHz and 2.6 GHz) that travel in the medium; when pulses encounter materials having different permittivities they are reflected, refracted or scattered back to the surface where a receiving antenna records the signal. Due to the frequency-dependent attenuation, lower frequencies lead to higher soil depth penetrations. However, higher frequencies provide higher resolutions. Two classes of GPR methods to estimate soil moisture can be distinguished: single-antenna separation methods (from scattering objects and traditional GPR sections), and different antenna separation methods (which require multiple measurements). For a comprehensive presentation of each single method, Huisman et al. [26] provide a detailed review.

Considering that the use of GPR systems is difficult in uneven irregular areas, an application is here presented in the plain permanent meadow case study. A commercial PulseEkko Pro radar system (Sensors and Software Inc., Mississauga, ON, Canada) with 100 MHz antennas has been used. For more details please refer to Rossi et al. [32].

3. Results: Testing Sites Description

In this section we describe the main characteristics of the three experimental sites together with the field campaigns layouts and the measurement methods applied in each site. All sites are located in Northwestern Italy. However, they are representative of very different conditions: the first is a permanent meadow located in a plain, while the second is a meadow located in a sloping mountain abandoned pasture; finally, the third is a vineyard in a hilly zone. A synoptic scheme of the different equipment used in the different testing sites is given in Table 1.

Table 1. Synoptic table summarizing the different equipment used in the different testing sites.

	Plain [Permanent Meadow]		Mountain [Permanent Meadow]		Hill [Vineyard]	
	Long Time Surveys	Single Campaigns	Long Time Surveys	Single Campaigns	Long Time Surveys	Single Campaigns
Time Domain Reflectometry (TDR)	√	√		√		√
Soil moisture sensors directly connected to data loggers			√		√	
Electromagnetic Induction (EMI)				√		√
Electrical Resistivity Tomography (ERT)		√		√		
Ground Penetrating Radar (GPR)		√				

Authors make the datasets collected within this work available. Readers or researchers interested in receiving and/or analysing one or more dataset shown in this work must address their specific request to the corresponding author.

3.1. Plain Permanent Meadow

Site and instrumental description. The experimental site is located in Grugliasco (Torino), in the northwestern part of the Po Plain, Italy (45°03′ N, 7°35′ E) at 290 m a.s.l. (Figure 1). In this area, rainfall climatology is characterized by two maxima, respectively in spring (April–May) and fall (October–November), and by a relatively dry winter and summer [107,108]. The vegetation growing season lasts, approximately, from late March to mid-October. The monitored area is 1500 m^2, constituted by a permanent meadow. The soil is loamy–sand to sandy (≈80% of sand [32], which progressively rises to ≈95% below the 1st meter), without any gravel, and with a slope of about 1%. These last conditions are particularly suitable for TDR and GPR applications for both practical operation simplification (e.g., TDR probes insertion; GPR surveying process) and porous medium characteristics (i.e., low signal attenuation) e.g., [109,110].

Figure 1. Experimental site located in Grugliasco (Torino–Italy) and a picture shot in April 2005.

Long-term soil moisture measurements. The soil water content monitoring station is based on a TDR Tektronix 1502C and a personal computer controlling 11 multiplexers connected to 160 probes vertically inserted into the soil. The great number of installed probes was conceived to consent the calibration of the 3D numerical codes described by Paniconi et al. [111]. The sandy soil and the absence of stones allowed the vertical insertion without disturbance and/or induced lack of contact. Probes are made out of two parallel stainless steel rods (diameter 6 mm) with the following lengths: 150, 300, 600,

1000 and 2000 mm. To prevent geometrical probes = deformations and keep rods parallel during the insertion process, a steel guide was used. The TDR probes' layout in the experimental field is depicted in Figure 2.

The measurements started in 1997 and, since summer 2005, the TDR measurements have been automatic. The TDR signals are sampled and acquired using WinTDR software [112]. Volumetric soil water content is computed using the composite dielectric approach described by Roth et al. [68]. This relation has been validated by comparison with gravimetric sample measurements in the oven, obtaining an average error of 2%. The sampling time interval between TDR measurements can vary from one hour (during and after rainfall events, or during intense exfiltration periods) up to one day. In the same field, a meteorological station collects meteorological data at hourly intervals.

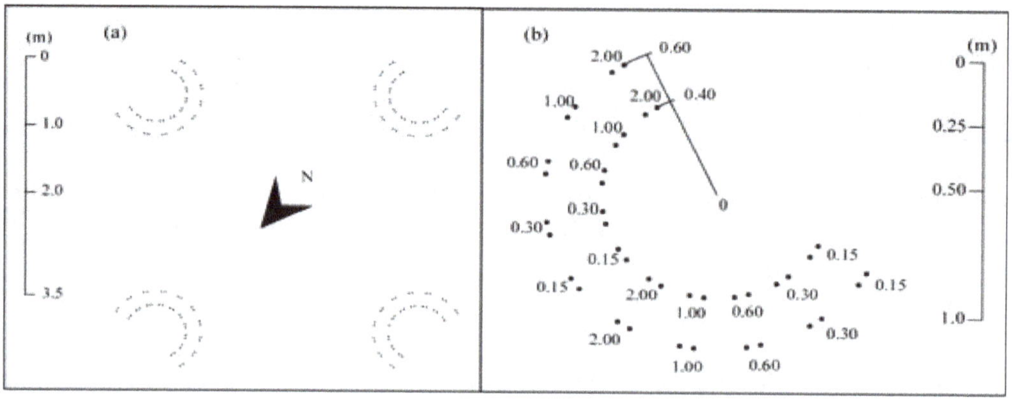

Figure 2. Grugliasco site. (**a**) Layout of the TDR probes in the experimental field; and (**b**) detail of the disposition of the probes of different lengths within one of the four groups.

ERT, GPR and TDR single measurement campaign. To evaluate the soil moisture evolution along time and space, on 28 August 2009 an infiltration experiment (over a rectangular area of 18 m by 2.6 m) was performed at the Grugliasco site. The scheme and a picture of the experimental setup is shown in Figure 3 (for details, see Rossi et al. [32]).

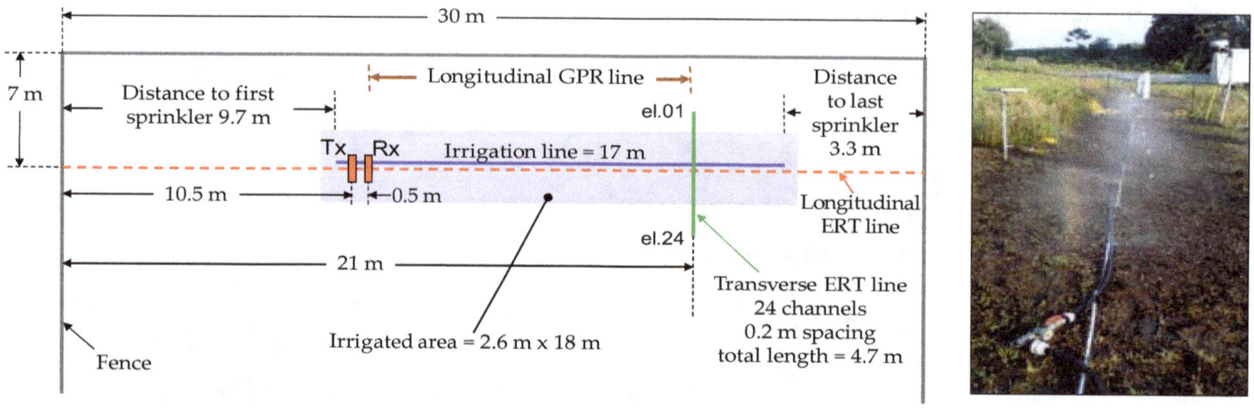

Figure 3. Grugliasco site, August 2009. Scheme of the irrigation experiment setup and a picture taken during the experiment.

The irrigation was performed in three steps: about 2.5 m^3 of water were distributed during the first irrigation step and about 1.5 m^3 during the others. ERT acquisitions occurred during irrigation breaks and with a precise scheduled time. Eight TDR probes with a length of 0.30 and 2.00 m were also vertically inserted into the soil, along the sprinkler line centred on the ERT profile. The absence of ponding was guaranteed by ensuring that the irrigation intensity was always lower than the soil infiltration capacity.

A longitudinal ERT line was acquired along the transect—see Figure 3 for the location—and measurements were taken only before and after irrigation. The line is composed of 48 electrodes with 0.6 m spacing, for a total length of 28.2 m. A dipole–dipole skip zero scheme was acquired, e.g., [113]

3.2. Mountain Permanent Meadow

Site and instrumental description. The Cogne Valley is a side valley of Valle d'Aosta (Northwestern Italy). The experimental site (45°36′ N, 7°21′ E) is located at 1730 m a.s.l. on a 26° slope facing south–southeast (169°)—Figure 4. The vegetation is characterized by herbaceous and shrub components typical of degraded pastures at high altitudes (e.g., *Hippophaë rhamnoides* L.), therefore it is representative of wide areas in the Alps (here, the pastures were abandoned due to the slope and low accessibility). This region is characterized by rainfall occurring mainly in the spring and autumn, with an average of 650 mm per year. The average annual temperature is about 4 °C. The soil depth ranges from 0.4 to 1.5 m, with a sandy loam texture and some gravel (≈73% of sand).

The experimental site is characterized by intense direct solar radiation due to its aspect, resulting in large daily temperature variations. The incident radiation accelerates snow melt and the drying of soil (compared to the plain).

Figure 4. Experimental site located in Cogne (Aosta–Italy) and a picture shot in October 2016.

Long-term soil moisture measurements. Soil water content is measured with three CS616 sensors (Campbell) connected to a CR1000 Data logger. In 2010, data were collected hourly by one probe horizontally inserted at 0.1 m depth. In 2015 two more CS616 probes horizontally inserted into the soil at 0.2 and 0.4 m depths were installed. The gravimetric calibration method and temperature correction were adopted.

EMI, ERT and TDR measurement campaigns. EMI surveys aiming at mapping the spatial and temporal variability of soil resistivity along the steep mountain meadow testing site were performed in 2010, on the 26 September and the 23 October.

Data were collected using a GF Instruments CMD1 in high penetration mode (hence the estimated investigation depth ranges from the surface down to −1.5 m). During the same investigation dates, an ERT survey was performed along a transect located on the slope. A dipole–dipole skip zero acquisition scheme, with full reciprocal acquisition, was adopted. Data were inverted using an Occam inversion approach, as implemented in the Profile R/R2/R3 software package, accounting for the error level estimated from the data themselves [114]; here 5% was chosen as a relative error. During the 26 September 2010 EMI and ERT monitoring campaigns, soil moisture values were also measured by vertically inserted TDR probes. Two transects were investigated: the first was realized with measurements at the beginning, at the centre and at the end of the ERT transect (0.75 m long TDR probes vertically inserted into the soil); the second was more superficial (0.15 and 0.30 m long TDR probes vertically inserted into the soil), conducted perpendicularly to the ERT profile.

3.3. Hilly Vineyard

Site and instrumental description. The "Tenuta Cannona" (44°40′ N, 8°37′ E, 296 m a.s.l.) is located in Carpeneto (AL), 85 km southeast of Torino, Northwest Italy (Figure 5). The area is in the Alto Monferrato hilly region, which is a valuable vine-growing and DOC (Controlled Designation of Origin) wine production area. The soil has a clay to clay–loam texture (≈35% of clay and ≈40% of silt). The climate presents an average annual precipitation of 965 mm (based on the Ovada meteorological station data), mainly concentrated from October to March. Mean annual air temperature measured at the experimental site in the period 2000–2013 was 13 °C and the average annual precipitation was 849 mm [115,116].

Figure 5. Experimental site located in Carpeneto (Alessandria–Italy) and a picture shot in March 2016.

Long-term soil moisture measurements. A continuous monitoring was performed starting from August 2011 to provide information about the space–time variability of hydrological processes with different cultivation techniques. The test sites, which include conventional tillage ("TI") and grass-covered soil ("GC"), are adjacent to each other on a parcel located on a hill with a SE aspect and average slope of 15%. Each site includes six vine rows aligned along the slope, where the vines are spaced 1.0 m along the row and there is 2.75 m between the rows (Figure 6).

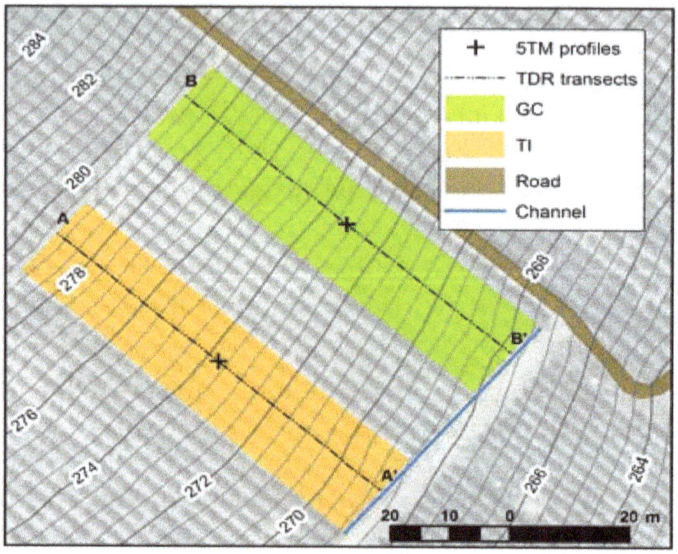

Figure 6. Carpeneto site. Plot of the experimental field. Grass Cover (GC) and Conventional Tillage (TI) represent the two investigated cultivation techniques. The two dotted lines A–A′ and B–B′ represent the investigated TDR transects.

Soil moisture data were collected by measuring the soil dielectric constant using four series of 5TM capacitance probes. Sensors have been horizontally inserted both in the central part of the inter-row (no soil compaction due to the passage of tractors—"NT") and in the track position ("T"), affected by the passage of tractor wheels. During the monitored period (winter 2015–2016), the soil was never frozen.

As regards to the sensors, we followed the calibration suggested by Decagon Devices which apply the Starr and Paltineanu approach [117], and resulting in an accuracy of ± 0.03 m^3/m^3 ($\pm 3\%$ VWC). Measurements are recorded and stored by a Decagon EM50 Data logger every 60 min.

EMI and TDR combined campaign. In this field campaign, we used the EMP400 multi-frequency EMI sensor, which can collect data up to three frequencies from 1 to 16 kHz, in horizontal dipole orientation (HDO) and vertical dipole orientation (VDO), with a coil spacing equal to 1.22 m. In this study, the EMP400 was used in VDO operating mode, with selected frequencies of 5, 10, and 15 kHz. The EMP400 is very sensitive at greater depths because of the larger coil distance. The EMP400, equipped with an integrated WAAS GPS, was carried by hand.

The TDR soil moisture was detected by a TDR Tektronix 1502C with three probes (made out of two parallel stainless steel rods with diameter 6 mm) vertically inserted into the soil from the surface down to the following depths: 0–0.15 m, 0–0.30 m, and 0–0.75 m. Measurements were performed every 10 m along two transects both in conventional tillage (TI), and in grass cover (GC) rows.

4. Results: Measurement Campaigns and Surveys Outputs

4.1. Plain Permanent Meadow

Long-term soil moisture measurements. The long-time survey depicted in Figure 7 shows the soil moisture data from 1 January 2004 to 31 December 2006 at the following vertical depths: 0–0.3, 0–0.6, and 0–1.0 m. Values are obtained through the arithmetic average of all the probes with the same rod lengths (namely, 16 probes per each considered length).

During the whole of 2004, until summer 2005, measurements were collected on a weekly base. Despite the general agreement with the precipitation data, the weekly sampling time is not exhaustive in terms of hydrological processes description (especially with coarse texture soils characterized by fast dynamics). Under such condition, minor meteorological events characterized by low magnitudes, are often not detected by the soil moisture profiles. At the same time soil moisture peaks corresponding to high precipitation events appear often damped (i.e., rarely exceeding the 0.20 m^3/m^3). Starting from August 2005, data have been automatically collected at hourly steps. With this sampling interval, minor events are also detected, and soil moisture peaks are better described, arriving to reach at the 0.30 m^3/m^3 threshold (which can be considered close to saturation for the investigated sandy soil).

With reference to the adopted experimental set-up, characterized by TDR probes vertically inserted into the soil, all the different lengths of probes react to meteorological forcing. Obviously this reaction is inversely proportional to the investigated soil volume/layer in relation to the magnitude of the events (both in terms of wetting and drying). Accordingly, the most intense reaction is observable in the data collected by the shortest probes (in Figure 7 represented by the 0.30 m lengths), which investigate the shallowest layer. Moreover, during the winter months, the presence of frozen topsoil can lead to TDR measurement complications due to the TDR's inability to detect simultaneously liquid and ice contents in frozen soils within the bandwidth included between a few megahertz and 1.5 GHz [118,119]. These complications, if not properly considered, can cause measurement errors proportional to the soil depth reached by the frost in relation to the depth investigated by the probes.

ERT, GPR and TDR single measurement campaign. Figure 8 shows how visible is the patch of soil where irrigation had taken place with respect to the non-irrigated zone. However, at this measurement scale, it is hard to identify the depth of infiltration as the resolution cannot be any better (and is actually worse) than the electrode spacing (0.6 m in this case).

In order to capture in more detail the changes in moisture content with depth, ERT measurements were collected also using 24 electrodes (spaced 20 cm), set on a transect perpendicular to the sprinkler's line over a total length of 4.6 m. Surface data were obtained with a Syscal-Pro resistivimeter (IRIS Instruments, Orleans, France) and a dipole–dipole skip zero (dipoles with minimal distance equal to one electrode spacing) acquisition scheme. Time-lapse measurements were taken periodically, using a dipole–dipole skip 0 scheme and full acquisition of reciprocals to estimate the data error level [96,120,121]. Consistently, the data inversion used an Occam inversion approach, as implemented in the Profile R/R2/R3 software package, accounting for the error level estimated from the data themselves [114]. Results are shown in Figures 9–11.

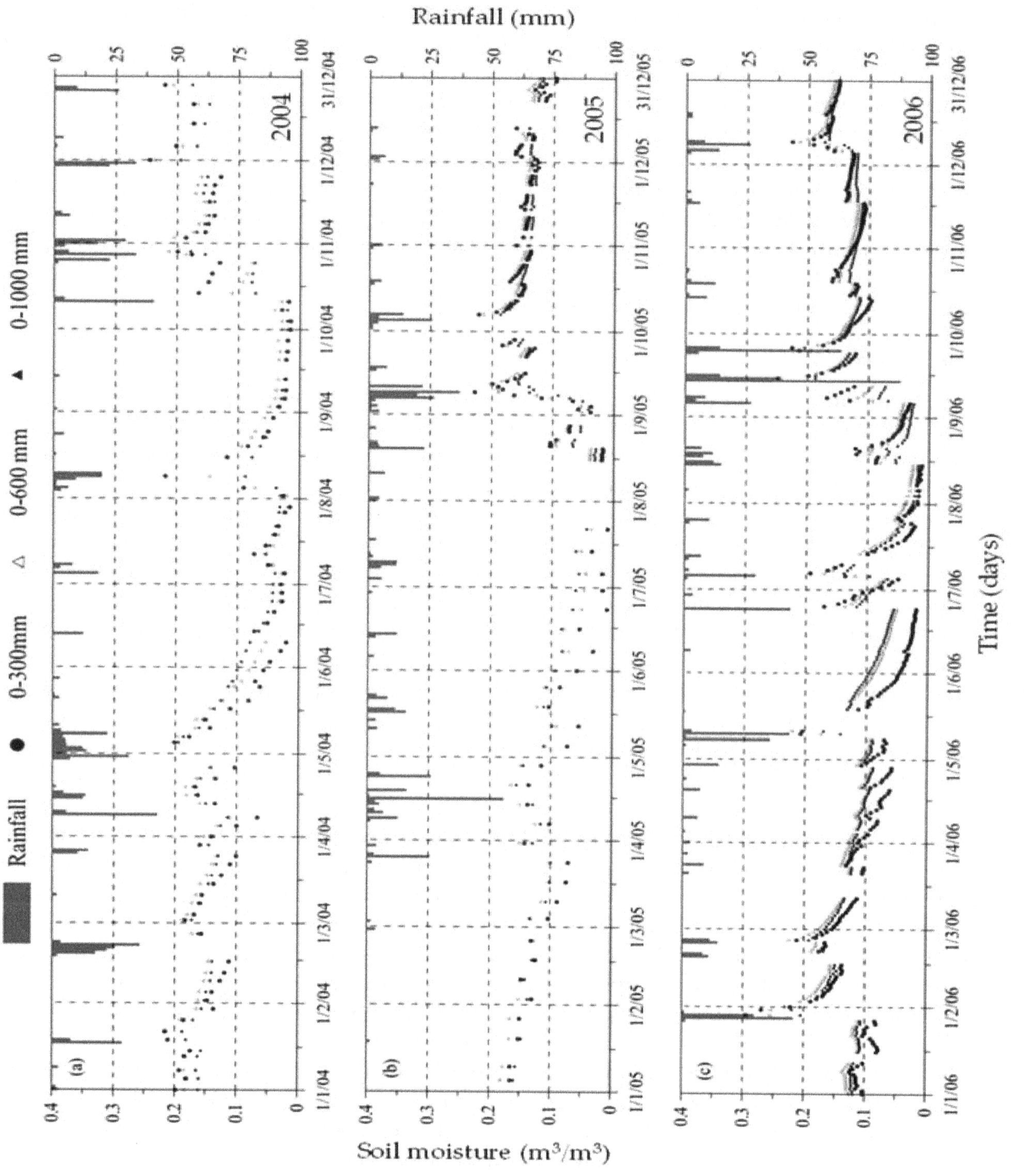

Figure 7. Grugliasco site, 2004–2006. Daily soil water content and rainfall data within the layers 0–0.3, 0–0.6, and 0–1.0 m from 1 January 2004 to 31 December 2006. Since 2005 data have been automatically sampled at hourly step intervals.

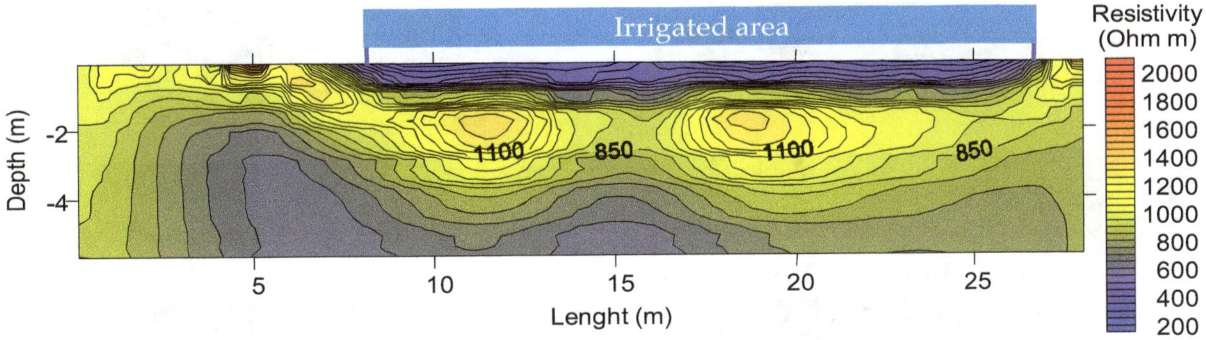

Figure 8. Grugliasco site, 28 August 2009 (hour 17:30). Absolute ERT image at the end of irrigation along the longitudinal ERT line (see Figure 3 for location).

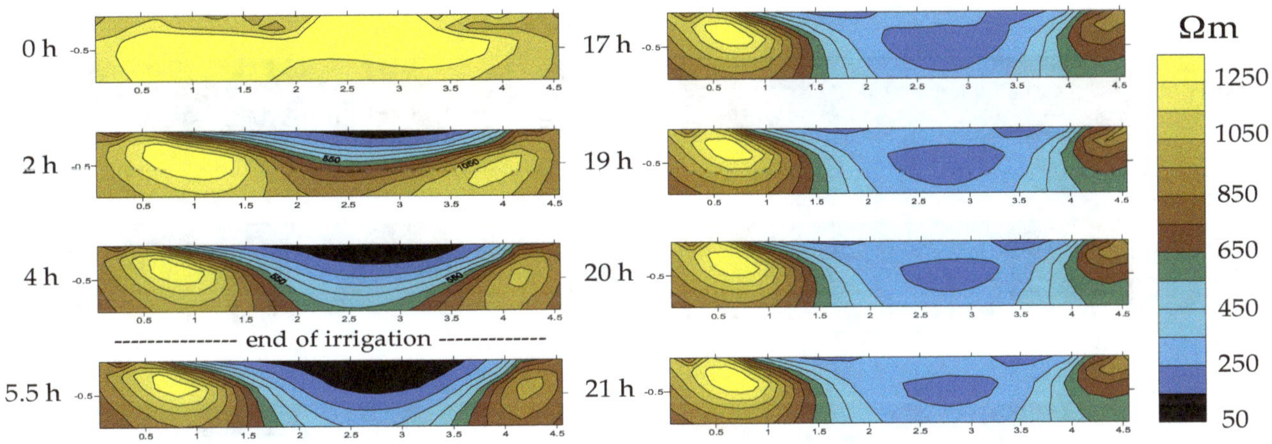

Figure 9. Grugliasco site, August 2009. Absolute ERT images along the transverse line (see Figure 3 for location).

Figure 9 shows the time evolution of absolute resistivity during and after irrigation. The changes in resistivity are particularly clear as the soil profile was initially very dry (at the end of August). The same patterns are apparent, with a more distinctive character, in the relative time-lapse images shown in Figure 10. Here a comparison is also made between a simple approach, where ratios of the absolute images in Figure 9 are taken pixel by pixel, and the more accurate approach based upon ratio inversion, e.g., [96]. In this specific case the two approaches yield practically the same results as, given the geometry of the system and the corresponding resistivity variations over time (the ERT line is transverse to a long stripe where resistivity changes in a fairly homogeneous manner), no 3D effect is present. Note that in general a simple pixel-by-pixel ratio of absolute images does not lead to acceptable results, as the absolute images are 2D inversions in what is actually a 3D space, and ratio or difference inversions shall be used [122].

Finally, Figure 11 shows the estimated soil moisture patterns in the ERT section, derived from converting resistivity into moisture content by means of a laboratory-calibrated Archie's law [86]. The infiltration front estimated by GPR measurements is also depicted (red dotted lines).

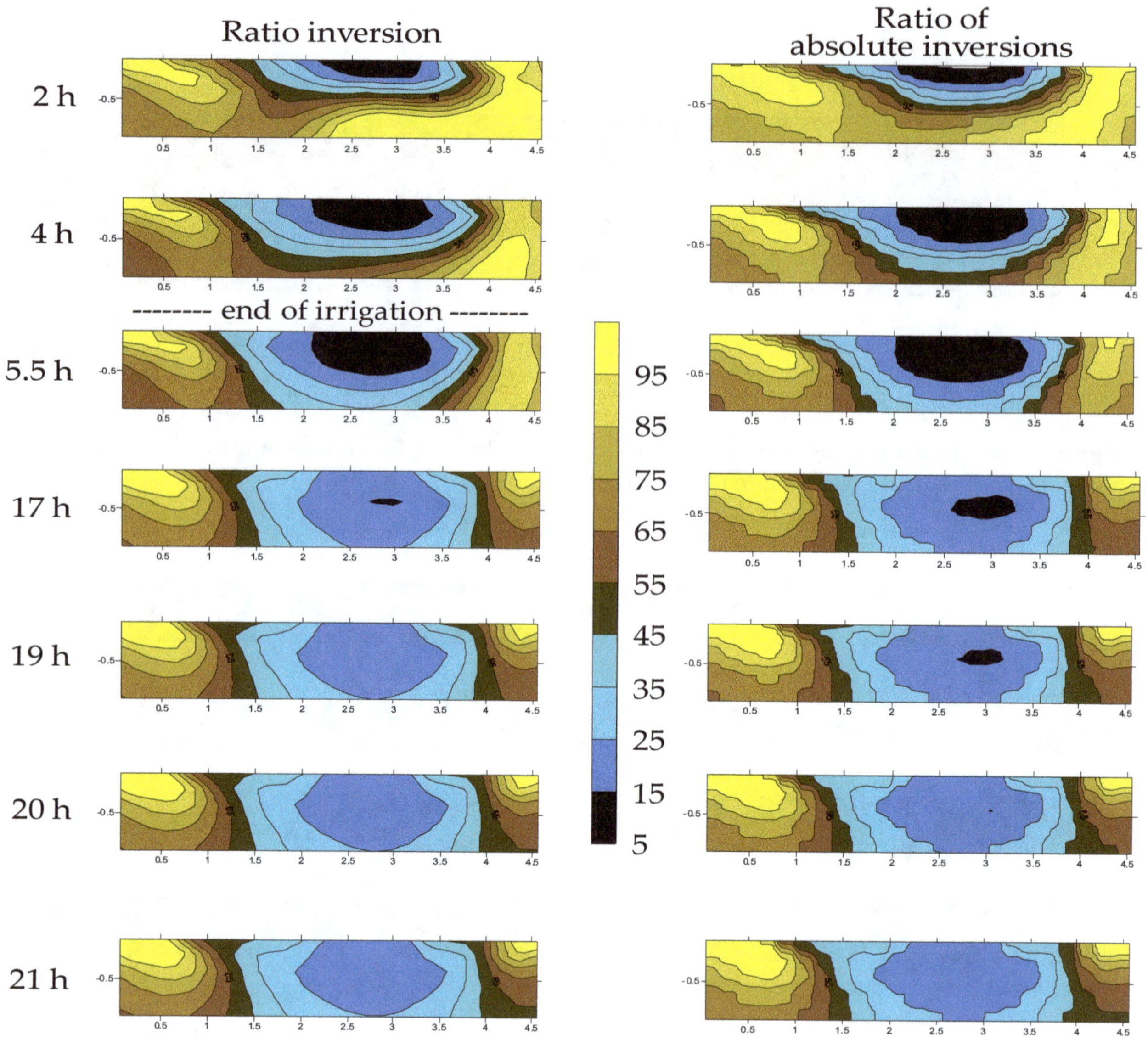

Figure 10. Grugliasco site, August 2009. Relative ERT images along the transverse line (see Figure 3 for location). Both results of ratio inversion and ratios of absolute inversion are shown as percentage of the background (pre-irrigation) values. In this case the two approaches yield very similar results, as the resistivity changes occur homogeneously across the measurement line (no 3D effects).

The performed TDR measurements are depicted in Figure 12 as a function of time and depth. The data are represented as averaged values together with their standard deviations. Initial conditions are dry and homogeneous with soil water contents about 0.08–0.10 m^3/m^3 along all the investigated soil profiles. After the first irrigation step the surface layer immediately reaches values close to saturation (about 0.30 m^3/m^3), but variability is evident between measurements despite the uniform testing conditions. In the following irrigation steps the saturation conditions persist in the surface layer investigated but the variability between measurements progressively decreases. Considering the whole profile, the soil moisture gradually increases along with the irrigation volumes. The following day, starting with the early morning, conditions changed considerably. At the surface, the soil moisture was more uniform and decreased from saturation conditions down to the field capacity (around 0.17 m^3/m^3), while, considering the whole depth investigated, conditions remain the same. All these considerations agree and integrate the information provided by the ERT imaging.

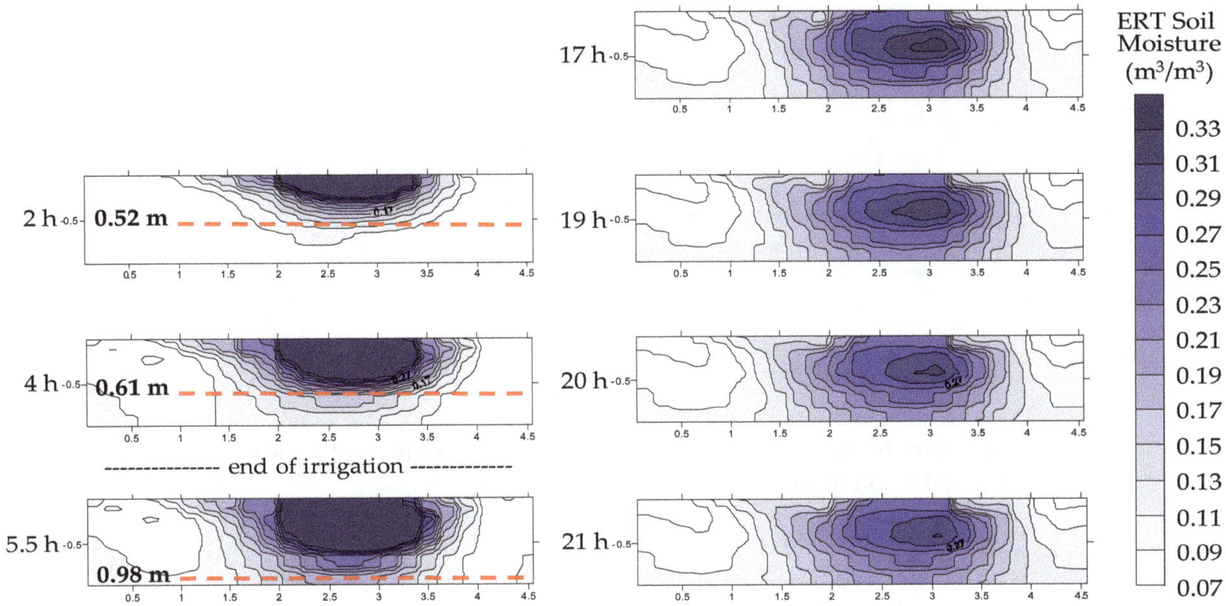

Figure 11. Grugliasco site, August 2009. Moisture content at different times after irrigation, estimated from ERT imaging and laboratory calibrated Archie's law results. A comparison is made against the depth of the infiltration front estimated by GPR measurements (red dotted lines) [32].

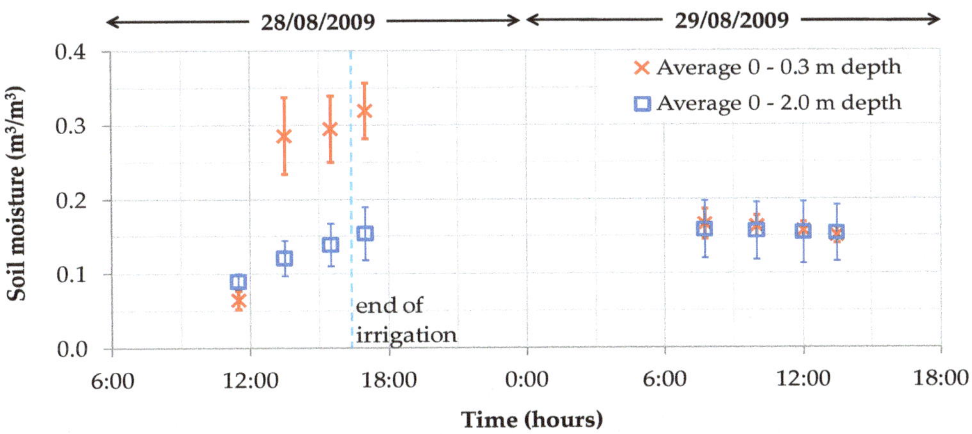

Figure 12. Grugliasco site, August 2009. Moisture contents and their standard deviation at different times after irrigation, computed from TDR measurements.

Usability of TDR for deep soil measurement campaigns. A further interesting application for a low-attenuation porous medium (such as the Grugliasco sandy soil case) is represented by the suitability of TDR technique to investigate very thick layers (down to 4 m). Despite the complications connected with the probes' insertion (where an insertion guide apparatus is essential to prevent geometrical probe deformations), Figure 13 illustrates the waveform detected during a single measurement campaign with soil moisture ranging between 0.11 and 0.14 m^3/m^3 along the profile. Only the surface layer (0–0.5 m depth) shows a higher value, around 0.22 m^3/m^3. Figure 13 highlights that retrieved waveforms are really clear, even for 4 m long probes, and signal reflections are evident and easy toanalyse.

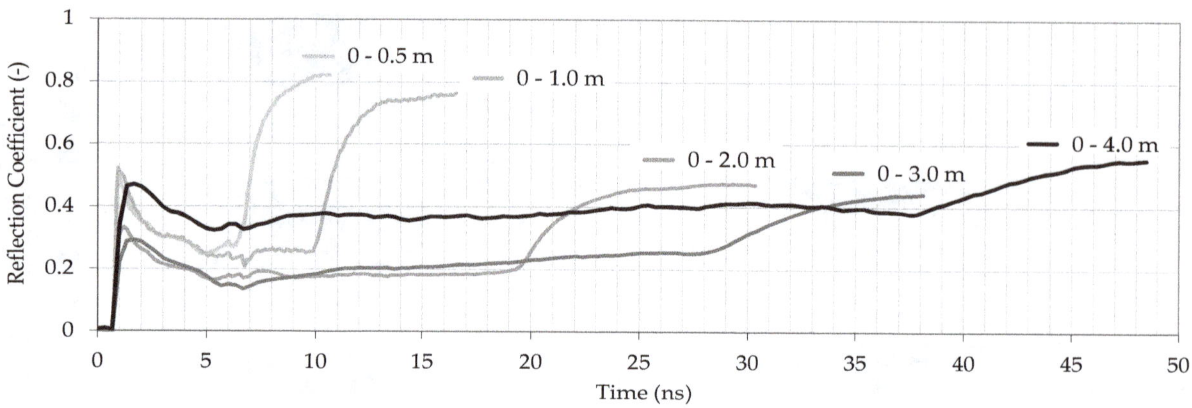

Figure 13. Grugliasco site, 20 May 2010. Waveforms detected with different TDR waveguides lengths vertically inserted into the soil from the surface down to 4 m depth.

4.2. Mountain Permanent Meadow

Long-term soil moisture measurements. In the graph below (Figure 14), the soil volumetric water content—averaged every 30 min—together with rainfall data—collected by the ARPA VdA meteorological station located nearby—are plotted on an hourly base from 1 January to 31 December 2016.

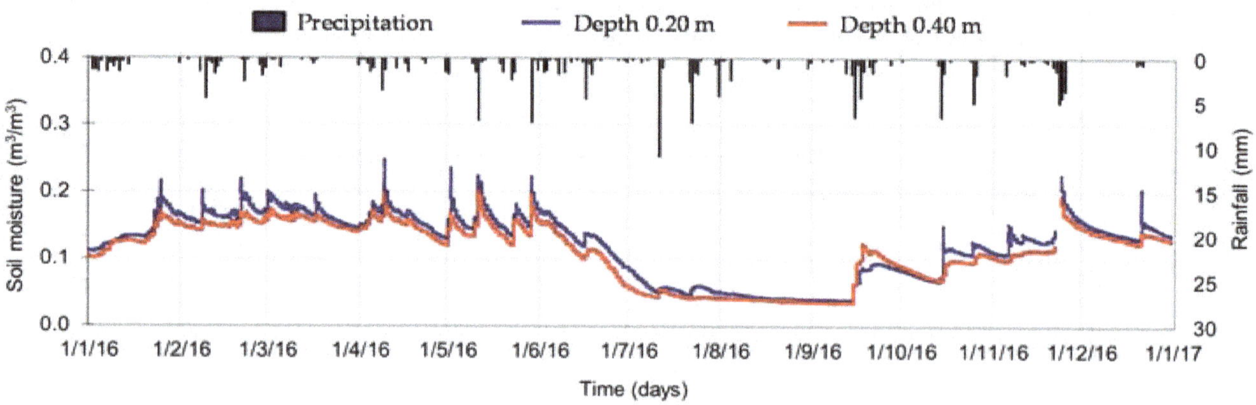

Figure 14. Cogne site, January–December 2016. Precipitation and soil moisture measured at 0.20 and 0.40 m depths.

The volumetric soil moisture collected in the Gimillan site (Figure 4) during the whole year 2016 never exceeded 0.3 m^3/m^3. This threshold can be considered to be in agreement with the literature data for coarse texture soils. Interesting data can be observed at the beginning of the year. In this period the soil is frozen, with a low amount of liquid water content. Then, in the second half of January, a peak occurs without any surveyed rainfall. This behaviour can be ascribed to the soil water (and snow) melting due to the temperature increase, especially considering the south–southeast aspect. Other interesting data can be provided during the summer (in particular from July to the first half of September). In this period, soil moisture is characterized by progressively lower values, with a minimum situated around 0.05 m^3/m^3. In this dry period the correspondence between the occurrence of rainfalls and soil moisture increase is not so evident. This behaviour might be attributed to the effect of the dense vegetation coverage (withered by the water scarcity), driving to interception rather than infiltration into the soil.

With reference to the soil moisture dynamics, fast-acting peaks of soil water increase/decrease can be observed. This attitude can mainly be attributed to the following factors: (1) hydrological properties of the coarse soil texture (e.g., fast infiltration processes; limited water retention/storage);

(2) morphological characteristics of the site, namely the SSE slope aspect, which implies accelerated atmospheric and evaporation dynamics (e.g., solar radiation incidence; mountain/valley breezes). Notwithstanding the abovementioned aspects, it is interesting to highlight that during the surveyed period, the soil moisture values detected at 0.20 m depth were higher than the 0.40 m depth ones, probably due to the dense vegetation and the spatial variability. However, the time dynamic is more enhanced at 20 cm depth.

EMI, ERT and TDR measurement campaigns. EMI surveys at this site are particularly complex given the steep slope and the shrubs (see Figure 4), thus spatial sampling is different at the two dates (as seen in the extent of the mapped areas represented in Figure 15). The most evident feature refers to the strong difference in mean electrical resistivity, with October being much drier than September. Here, the irregular morphology characterized by depressions and bumps, together with the variability of soil and soil-cover vegetation, led to heterogeneous dynamics in time and space, more evident in drier conditions.

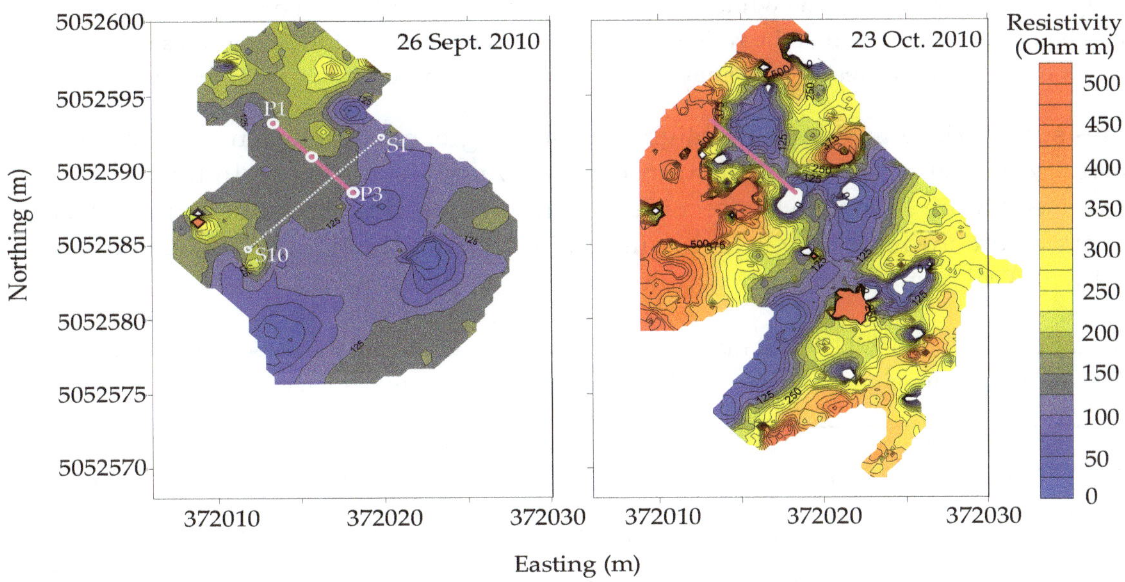

Figure 15. Cogne site, 2010. EMI maps, shown as electrical resistivity values. The strong change between 26 September (more conductive) and 23 ctober (more resistive) is remarkable. The maps represent the mean electrical resistivity from the ground surface to a depth estimated around 1.5 m. The purple line shows the location of the ERT line (results in Figure 16). The white dots represent the two TDR transects: P indicates the progressive 0.75 m depth measurements; S indicates the progressive 0.15 and 0.30 m depth measurements.

It is very informative to compare EMI maps with localized ERT profiles (Figure 16). The latter are taken in reference to the purple line marked in Figure 13. Considering that the values in the EMI maps are obtained in a configuration that gives one value somehow averaged from the ground surface down to 1.5 m depth, i.e., practically involving the entire thickness shown by ERT sections in Figure 16, and given the strong resistivity contrast (both in September and October), this average only has an indicative meaning. Exploring the ERT transects output, note how at both times the shallow soil layer (around 0.5 m thick) is much more resistive than the deeper subsoil. This leads to a very heterogeneous soil profile, so that the EMI maps in Figure 15 only provide a rough mean of the electrical properties averaged over very different values. The ERT images show, consistently with the EMI maps, that in October the site was much drier than in September. More advanced approaches using multi-coil/multi-frequency EMI and depth inversion are, in general, strongly recommended, e.g., [91].

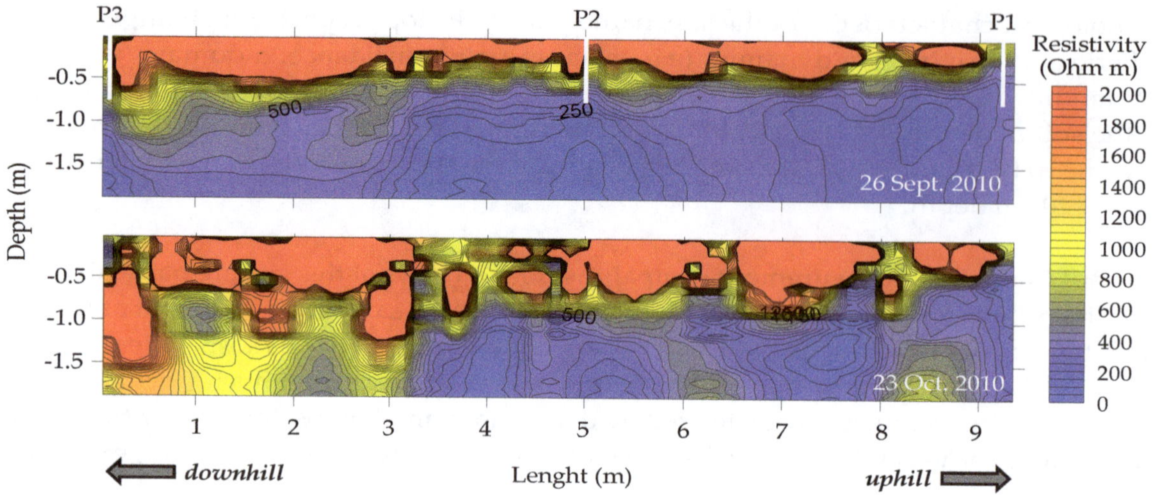

Figure 16. Cogne site, 2010. ERT profiles acquired at the two survey dates along a short profile (see Figure 15 for location). The white lines identified by the P indicate the three 0.75 m depths TDR measurement points. Note that the resistivity scale is much larger in the ERT profiles than in the EMI maps, in order to give a clearer presentation of the variations of resistivity with depth.

Values summarized in Table 2 highlight a general agreement between TDR water content and ERT resistivity behaviour along all the transect ("P" values—see Figure 16 for the exact location). "S" values—see Figure 15 for the transect identification—confirm the water content decrease in the upper layers where the resistivity is higher (as easily seen also in Figure 16) but a strong variability connected with the irregular micro morphology of each measurement point become evident.

Table 2. Cogne site, 2010. TDR soil moisture values measured during the 26 September 2010 EMI and ERT monitoring campaigns. The two TDR transect alignments are depicted in Figure 15.

Probe Code	Transect "P"			Transect "S"														
	P1	*P2*	*P3*	*S1*	*S2*		*S3*	*S4*		*S5*	*S6*		*S7*	*S8*	*S9*		*S10*	
Depth from surface (m)		0.75		0.15	0.15	0.30	0.15	0.15	0.30	0.15	0.15	0.30	0.15	0.15	0.30	0.15	0.15	0.30
Soil moisture (m^3/m^3)	0.41	0.37	0.38	0.33	0.33	0.31	-	0.39	0.22	0.31	0.41	0.40	0.33	0.35	0.34	0.32	0.32	0.27

During the 23 October 2010 EMI and ERT monitoring campaigns, no TDR specific transects were surveyed, however, the CS616 0.10 m depth probe provided the hourly dataset depicted in Figure 15. Despite the data gaps, Figure 17 provides interesting information both on the soil moisture dynamics during the days prior to the measurement, and to the absolute measured soil water content (in agreement with the expected values).

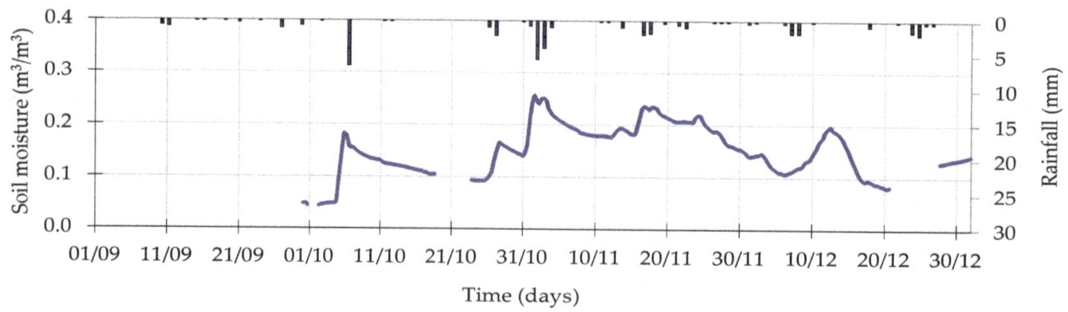

Figure 17. Cogne site, September–December 2010. Precipitation and soil moisture at 0.10 m depth.

4.3. *Hilly Vineyard*

Long-term soil moisture measurements. A plot of the soil moisture data detected at different depths (namely 0.1, 0.2, 0.3 and 0.4 m below the surface) is depicted in Figure 18, together with the rainfall data.

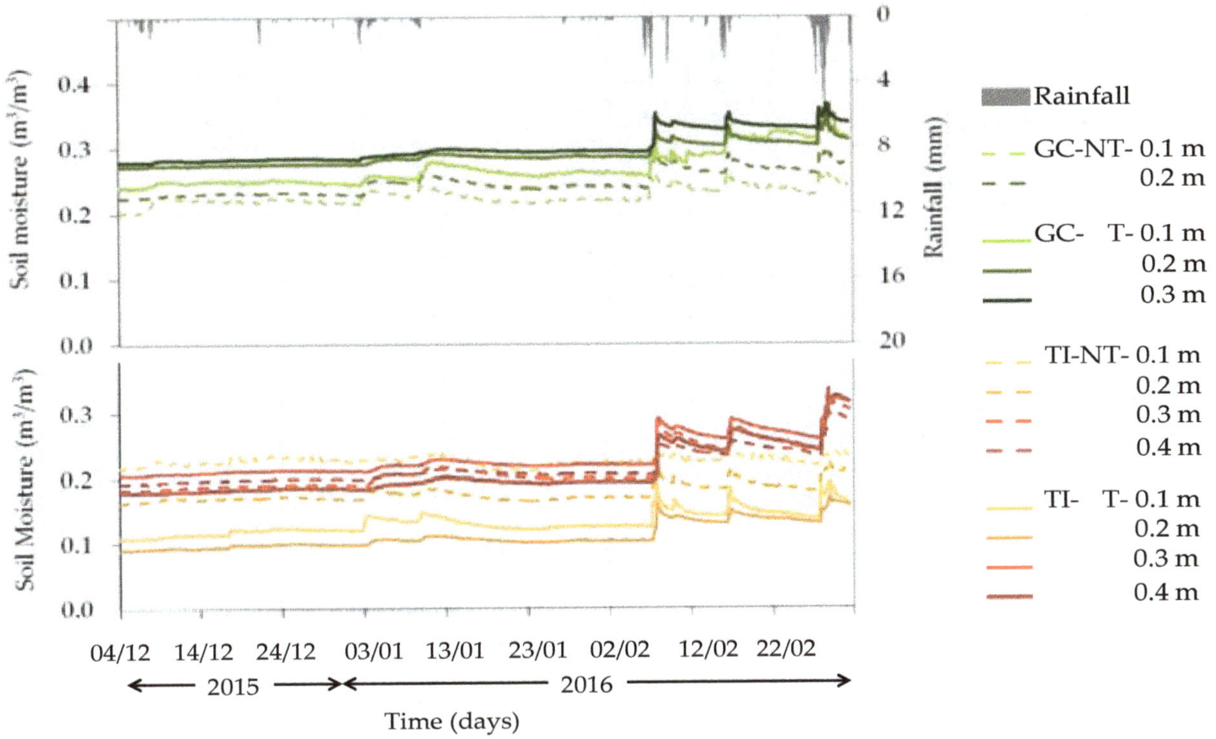

Figure 18. Carpeneto site, winter 2015–2016. Rainfall and soil moisture data collected with an hourly time step (from 0.1 to 0.4 m soil depths). Green lines refer to the Grass Cover condition (GC); red lines refer to the Conventional Tillage condition (TI). Dotted lines refer to the "No Track" condition (NT); solid lines refer to the "Track" condition (T).

The results of the long-term survey (Figure 18) suggest that, under conventional tillage conditions, surface soil moisture in the track position was usually lower than in the central no-track zone. In agreement with Leonard and Andrieux [123] (who found that the major causes in variability of infiltration rates in vineyards are referred to the history of cultivation and the structure of the first soil centimetres), the monitored behaviour can mainly be ascribed to both the agricultural treatment and the soil compaction.

In the grass-covered plot, the soil moisture in the track position was slightly higher than in the central no-track zone along the whole monitoring period. As already observed by Ferrero et al. [124], the tractor traffic in vineyards has a great influence on the spatial variability of soil physical properties, which are strictly related to the topsoil water content. The recurrence of tillage could temporarily decrease this effect, but affects the variability of soil properties over a long period (for more details, see Biddoccu et al. [97]).

EMI and TDR combined campaign. The following Figure 19 show the result obtained the 21 March 2016, from the top of the hill down to the bottom (the last meters of the transects were excluded to avoid measurement distortions due to the presence of a steel grid used to survey the runoff erosion). EMI data were collected in all the rows, starting from the right side (row 18) and moving to the left side (row 0).

In Figure 19, the dots represent the TDR measurement sites along the two transects. The TDR output values are depicted in Figure 20.

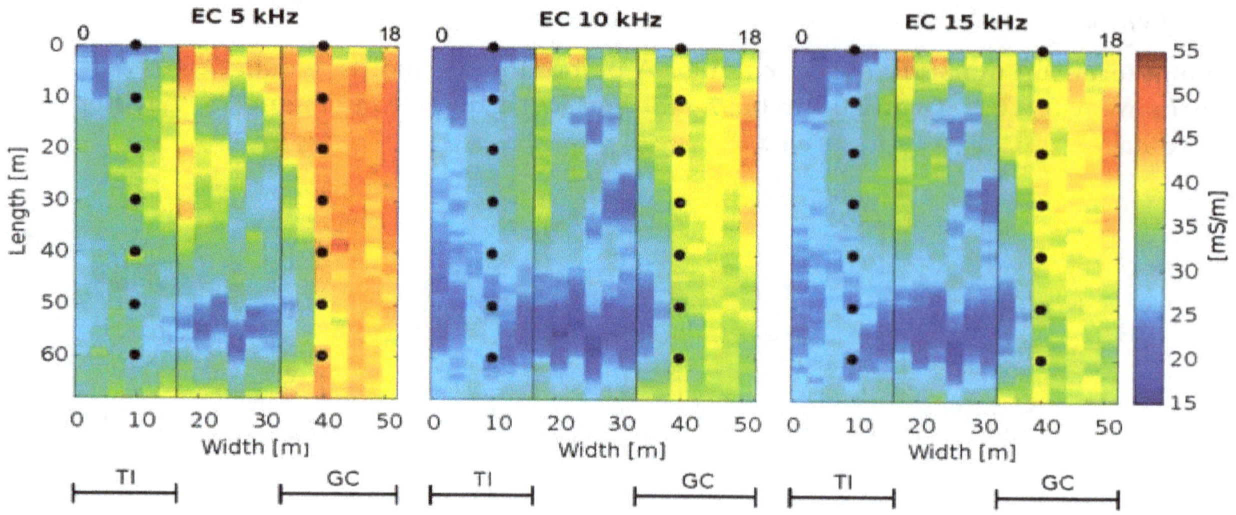

Figure 19. Carpeneto site, 21 March 2016. Electrical conductivity monitored by the EMI equipment and TDR probes locations (black dots) in the testing plot. The three images represent lower depths of investigation, from right to left.

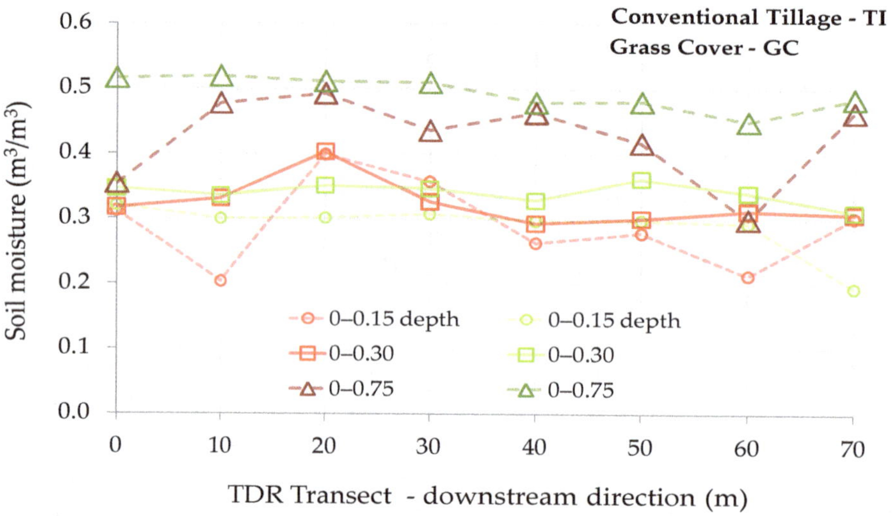

Figure 20. Carpeneto site, 21 March 2016. Soil water content monitored along the two TDR transects (conventional tillage and grass cover) depicted in Figure 19.

The correlation between EC_a and soil moisture is evident (Figures 19 and 20). The results of the EMI measurement campaign show EC_a values along the TI rows lower than those measured along the GC rows. The high value of EC_a along the GC rows can be ascribed to the soil water content, which is higher than in the TI rows, especially in the deeper layers (Figure 20). Moreover, EC_a measured at 5 kHz (highest depth of investigation) is higher than EC_a measured at 10 and 15 kHz, and is in agreement with the soil water content measured by TDR (Figure 20) and also with the long-term measurements (Figure 18). However, the low EC_a values at the bottom of the TI rows (corresponding to the 60 m coordinate) do not completely agree with the TDR water content value: at this point, notwithstanding the lowest TI soil moisture, the EC_a values are not as low as the values detected in similar conditions. Being in the tilled area, this behaviour might be ascribed to the deposition of fine particles eroded from the top toward the bottom.

5. Discussion

Through a comprehensive analysis of the presented case studies, it is possible to evaluate the application pros and cons of the different methods used.

The TDR technique offers wide flexibility in different contexts for both single campaigns and long-term surveys, with excellent temporal and spatial resolution attainable through the automation of measurements. Measurements are rapid, non-destructive and simple to obtain. The higher the number of probes to be acquired simultaneously, the more complex the system will have to be (eventually adopting a multiplexed system). Depending on the monitoring needs, specific experimental layouts can be adopted with vertical and/or horizontal probes of different lengths. In ideal conditions (i.e., low-attenuation porous media with no coarse skeletal material—stones/rocks) TDR probes can be vertically inserted down to unusual depths (e.g., 4 m, as demonstrated in the plain permanent meadow case study) to explore the soil water content/behaviour below the most commonly monitored layers. For these last applications, the adoption of a steel guide during the insertion process is strongly recommended to maintain rods in parallel and prevent geometrical probe deformations. Due to the TDR's inability to detect liquid and ice contents simultaneously (within a bandwidth between a few megahertz and 1.5 GHz), in the case of frozen soils additional measurements with lower frequencies (between 0.1 and 200 kHz [118,119]) would be necessary to avoid underestimations or errors.

Low-frequency sensors directly connected to data loggers combine good accuracy and reliability with a lower cost (and can be used in place of TDR). By their nature they need specific calibration and they suffer from some restrictions related to the constructive features (e.g., probe lengths; more fragility—depending on the types and models), but they are easy to use and efficient in the most common soil moisture investigations.

To determine the spatial distribution of soil moisture, whereas TDR evaluates the specific water content analysing the full signal response (and relating it to the dielectric properties), the ERT system determines the two-dimensional electric resistivity distribution (from surface-based geoelectric measurements and subsequent data inversion). A combination of both methods would lead to better results (e.g., for calibrating the ERT approach). However, determining the soil water content from the resulting electrical resistivity values is more difficult than in the case of the dielectric coefficients using TDR. The inversion procedure can produce mass balance errors, due to a rapid decrease of ERT resolution with depth [40]. This problem may be solved through a coupled hydro-geophysical approach [40,101]. However, if only qualitative assessment of spatial distribution and time variations of soil moisture content are sufficient, an uncoupled approach can be used [32], based on the following steps: (1) inversion of geophysical field data gives the spatial distribution of electrical resistivity; (2) application of a petro-physical relationship to obtain an estimation of moisture content distribution from the electrical resistivity.

It must be noted that electrical resistivity and electrical conductivity, as measured by EMI, are just reciprocal quantities. Thus ERT and EMI measure the same physical parameter, and can be effectively used to complement each other: EMI gives quicker information over large areas, while ERT is more effective at in-depth investigations. In the hilly vineyard test case, for example, EMI measurements allow for identifying differences in EC_a (and thus soil moisture) between the two treatments and in relation to depth, with results that are comparable with the more accurate (and time-consuming) TDR measurements. In the permanent mountain meadow, since EMI measurements are fast (and not invasive), it was possible to map the EC_a spatial variations over the whole field-size plot (also in complex logistical conditions). Some discordances between EMI and TDR measurements could be ascribed to local conditions that can affect TDR local measurements. When possible, a combined EMI and TDR campaign allows us to combine accuracy with wider spatial representativeness of measurements than only TDR campaigns.

Without considering the borehole applications (which have not been analysed in this work), GPR can be used to perform specific campaigns aimed at obtaining soil sub-surface "images" (typically in the range of sub-meters up to tens of meters) referred to the interaction between the transmitted

EM energy and the spatial variation in the complex, frequency-dependent EM properties of the earth materials in the subsurface [125]. For small-scale, site-specific experiments, GPR can be used to monitor the movement of water into the—and through the—subsurface (such as the depth of the infiltration front estimated by GPR measurements in the plain permanent meadow case study). However, the use of GPR systems remains difficult in uneven irregular areas due to the need to drag the instrument to keep it in contact with the soil surface (such as at the permanent mountain meadow testing site). Limitations are also acknowledged in clay-laden soils (such as the hilly vineyard testing site), where the signal penetration into the soil is strongly reduced by the high electrical conductivity.

6. Conclusions

Soil moisture is highly variable at both the spatial and the temporal scale, and soil moisture field measurements are recognized to be fundamental in the integration of in situ, satellite and modelled data. For these reasons, in our work we provide a synoptic view of techniques, supported by case studies, to show that also in very different field conditions (in terms of altitude, land use, and soil type, namely a plain, a mountain meadow and a hilly vineyard) it is possible to estimate time and spatially resolved soil moisture by the same combination of instruments: contact-based methods (i.e., Time Domain Reflectometry—TDR, and two low-frequency probes) for the time resolved, and minimally invasive hydro-geophysical methods (i.e., Electromagnetic Induction—EMI, Ground Penetrating Radar—GPR, and the Electrical Resistivity Tomography—ERT) for the spatially resolved. By doing so, soil moisture dynamics determined by soil's heterogeneity and meteorological events can be efficiently observed and measured. In particular, while plot-scale methods—like EMI, GPR and ERT—provide efficient spatial surveys, local-scale methods—like TDR and other lower-cost soil moisture sensors—provide efficient soil moisture monitoring across time (and space—where properly structured). With less accuracy, and with smaller volume support, the low-frequency sensors can also be used in place of TDR.

An important key future challenge might be represented by the creation of a framework in which the information contained in the local and plot observations is combined with model predictions of soil moisture dynamics, at different spatial scales. Nowadays, this integrated approach would represent the best method to achieve the great potential, especially in hydrological applications, of soil moisture studies.

Acknowledgments: This work has been partially funded by the EU-funded H2020 Project "*ECOPOTENTIAL: Improving Future Ecosystem Benefits through Earth Observations*" (http://www.ecopotential-project.eu) and the Italian National Research Programme 2011–2013 by The Project of Interest "*NextData: a national system for the retrieval, storage, access and diffusion of environmental and climate data from mountain and marine areas*" (http://www.nextdataproject.it/?q=en). The "Agrion—Centro sperimentale per la Vitivinicoltura" (which managed the "Tenuta Cannona" vineyards), is also gratefully acknowledged.

Author Contributions: All co-authors, each within his competence, actively participated in the field campaigns, data processing, and data analysis. Stefano Ferraris is the Grugliasco and Cogne experimental sites scientific coordinator. Eugenio Cavallo coordinated the experiments in Tenuta Cannona. Giulia Raffelli, Stefano Ferraris and Maurizio Previati conceived and designed the paper. The manuscript was written by Maurizio Previati, Giulia Raffelli, Davide Canone, Stefano Ferraris and Giorgio Cassiani. Maurizio Previati, Ivan Bevilacqua, Giulia Raffelli, Giorgio Cassiani and Stefano Ferraris contributed to the figures and the manuscript editing.

References

1. Lin, H. Temporal stability of soil moisture spatial pattern and subsurface preferential flow pathways in the shale hills catchment. *Vadose Zone J.* **2006**, *5*, 317–340. [CrossRef]
2. Tromp-van Meerveld, H.J.; McDonnell, J.J. On the interrelations between topography, soil depth, soil moisture, transpiration rates and species distribution at the hillslope scale. *Adv. Water Resour.* **2006**, *29*, 293–310. [CrossRef]

3. Teuling, A.J.; Hupet, F.; Uijlenhoet, R.; Troch, P.A. Climate variability effects on spatial soil moisture dynamics. *Geophys. Res. Lett.* **2007**, *34*. [CrossRef]
4. Vereecken, H.; Huisman, J.A.; Bogena, H.; Vanderborght, J.; Vrugt, J.A.; Hopmans, J.W. On the value of soil moisture measurements in vadose zone hydrology: A review. *Water Resour. Res.* **2008**, *44*. [CrossRef]
5. Vereecken, H.; Huisman, J.A.; Pachepsky, Y.; Montzka, C.; Van Der Kruk, J.; Bogena, H.; Vanderborght, J. On the spatio-temporal dynamics of soil moisture at the field scale. *J. Hydrol.* **2014**, *516*, 76–96. [CrossRef]
6. Ceaglio, E.; Mitterer, C.; Maggioni, M.; Ferraris, S.; Segor, V.; Freppaz, M. The role of soil volumetric liquid water content during snow gliding processes. *Cold Reg. Sci. Technol.* **2017**, *136*, 17–29. [CrossRef]
7. Bomblies, A. Modeling the role of rainfall patterns in seasonal malaria transmission. *Clim. Chang.* **2012**, *112*, 673–685. [CrossRef]
8. Previati, M.; Bevilacqua, I.; Canone, D.; Ferraris, S.; Haverkamp, R. Evaluation of soil water storage efficiency for rainfall harvesting on hillslope micro-basins built using time domain reflectometry measurements. *Agric. Water Manag.* **2010**, *97*, 449–456. [CrossRef]
9. Canone, D.; Previati, M.; Bevilacqua, I.; Salvai, L.; Ferraris, S. Field measurements based model for surface irrigation efficiency assessment. *Agric. Water Manag.* **2015**, *156*, 30–42. [CrossRef]
10. Canone, D.; Previati, M.; Ferraris, S. Evaluation of Stemflow Effects on the Spatial Distribution of Soil Moisture Using TDR Monitoring and an Infiltration Model. *J. Irrig. Drain. Eng. ASCE* **2017**, *143*, 04016075. [CrossRef]
11. Robinson, D.A.; Campbell, C.S.; Hopmans, J.W.; Hornbuckle, B.K.; Jones, S.B.; Knight, R.; Ogden, F.; Selker, J.; Wendroth, O. Soil moisture measurement for ecological and hydrological watershed-scale observatories: A review. *Vadose Zone J.* **2008**, *7*, 358–389. [CrossRef]
12. Fares, A.; Temimi, M.; Morgan, K.; Kelleners, T.J. In-situ and remote soil moisture sensing technologies for vadose zone hydrology. *Vadose Zone J.* **2013**, *12*. [CrossRef]
13. Brocca, L.; Ciabatta, L.; Massari, C.; Camici, S.; Tarpanelli, A. Soil Moisture for Hydrological Applications: Open Questions and New Opportunities. *Water* **2017**, *9*, 140. [CrossRef]
14. Romano, N. Soil moisture at local scale: Measurements and simulations. *J. Hydrol.* **2014**, *516*, 6–20. [CrossRef]
15. Topp, G.C.; Davis, J.L.; Annan, A.P. Electromagnetic determination of soil water content: Measurements in coaxial transmission lines. *Water Resour. Res.* **1980**, *16*, 574–582. [CrossRef]
16. Topp, G.C. State of the art of measuring soil water content. *Hydrol. Process.* **2003**, *17*, 2993–2996. [CrossRef]
17. Robinson, D.A.; Jones, S.B.; Wraith, J.M.; Or, D.; Friedman, S.P. A review of advances in dielectric and electrical conductivity measurement in soils using time domain reflectometry. *Vadose Zone J.* **2003**, *2*, 444–475. [CrossRef]
18. Robinson, D.A.; Jones, S.B.; Blonquist, J.M.; Friedman, S.P. A physically derived water content/permittivity calibration model for coarse-textured, layered soils. *Soil Sci. Soc. Am. J.* **2005**, *69*, 1372–1378. [CrossRef]
19. Kizito, F.; Campbell, C.S.; Campbell, G.S.; Cobos, D.R.; Teare, B.L.; Carter, B.; Hopmans, J.W. Frequency, electrical conductivity and temperature analysis of a low-cost capacitance soil moisture sensor. *J. Hydrol.* **2008**, *352*, 367–378. [CrossRef]
20. Blonquist, J.M.; Jones, S.B.; Robinson, D.A.; Rasmussen, V.P.; Or, D. Standardizing characterization of electromagnetic water content sensors. *Vadose Zone J.* **2005**, *4*, 1048–1058. [CrossRef]
21. Bogena, H.R.; Huisman, J.A.; Oberdorster, C.; Vereecken, H. Evaluation of a low-cost soil water content sensor for wireless network applications. *J. Hydrol.* **2007**, *344*, 32–42. [CrossRef]
22. Rosenbaum, U.; Huisman, J.A.; Weuthen, A.; Vereecken, H.; Bogena, H.R. Sensor-to-sensor variability of the ECH_2O EC-5, TE, and 5TE sensors in dielectric liquids. *Vadose Zone J.* **2010**, *9*, 181–186. [CrossRef]
23. Baker, J.M.; Allmaras, R.R. System for automating and multiplexing soil moisture measurement by time-domain reflectometry. *Soil. Sci. Soc. Am. J.* **1990**, *54*, 1–6. [CrossRef]
24. Heimovaara, T.J.; Bouten, W. A computer-controlled 36-channel time domain reflectometry system for monitoring soil water contents. *Water Resour. Res.* **1990**, *26*, 2311–2316.
25. Dobriyal, P.; Qureshi, A.; Badola, R.; Hussain, S.A. A review of the methods available for estimating soil moisture and its implications for water resource management. *J. Hydrol.* **2012**, *458*, 110–117. [CrossRef]
26. Huisman, J.A.; Hubbard, S.S.; Redman, J.D.; Annan, A.P. Measuring soil water content with ground penetrating radar. *Vadose Zone J.* **2003**, *2*, 476–491. [CrossRef]
27. Serbin, G.; Or, D. Ground-penetrating radar measurement of crop and surface water content dynamics. *Remote Sens. Environ.* **2005**, *96*, 119–134. [CrossRef]

28. Lambot, S.; Weihermüller, L.; Huisman, J.A.; Vereecken, H.; Vanclooster, M.; Slob, E.C. Analysis of air-launched ground-penetrating radar techniques to measure the soil surface water content. *Water Resour. Res.* **2006**, *42*. [CrossRef]
29. Lambot, S.; Slob, E.; Chavarro, D.; Lubczynski, M.; Vereecken, H. Measuring soil surface water content in irrigated areas of southern Tunisia using full-waveform inversion of proximal GPR data. *Near Sur. Geophys.* **2008**, *6*, 403–410. [CrossRef]
30. Busch, S.; Van der Kruk, J.; Bikowski, J.; Vereecken, H. Quantitative conductivity and permittivity estimation using full-waveform inversion of onground GPR data. *Geophysics* **2012**, *77*, H79–H91. [CrossRef]
31. Klotzsche, A.; Van der Kruk, J.; Linde, N.; Doetsch, J.; Vereecken, H. 3-D characterization of high-permeability zones in a gravel aquifer using 2-D crosshole GPR full-waveform inversion and waveguide detection. *Geophys. J. Int.* **2013**, *195*, 932–944. [CrossRef]
32. Rossi, M.; Manoli, G.; Pasetto, D.; Deiana, R.; Ferraris, S.; Strobbia, C.; Cassiani, G. Coupled inverse modeling of a controlled irrigation experiment using multiple hydro-geophysical data. *Adv. Water Resour.* **2015**, *82*, 150–165. [CrossRef]
33. Corwin, D.L.; Lesch, S.M. Apparent soil electrical conductivity measurements in agriculture. *Comput. Electron. Agric.* **2005**, *46*, 11–43. [CrossRef]
34. Robinson, D.A.; Lebron, I.; Kocar, B.; Phan, K.; Sampson, M.; Crook, N.; Fendorf, S. Time-lapse geophysical imaging of soil moisture dynamics in tropical deltaic soils: An aid to interpreting hydrological and geochemical processes. *Water Resour. Res.* **2009**, *45*. [CrossRef]
35. Cassiani, G.; Ursino, N.; Deiana, R.; Vignoli, G.; Boaga, J.; Rossi, M.; Ludwig, R. Non-invasive monitoring of soil static characteristics and dynamic states: A case study highlighting vegetation effects on agricultural land. *Vadose Zone J.* **2012**, *11*. [CrossRef]
36. Brovelli, A.; Cassiani, G. Effective permittivity of porous media: A critical analysis of the complex refractive index model. *Geophys. Prospect.* **2008**, *56*, 715–727. [CrossRef]
37. Brovelli, A.; Cassiani, G. Combined estimation of effective electrical conductivity and permittivity for soil monitoring. *Water Resour. Res.* **2011**, *47*. [CrossRef]
38. Loke, M.H.; Chambers, J.E.; Rucker, D.F.; Kuras, O.; Wilkinson, P.B. Recent developments in the direct-current geoelectrical imaging method. *J. Appl. Geophys.* **2013**, *95*, 135–156. [CrossRef]
39. Vanderborght, J.; Huisman, J.A.; Van der Kruk, J.; Vereecken, H. Geophysical methods for field-scale imaging of root zone properties and processes. In *Soil–Water–Root Processes: Advances in Tomography and Imaging*; Anderson, S.H., Hopmans, J.W., Eds.; SSSA: Madison, WI, USA, 2013; pp. 247–281.
40. Manoli, G.; Rossi, M.; Pasetto, D.; Deiana, R.; Ferraris, S.; Cassiani, G.; Putti, M. An iterative particle filter approach for coupled hydro-geophysical inversion of a controlled infiltration experiment. *J. Comput. Phys.* **2015**, *283*, 37–51. [CrossRef]
41. Hornbuckle, B.K.; England, A.W.; De Roo, R.D.; Fischman, M.A.; Boprie, D.L. Vegetation canopy anisotropy at 1.4 GHz. *IEEE Trans. Geosci. Remote Sens.* **2003**, *41*, 2211–2223. [CrossRef]
42. Saleh, K.; Wigneron, J.P.; Waldteufel, P.; De Rosnay, P.; Schwank, M.; Calvet, J.C.; Kerr, Y.H. Estimates of surface soil moisture under grass covers using L-band radiometry. *Remote Sens. Environ.* **2007**, *109*, 42–53. [CrossRef]
43. Hong, S.; Shin, I. A physically-based inversion algorithm for retrieving soil moisture in passive microwave remote sensing. *J. Hydrol.* **2011**, *405*, 24–30. [CrossRef]
44. Kornelsen, K.C.; Coulibaly, P. Advances in soil moisture retrieval from synthetic aperture radar and hydrological applications. *J. Hydrol.* **2013**, *476*, 460–489. [CrossRef]
45. Reigber, A.; Scheiber, R.; Jager, M.; Prats-Iraola, P.; Hajnsek, I.; Jagdhuber, T.; Horn, R. Very-high-resolution airborne synthetic aperture radar imaging: Signal processing and applications. *Proc. IEEE* **2013**, *101*, 759–783. [CrossRef]
46. Zreda, M.; Desilets, D.; Ferré, T.P.A.; Scott, R.L. Measuring soil moisture content non-invasively at intermediate spatial scale using cosmic-ray neutrons. *Geophys. Res. Lett.* **2008**, *35*. [CrossRef]
47. Zreda, M.; Shuttleworth, W.J.; Zeng, X.; Zweck, C.; Desilets, D.; Franz, T.; Rosolem, R. COSMOS: The COsmic-ray soil moisture observing system. *Hydrol. Earth Syst. Sci.* **2012**, *16*, 4079–4099. [CrossRef]
48. Rivera Villarreyes, C.A.; Baroni, G.; Oswald, S.E. Integral quantification of seasonal soil moisture changes in farmland by cosmic-ray neutrons. *Hydrol. Earth Syst. Sci.* **2011**. [CrossRef]

49. Bogena, H.R.; Huisman, J.A.; Baatz, R.; Hendricks Franssen, H.J.; Vereecken, H. Accuracy of the cosmic-ray soil water content probe in humid forest ecosystems: The worst case scenario. *Water Resour. Res.* **2013**, *49*, 5778–5791. [CrossRef]
50. Zhu, Z.; Tan, L.; Gao, S.; Jiao, Q. Observation on soil moisture of irrigation cropland by cosmic-ray probe. *IEEE Geosci. Remote Sens. Lett.* **2015**, *12*, 472–476. [CrossRef]
51. Shin, Y.; Mohanty, B.P. Development of a deterministic downscaling algorithm for remote sensing soil moisture footprint using soil and vegetation classifications. *Water Resour. Res.* **2013**, *49*, 6208–6228. [CrossRef]
52. Ines, A.V.M.; Mohanty, B.P.; Shin, Y. An unmixing algorithm for remotely sensed soil moisture. *Water Resour. Res.* **2013**, *49*, 408–425. [CrossRef]
53. Peng, J.; Loew, A.; Zhang, S.; Wang, J.; Niesel, J. Spatial downscaling of satellite soil moisture data using vegetation temperature condition index. *IEEE Trans. Geosci. Remote Sens.* **2016**, *54*, 558–566. [CrossRef]
54. Peng, J.; Niesel, J.; Loew, A. Evaluation of soil moisture downscaling using a simple thermal-based proxy: The REMDHUS network (Spain) example. *Hydrol. Earth Syst. Sci.* **2015**, *19*, 4765–4782. [CrossRef]
55. Calvet, J.-C.; Noilhan, J. From near-surface to root-zone soil moisture using year-round data. *J. Hydrometeorol.* **2000**, *1*, 393–411. [CrossRef]
56. Sabater, J.M.; Jarlan, L.; Calvet, J.C.; Bouyssel, F.; De Rosnay, P. From near-surface to root-zone soil moisture using different assimilation techniques. *J. Hydrometeorol.* **2007**, *8*, 194–206. [CrossRef]
57. Albergel, C.; Rudiger, C.; Pellarin, T.; Calvet, J.-C.; Fritz, N.; Froissard, F.; Suquia, D.; Petitpa, A.; Piguet, B.; Martin, E. From near-surface to root-zone soil moisture using an exponential filter: An assessment of the method based on in-situ observations and model simulations. *Hydrol. Earth Syst. Sci.* **2008**, *12*, 1323–1337. [CrossRef]
58. Manfreda, S.; Brocca, L.; Moramarco, T.; Melone, F.; Sheffield, J. A physically based approach for the estimation of root-zone soil moisture from surface measurements. *Hydrol. Earth Syst. Sci.* **2014**, *18*, 1199–1212. [CrossRef]
59. Tebbs, E.; Gerard, F.; Petrie, A.; DeWitte, E. Emerging and Potential Future Applications of satellite-Based Soil Moisture products. In *Satellite Soil Moisture Retrievals: Techniques and Applications*; Petropoulos, G.P., Srivastava, P., Kerr, Y., Eds.; Elsevier: Amsterdam, The Netherlands, 2016; Volume 19, pp. 379–400.
60. Peng, J.; Loew, A.; Merlin, O.; Verhoest, N.E.C. A review of spatial downscaling of satellite remotely sensed soil moisture. *Rev. Geophys.* **2017**, *55*, 341–366. [CrossRef]
61. Mohanty, B.P.; Cosh, M.H.; Lakshmi, V.; Montzka, C. Soil moisture remote sensing: State-of-the-science. *Vadose Zone J.* **2017**, *16*. [CrossRef]
62. Desilets, D.; Zreda, M. Footprint diameter for a cosmic-ray soil moisture probe: Theory and Monte Carlo simulations. *Water Resour. Res.* **2013**, *49*. [CrossRef]
63. Franz, T.E.; Zreda, M.; Ferré, T.P.A.; Rosolem, R.; Zweck, C.; Stillman, S.; Zeng, X.; Shuttleworth, W.J. Measurement depth of the cosmic ray soil moisture probe affected by hydrogen from various sources. *Water Resour. Res.* **2012**, *48*, W08515. [CrossRef]
64. Köhli, M.; Schrön, M.; Zreda, M.; Schmidt, U.; Dietrich, P.; Zacharias, S. Footprint characteristics revised for field-scale soil moisture monitoring with cosmic-ray neutrons. *Water Resour. Res.* **2015**, *51*, 5772–5790. [CrossRef]
65. Hallikainen, M.T.; Ulaby, F.T.; Dobson, M.C.; El-Rayes, M.A.; Wu, L.K. Microwave dielectric behavior of wet soil-part 1: Empirical models and experimental observations. *IEEE Trans. Geosci. Remote Sens.* **1985**, *1*, 25–34. [CrossRef]
66. Evett, S.R. Soil water measurement by time domain reflectometry. In *Encyclopedia of Water Science*; United States Department of Agriculture (USDA): Bushland, TX, USA, 2003; pp. 894–898.
67. Topp, G.C.; Ferré, T.P.A. Methods for measurement of soil water content: Time domain reflectometry. In *Methods of Soil Analysis, Part 4*; Dane, J.H., Topp, G.C., Eds.; SSSA Book Series No. 5; Soil Science Society of America: Madison, WI, USA, 2002; pp. 434–446.
68. Roth, K.; Schulin, R.; Flühler, H.; Attinger, W. Calibration of time domain reflectometry for water content measurement using a composite dielectric approach. *Water Resour. Res.* **1990**, *26*, 2267–2273. [CrossRef]
69. Herkelrath, W.N.; Hamburg, S.P.; Murphy, F. Automatic, real-time monitoring of soil moisture in a remote field area with time domain reflectometry. *Water Resour. Res.* **1991**, *27*, 857–864. [CrossRef]
70. Topp, G.C.; Davis, J.L.; Annan, A.P. Electromagnetic determination of soil water content using TDR: I. Applications to wetting fronts and steep gradients. *Soil Sci. Soc. Am. J.* **1982**, *46*, 672–678. [CrossRef]

71. Topp, G.C.; Davis, J.L.; Annan, A.P. Electromagnetic determination of soil water content using TDR: II. Evaluation of installation and configuration of parallel transmission lines. *Soil Sci. Soc. Am. J.* **1982**, *46*, 678–684. [CrossRef]
72. Zegelin, S.J.; White, I.; Jenkins, D.R. Improved field probes for soil water content and electrical conductivity measurement using time domain reflectometry. *Water Resour. Res.* **1989**, *25*, 2367–2376. [CrossRef]
73. Heimovaara, T.J. Design of triple-wire time domain reflectometry probes in practice and theory. *Soil Sci. Soc. Am. J.* **1993**, *57*, 1410–1417. [CrossRef]
74. Ferré, P.A.; Rudolph, D.L.; Kachanoski, R.G. Spatial averaging of water content by time domain reflectometry: Implications for twin rod probes with and without dielectric coatings. *Water Resour. Res.* **1996**, *32*, 271–279. [CrossRef]
75. Ferré, P.A.; Knight, J.H.; Rudolph, D.L.; Kachanoski, R.G. The sample areas of conventional and alternative time domain reflectometry probes. *Water Resour. Res.* **1998**, *34*, 2971–2979. [CrossRef]
76. Canone, D.; Previati, M.; Ferraris, S.; Haverkamp, R. A new coaxial time domain reflectometry probe for water content measurement in forest floor litter. *Vadose Zone J.* **2009**, *8*, 363–372. [CrossRef]
77. Adelakun, I.A.; Ranjan, R.S. Design of a Multilevel TDR Probe for Measuring Soil Water Content at Different Depths. *Trans. ASABE* **2013**, *56*, 1451–1460. [CrossRef]
78. Nissen, H.H.; Ferré, T.P.A.; Moldrup, P. Sample area of two- and three-rod time domain reflectometry probes. *Water Resour. Res.* **2003**, *39*, 1289. [CrossRef]
79. Vaz, C.M.; Jones, S.; Meding, M.; Tuller, M. Evaluation of standard calibration functions for eight electromagnetic soil moisture sensors. *Vadose Zone J.* **2013**, *12*. [CrossRef]
80. Campbell, C.S.; Campbell, G.S.; Cobos, D.R.; Bissey, L.L. Calibration and Evaluation of an Improved Low-Cost Soil Moisture Sensor. 2009. Available online: http://www.decagon.com (accessed on 12 September 2017).
81. Logsdon, S.D.; Hernandez-Ramirez, G.; Hatfield, J.L.; Sauer, T.J.; Prueger, J.H.; Schilling, K.E. Soil water and shallow groundwater relations in an agricultural hillslope. *Soil Sci. Soc. Am. J.* **2009**, *73*, 1461. [CrossRef]
82. Parsons, L.R.; Bandaranayake, W.M. Performance of a new capacitance soil moisture probe in a sandy soil. *Soil Sci. Soc. Am. J.* **2009**, *73*, 1378–1385. [CrossRef]
83. Evett, S.R.; Schwartz, R.C. Discussion of "Soil Moisture Measurements: Comparison of Instrumentation Performances" by Ventura Francesca, Facini Osvaldo, Piana Stefano, and Rossi Pisa Paola. *J. Irrig. Drain. Res.* **2011**, *137*, 466–468. [CrossRef]
84. Sakaki, T.; Limsuwat, A.; Illangasekare, T.H. A simple method for calibrating dielectric soil moisture sensors: Laboratory validation in sands. *Vadose Zone J.* **2011**, *10*, 526–531. [CrossRef]
85. Friedman, S.P. Soil properties influencing apparent electrical conductivity: A review. *Comput. Electron. Agric.* **2005**, *46*, 45–70. [CrossRef]
86. Archie, G.E. The electrical resistivity log as an aid in determining some reservoir characteristics. *Trans. AIME* **1942**, *146*, 54–62. [CrossRef]
87. Cassiani, G.; Boaga, J.; Rossi, M.; Fadda, G.; Putti, M.; Majone, B.; Bellin, A. Soil-plant interaction monitoring: Small scale example of an apple orchard in Trentino, North-Eastern Italy. *Sci. Total Environ.* **2016**, *543*, 851–861. [CrossRef] [PubMed]
88. Ortuani, B.; Chiaradia, E.A.; Priori, S.; L'Abate, G.; Canone, D.; Comunian, A.; Giudici, M.; Mele, M.; Facchi, A. Mapping soil water capacity through EMI survey to delineate site specific management units within an irrigated field. *Soil Sci.* **2016**, *181*, 252–263. [CrossRef]
89. Deidda, G.P.; Fenu, C.; Rodriguez, G. Regularized solution of a nonlinear problem in electromagnetic sounding. *Inverse Probl.* **2014**, *30*, 125014. [CrossRef]
90. Díaz De Alba, P.; Rodriguez, G. Regularized inversion of multi-frequency EM data in geophysical applications. In *Trends in Differential Equations and Applications*; Ortegón, F., Gallego, M.V., Redondo, N., Rodríguez Galván, J.R., Eds.; SEMA SIMAI Springer Series; Springer: Cham, Switzerland, 2016; Volume 8, pp. 357–369.
91. Shanahan, P.W.; Binley, A.; Whalley, W.R.; Watts, C.W. The Use of Electromagnetic Induction to Monitor Changes in Soil Moisture Profiles beneath Different Wheat Genotypes. *Soil Sci. Soc. Am. J.* **2015**, *79*, 459–466. [CrossRef]
92. Boaga, J. The use of FDEM in hydrogeophysics: A review. *J. Appl. Geophys.* **2017**, *139*, 36–46. [CrossRef]
93. Everett, M.E.; Meju, M.A. Near-surface controlled-source electromagnetic induction: Background and recent advances. In *Hydrogeophysics*; Rubin, Y., Hubbard, S.S., Eds.; Springer: New York, NY, USA, 2005; pp. 157–183.

94. Binley, A.; Kemna, A. DC resistivity and induced polarization methods. In *Hydrogeophysics*; Rubin, Y., Hubbard, S.S., Eds.; Springer: New York, NY, USA, 2005; pp. 129–156.
95. Binley, A.M.; Cassiani, G.; Deiana, R. Hydrogeophysics—Opportunities and Challenges. *Boll. Geofis. Teor. Appl.* **2010**, *51*, 267–284.
96. Cassiani, G.; Bruno, V.; Villa, A.; Fusi, N.; Binley, A.M. A saline trace test monitored via time-lapse surface electrical resistivity tomography. *J. Appl. Geophys.* **2006**, *59*, 244–259. [CrossRef]
97. Constable, S.C.; Parker, R.L.; Constable, C.G. Occam's Inversion: A practical algorithm for generating smooth models from EM sounding data. *Geophysics* **1987**, *52*, 289–300. [CrossRef]
98. La Brecque, D.J.; Heath, G.; Sharpe, R.; Versteeg, R. Autonomous monitoring of fluid movement using 3-D electrical resistivity tomography. *J. Environ. Eng. Geophys.* **2004**, *9*, 167–176. [CrossRef]
99. Cassiani, G.; Godio, A.; Stocco, S.; Villa, A.; Deiana, R.; Frattini, P.; Rossi, M. Monitoring the hydrologic behaviour of a mountain slope via time-lapse electrical resistivity tomography. *Near Sur. Geophys.* **2009**, *7*, 475–486. [CrossRef]
100. Hinnell, A.C.; Ferré, T.P.A.; Vrugt, J.A.; Huisman, J.A.; Moysey, S.; Rings, J.; Kowalsky, M.B. Improved extraction of hydrologic information from geophysical data through coupled hydrogeophysical inversion. *Water Resour. Res.* **2010**, *46*. [CrossRef]
101. Beff, L.; Günther, T.; Vandoorne, B.; Couvreur, V.; Javaux, M. Three-dimensional monitoring of soil water content in a maize field using Electrical Resistivity Tomography. *Hydrol. Earth Syst. Sci.* **2013**, *17*, 595–609. [CrossRef]
102. Garré, S.; Coteur, I.; Wongleecharoen, C.; Kongkaew, T.; Diels, J.; Vanderborght, J. Non-invasive monitoring of soil water dynamics in mixed cropping systems: A case study in Ratchaburi Province, Thailand. *Vadose Zone J.* **2013**, *12*. [CrossRef]
103. Slater, L.; Binley, A.M.; Daily, W.; Johnson, R. Cross-hole electrical imaging of a controlled saline tracer injection. *J. Appl. Geophys.* **2000**, *44*, 85–102. [CrossRef]
104. Cassiani, G.; Binley, A. Modeling unsaturated flow in a layered formation under quasi-steady state conditions using geophysical data constraints. *Adv. Water Resour.* **2005**, *28*, 467–477. [CrossRef]
105. Linde, N.; Binley, A.; Tryggvason, A.; Pedersen, L.B.; Revil, A. Improved hydrogeophysical characterization using joint inversion of cross-hole electrical resistance and ground-penetrating radar traveltime data. *Water Resour. Res.* **2006**, *42*. [CrossRef]
106. Binley, A.; Cassiani, G.; Middleton, R.; Winship, P. Vadose zone flow model parameterisation using cross-borehole radar and resistivity imaging. *J. Hydrol.* **2002**, *267*, 147–159. [CrossRef]
107. Zuecco, G.; Borga, M.; Penna, D.; Canone, D.; Previati, M.; Ferraris, S. Towards improved understanding of land use effect on soil moisture variability: Analysis and modeling at the plot scale. *Procedia Environ. Sci.* **2013**, *19*, 456–464. [CrossRef]
108. Baudena, M.; Bevilacqua, I.; Canone, D.; Ferraris, S.; Previati, M.; Provenzale, A. Soil water dynamics at a midlatitude test site: Field measurements and box modeling approaches. *J. Hydrol.* **2012**, *414*, 329–340. [CrossRef]
109. Jones, S.B.; Wraith, J.M.; Or, D. Time domain reflectometry measurement principles and applications. *Hydrol. Processes* **2002**, *16*, 141–153. [CrossRef]
110. Doolittle, J.A.; Minzenmayer, F.E.; Waltman, S.W.; Benham, E.C.; Tuttle, J.W.; Peaslee, S.D. Ground-penetrating radar soil suitability map of the conterminous United States. *Geoderma* **2007**, *141*, 416–421. [CrossRef]
111. Paniconi, C.; Ferraris, S.; Putti, M.; Pini, G.; Gambolati, G. Three-dimensional numerical codes for simulating groundwater contamination: FLOW3D, flow in saturated and unsaturated porous media. In *Pollution Modeling*; CMP: Boston, MA, USA, 1994; Volume 1, pp. 149–156.
112. Or, D.; Jones, S.B.; VanSchaar, J.R.; Humphries, S.D.; Koberstein, R.L. WinTDR v.6.1: A Windows-based Time Domain Reflectometry Program for Measurement of Soil Water Content and Electrical Conductivity—User Manual. Utah State Univ. Soil Physics Group, Logan. 2004. Available online: https://psc.usu.edu/ou-files/wintdr/Introduction.pdf (accessed on 15 September 2017).
113. Ursino, N.; Cassiani, G.; Deiana, R.; Vignoli, G.; Boaga, J. Measuring and Modelling water related soil—Vegetation feedbacks in a fallow plot. *Hydrol. Earth Syst. Sci.* **2014**, *18*, 1105–1118. [CrossRef]
114. Binley, A. Resistivity Inversion Software. 2013. Available online: http://www.es.lancs.ac.uk/people/amb/Freeware/freeware.htm (accessed on 12 September 2017).
115. Biddoccu, M.; Ferraris, S.; Opsi, F.; Cavallo, E. Long-term monitoring of soil management effects on runoff

and soil erosion in sloping vineyards in Alto Monferrato (North-West Italy). *Soil Tillage Res.* **2016**, *155*, 176–189. [CrossRef]

116. Biddoccu, M.; Ferraris, S.; Pitacco, A.; Cavallo, E. Temporal variability of soil management effects on soil hydrological properties, runoff and erosion at the field scale in a hillslope vineyard, North-West Italy. *Soil Tillage Res.* **2017**, *165*, 46–58. [CrossRef]
117. Starr, J.L.; Paltineanu, I.C. Methods for Measurement of Soil Water Content: Capacitance Devices. In *Methods of Soil Analysis: Part 4 Physical Methods*; Dane, J.H., Topp, G.C., Eds.; Soil Science Society of America, Inc.: Madison, WI, USA, 2002; pp. 463–474.
118. Bittelli, M.; Flury, M.; Roth, K. Use of dielectric spectroscopy to estimate ice content in frozen porous media. *Water Resour. Res.* **2004**, *40*, W04212. [CrossRef]
119. He, H.; Dyck, M. Application of multiphase dielectric mixing models for understanding the effective dielectric permittivity of frozen soils. *Vadose Zone J.* **2013**, *12*, 12. [CrossRef]
120. Binley, A.; Ramirez, A.; Daily, W. Regularised image reconstruction of noisy electrical resistance tomography data. In Proceedings of the 4th Workshop of the European Concerted Action on Process Tomography, Bergen, Norway, 6–8 April 1995; pp. 401–410.
121. Monego, M.; Cassiani, G.; Deiana, R.; Putti, M.; Passadore, G.; Altissimo, L. Tracer test in a shallow heterogeneous aquifer monitored via time-lapse surface ERT. *Geophysics* **2010**, *75*, WA61–WA73. [CrossRef]
122. Daily, W.; Ramirez, A.; LaBrecque, D.; Nitao, J. Electrical resistivity tomography of vadose water movement. *Water Resour. Res.* **1992**, *28*, 1429–1442. [CrossRef]
123. Leonard, J.; Andrieux, P. Infiltration characteristics of soils in Mediterranean vineyards in Southern France. *Catena* **1998**, *32*, 209–223. [CrossRef]
124. Ferrero, A.; Usowicz, B.; Lipiec, J. Effects of tractor traffic on spatial variability of soil strength and water content in grass covered and cultivated sloping vineyard. *Soil Till. Res.* **2005**, *84*, 127–138. [CrossRef]
125. Robinson, D.A.; Binley, A.; Crook, N.; Day-Lewis, F.D.; Ferré, T.P.A.; Grauch, V.J.S.; Knight, R.; Knoll, M.; Lakshmi, V.; Miller, R.; et al. Advancing process-based watershed hydrological research using near-surface geophysics: A vision for, and review of, electrical and magnetic geophysical methods. *Hydrol. Process.* **2008**, *22*, 3604–3635. [CrossRef]

Permissions

All chapters in this book were first published in MDPI; hereby published with permission under the Creative Commons Attribution License or equivalent. Every chapter published in this book has been scrutinized by our experts. Their significance has been extensively debated. The topics covered herein carry significant findings which will fuel the growth of the discipline. They may even be implemented as practical applications or may be referred to as a beginning point for another development.

The contributors of this book come from diverse backgrounds, making this book a truly international effort. This book will bring forth new frontiers with its revolutionizing research information and detailed analysis of the nascent developments around the world.

We would like to thank all the contributing authors for lending their expertise to make the book truly unique. They have played a crucial role in the development of this book. Without their invaluable contributions this book wouldn't have been possible. They have made vital efforts to compile up to date information on the varied aspects of this subject to make this book a valuable addition to the collection of many professionals and students.

This book was conceptualized with the vision of imparting up-to-date information and advanced data in this field. To ensure the same, a matchless editorial board was set up. Every individual on the board went through rigorous rounds of assessment to prove their worth. After which they invested a large part of their time researching and compiling the most relevant data for our readers.

The editorial board has been involved in producing this book since its inception. They have spent rigorous hours researching and exploring the diverse topics which have resulted in the successful publishing of this book. They have passed on their knowledge of decades through this book. To expedite this challenging task, the publisher supported the team at every step. A small team of assistant editors was also appointed to further simplify the editing procedure and attain best results for the readers.

Apart from the editorial board, the designing team has also invested a significant amount of their time in understanding the subject and creating the most relevant covers. They scrutinized every image to scout for the most suitable representation of the subject and create an appropriate cover for the book.

The publishing team has been an ardent support to the editorial, designing and production team. Their endless efforts to recruit the best for this project, has resulted in the accomplishment of this book. They are a veteran in the field of academics and their pool of knowledge is as vast as their experience in printing. Their expertise and guidance has proved useful at every step. Their uncompromising quality standards have made this book an exceptional effort. Their encouragement from time to time has been an inspiration for everyone.

The publisher and the editorial board hope that this book will prove to be a valuable piece of knowledge for researchers, students, practitioners and scholars across the globe.

List of Contributors

Luca Furnari, Giuseppe Mendicino and Alfonso Senatore
Department of Environmental Engineering, University of Calabria, 87036 Rende (Cosenza), Italy

Abdullah A. Alsumaiei and Mosaed S. Alrashidi
Civil Engineering Department, College of Engineering and Petroleum (COEP), Khaldiya Campus, Kuwait University, Safat 13060, Kuwait

Yuelong Zhu
College of Computer and Information, Hohai University, Nanjing 210098, China

Wenpeng Wang
State Key Laboratory of Hydrology-Water Resources and Hydraulic Engineering, Nanjing 210098, China
College of Computer and Information, Hohai University, Nanjing 210098, China

Bo Liu and Yuanfang Chen
College of Hydrology and Water Resources, Hohai University, Nanjing 210098, China

Xu Zhao
College of Environment, Hohai University, Nanjing 210098, China

Aaron Boone
CNRM-GAME, UMR 3589, Météo-France, CNRS, 31057 Toulouse, France

Vincent Häfliger
CNRM-GAME, UMR 3589, Météo-France, CNRS, 31057 Toulouse, France
Meteorologisches Institut, Bonn Universität, 53113 Bonn, Germany

Eric Martin
CNRM-GAME, UMR 3589, Météo-France, CNRS, 31057 Toulouse, France
IRSTEA, UR RECOVER, 13100 Aix-en-Provence, France

Sophie Ricci
CECI, CERFACS-CNRS, 42 avenue G. Coriolis, 31057 Toulouse, France

Sylvain Biancamaria
CNRS, LEGOS, UMR 5566-CNRS-CNES-IRD-Université Toulouse III, 31057 Toulouse, France

Ichiro Fujita
Department of Civil Engineering, Graduate School of Engineering, Kobe University, Kobe 657-8501, Japan

Giovanna Grossi, Amerigo Lendvai and Roberto Ranzi
Department of Civil, Environmental, Architectural Engineering and Mathematics, Università degli Studi di Brescia-DICATAM, Via Branze, 42, 25123 Brescia BS, Italy

Giovanni Peretti
ARPA Lombardia–Centro Nivometeorologico di Bormio (SO), 23032 Bormio SO, Italy

Luca Brocca, Luca Ciabatta, Christian Massari and Stefania Camici
Research Institute for Geo-Hydrological Protection, National Research Council, 06128 Perugia, Italy

Angelica Tarpanelli
Research Institute for Geo-Hydrological Protection, National Research Council, 06128 Perugia, Italy
Research Institute for Geo-Hydrological Protection, National Research Council, Via Madonna Alta 126, 06128 Perugia, Italy

Aifeng Lv, Shaofeng Jia and Wenbin Zhu
Key Laboratory of Water Cycle and Related Land Surface Processes, Institute of Geographic Sciences and Natural Resources Research, CAS, Beijing 100101, China

Bo Qu
Key Laboratory of Water Cycle and Related Land Surface Processes, Institute of Geographic Sciences and Natural Resources Research, CAS, Beijing 100101, China
University of Chinese Academy of Sciences, Beijing 100049, China

Changming Liu and Kang Liang
Key Laboratory of Water Cycle and Related Land Surface Processes, Institute of Geographic Sciences and Natural Resources Research, Chinese Academy of Sciences, Beijing 100101, China

Yanzhong Li
Key Laboratory of Water Cycle and Related Land Surface Processes, Institute of Geographic Sciences and Natural Resources Research, Chinese Academy of Sciences, Beijing 100101, China
University of Chinese Academy of Sciences, Beijing 100101, China

Bahram Saghafian
Technical and Engineering Department, Science and Research Branch, Islamic Azad University, Tehran 1477893855, Iran

Rahman Davtalab
Department of Civil, Environmental and Construction Engineering, University of Central Florida, Orlando, FL 32818, USA

Arezoo Rafieeinasab
Department of Civil Engineering, The University of Texas at Arlington, 416 Yates St. Suite 425, Arlington, TX 76019, USA

M. Reza Ghanbarpour
Alberta Environment and Parks, Calgary, AB T2E 7L7, Canada

Tommaso Moramarco and Silvia Barbetta
Research Institute for Geo-Hydrological Protection, National Research Council, Via Madonna Alta 126, 06128 Perugia, Italy

Chee Loong Wong
Faculty of Civil Engineering, Universiti Teknologi Malaysia, Johor Bahru 81310, Malaysia
Department of Irrigation and Drainage, Jalan Sultan Salahuddin, Kuala Lumpur 50626, Malaysia

Juneng Liew
School of Environment and Natural Resource Sciences, Universiti Kebangsaan Malaysia, Bangi 43600, Malaysia

Zulkifli Yusop and Tarmizi Ismail
Faculty of Civil Engineering, Universiti Teknologi Malaysia, Johor Bahru 81310, Malaysia
Centre for Environmental Sustainability and Water Security, Research Institute for Sustainable Environment, Universiti Teknologi Malaysia, Johor Bahru 81310, Malaysia

Raymond Venneker
UNESCO-IHE Institute for Water Education, Westvest 7, Delft 2611 AX, The Netherlands

Stefan Uhlenbrook
World Water Assessment Programme (WWAP), UNESCO Villa La Colombella – Località di Colombella Alta, Perugia 06134, Italy

Giulia Raffelli, Maurizio Previati, Davide Canone, Davide Gisolo, Ivan Bevilacqua and Stefano Ferraris
Interuniversity Department of Regional and Urban Studies and Planning (DIST), Polytechnic and University of Torino, viale Mattioli, 39, 10125 Torino, Italy

Giorgio Capello, Marcella Biddoccu and Eugenio Cavallo
Institute for Agricultural and Earthmoving Machines, Italian National Research Council, Strada delle Cacce 73, 10135 Torino, Italy

Rita Deiana
Department of Cultural Heritage (dBC), University of Padova, 35127 Padova, Italy

Giorgio Cassiani
Department of Geosciences, University of Padova, 35127 Padova, Italy

Index

T

V

W

www.ingramcontent.com/pod-product-compliance
Lightning Source LLC
LaVergne TN
LVHW060518210926
848485LV00005B/1101